Emerging Memory Technologies

Yuan Xie
Editor

Emerging Memory Technologies

Design, Architecture, and Applications

Editor
Yuan Xie
Pennsylvania State University
University Park, PA
USA

ISBN 978-1-4419-9550-6 ISBN 978-1-4419-9551-3 (eBook)
DOI 10.1007/978-1-4419-9551-3
Springer New York Heidelberg Dordrecht London

Library of Congress Control Number: 2013948866

Printed on acid-free paper

Springer is part of Springer Science+Business Media (www.springer.com)

Contents

Chapter 1
Introduction

Yuan Xie

Abstract Emerging non-volatile memory (NVM) technologies, such as PCRAM and STT-RAM, are getting mature in recent years. These emerging NVM technologies have demonstrated great potentials to be the candidates for future computer memory architecture design. It is important for SoC designers and computer architects to understand the benefits and limitations of such emerging memory technologies, to improve the performance/power/reliability of future memory architectures. This chapter gives a brief introduction of these memory technologies, reviews recent advances in memory architecture design, discusses the benefits of using at various levels of memory hierarchy, and also reviews the mitigation techniques to overcome the limitations of applying such emerging memory technologies for future memory architecture design.

1.1 Introduction

In the modern computer architecture design, the instruction/data storage follows a hierarchical arrangement called **memory hierarchy**, which takes advantage of locality and performance of memory technologies. Memory hierarchy design is one of the key components in modern computer systems. The importance of the memory hierarchy increases with the advances in performance of the microprocessors. Traditional memory hierarchy design consists of embedded memory (such as SRAM and eDRAM) as on-chip caches, commodity DRAM as main memory, and magnetic hard disk drivers (HDD) as the storage. Recently, solid-state drives (SSD) based on NAND-flash memory have also gained the momentum as the replacement or cache for the traditional magnetic HDD. The closer the memory is placed to microprocessor, the faster latency and higher bandwidth are required, with the penalty of the smaller

Y. Xie (✉)
Pennsylvania State University, University Park, PA, USA
e-mail: yuanxie@cse.psu.edu

Y. Xie (ed.), *Emerging Memory Technologies*,
DOI: 10.1007/978-1-4419-9551-3_1,

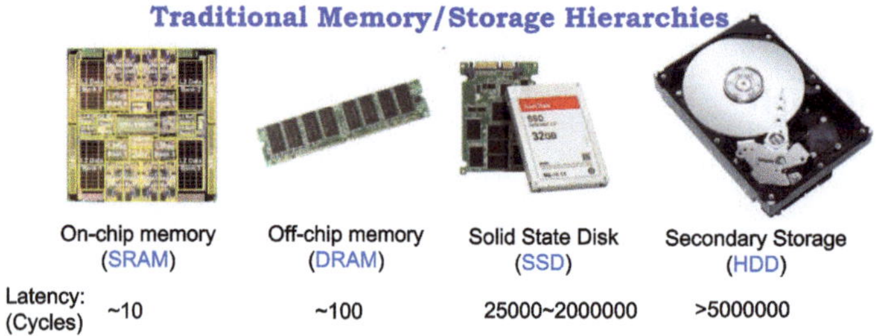

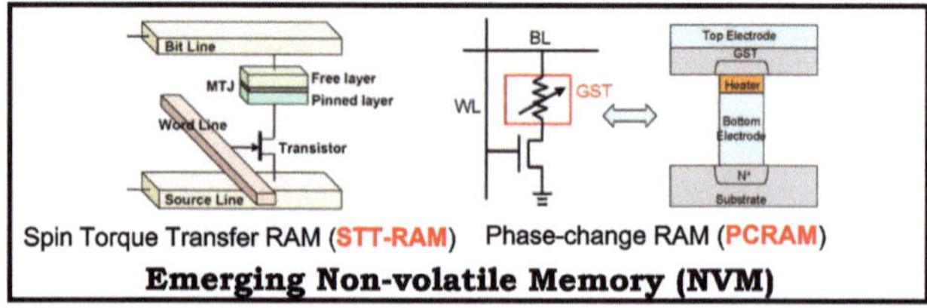

Fig. 1.1 What is the impact of emerging memory technologies on traditional memory/storage hierarchy design?

capacity. Figure 1.1 illustrates a typical memory hierarchy design, where each level of the hierarchy has the properties of smaller size, faster latency, and higher bandwidth than lower levels, with different memory technologies such as SRAM, DRAM, and magnetic hard disk drives (HDD).

Technology scaling of SRAM and DRAM (which are the common memory technologies used in traditional memory hierarchy) are increasingly constrained by fundamental technology limits. In particular, the increasing leakage power for SRAM/DRAM and the increasing refresh dynamic power for DRAM have posed challenges for circuit/architecture designers for future memory hierarchy design.

Recently, emerging memory technologies (such as Spin Torque Transfer RAM(STT-MRAM), Phase-change RAM (PCRAM), and Resistive RAM (ReRAM)), are being explored as potential alternatives of existing memories in future computing systems. Such emerging non-volatile memory (NVM) technologies combine the speed of SRAM, the density of DRAM, and the non-volatility of Flash memory, and hence, become very attractive as the alternatives for future memory hierarchy. It is anticipated that these NVM technologies will break important ground and move closer to market very rapidly.

Simply using new technologies as replacements of existing hierarchy may not be the most desirable approach. For example, using high-density STT-RAM to replace SRAM as on-chip cache can reduce the cache miss rate due to larger capacity and improve performance, on the other hand, the longer write latency for STT-RAM could hurt the performance for write-intensive applications; Also, using high density memory as an extra level of on-chip cache will reduce CPU requests to the traditional, off-package DRAM and thus reduce the average memory access time. However, to manage this large cache, a substantial amount of space on the CPU chip needs to be taken up by tags and logics, which could be used to increase the size of the next lower level cache. Moreover, trends toward Many-core and System-on-Chip may introduce the need and opportunity for new memory architectures. Consequently, as such emerging memory technologies are getting mature, it is important for SoC designers and computer architects to understand the benefits and limitations for better utilizing them to improve the performance/power/reliability of future computer architecture. Specifically, designers need to seek the answers to the following questions:

- How to model such emerging NVM technologies at the architectural level?
- What will be the impacts of such NVMs on the future memory hierarchy? What will be the novel architectures/applications?
- What are the limitations to overcome for such a new memory hierarchy?

This book includes 11 chapters that try to answer the questions mentioned above. These chapters cover different perspectives related to the modeling, design, and architectures of using the emerging memory technologies. We expect that this book can serve as a catalyst to accelerate the adoption of such emerging memory technologies for future computer system design from architecture and system design perspectives.

1.2 Preliminary on Emerging Memory Technologies

Many promising emerging memory technology candidates, such as Phase-Change RAM (PCRAM), Spin Torque Transfer Magnetic RAM (STT-RAM), Resistive RAM (ReRAM), and Memristor, have gained substantial attentions and are being actively pursued by industry [1]. In this section we will briefly describe the fundamentals of these promising emerging memory technologies to be surveyed in this paper, namely, the STT-RAM, the PCRAM, the ReRAM, and Memristor.

STT-RAM is a new type of Magnetic RAM (MRAM) [1], which features non-volatility, fast writing/reading speed (<10 ns), high programming endurance ($>10^{15}$ cycles) and zero standby power [1]. The storage capability or programmability of MRAM arises from magnetic tunneling junction (MTJ), in which a thin tunneling dielectric, e.g., MgO, is sandwiched by two ferromagnetic layers, as shown in Fig. 1.1. One ferromagnetic layer ("pinned layer") is designed to have its magnetization pinned, while the magnetization of the other layer ("free layer") can be flipped by a write event. An MTJ has a low (high) resistance if the magnetizations of the free layer and the pinned layer are parallel (anti-parallel). Prototyping STT-RAM chips have been demonstrated recently by various companies and research groups [2, 3]. Commercial MRAM products have been launched by companies like Everspin and NEC.

PCRAM technology is based on a chalcogenide alloy (typically, Ge_2–Sb_2–Te_5, GST) material) [1, 4]. The data storage capability is achieved from the resistance differences between an amorphous (high-resistance) and a crystalline (low-resistance) phase of the chalcogenide-based material. In SET operation, the phase change material is crystallized by applying an electrical pulse that heats a significant portion of the cell above its crystallization temperature. In RESET operation, a larger electrical current is applied and then abruptly cut off in order to melt and then quench the material, leaving it in the amorphous state. PCRAM has shown to offer compatible integration with CMOS technology, fast speed, high endurance, and inherent scaling of the phase-change process at 22-nm technology node and beyond [5]. Compared to STT-RAM, PCRAM is even denser with an approximate cell area of $6 \sim 12F^2$ [1], where F is the feature size. In addition, phase change material has a key advantage of the excellent scalability within current CMOS fabrication methodology, with continuous density improvement. Many PCRAM prototypes have been demonstrated in the past years by companies like Hitachi [6], Samsung [7], STMicroelectronics [8], and Numonyx [9].

Resistive RAM (ReRAM) and Memristor

ReRAM memory stores the data as two (single-level cell, or SLC) or more resistance states (multi-level cell, or MLC) of the resistive switch device (RSD). Resistive switching in transition metal oxides was discovered in thin NiO film decades ago. From then, a large variety of metal-oxide materials have been verified to have resistive switching characteristics, including TiO_2, NiO_x, Cr-doped $SrTiO_3$, PCMO, CMO [10], etc. Based on the storage mechanisms, ReRAM materials can be cataloged as filament-based, interface-based, programmable-metallization-cell (PMC), etc. Based on the electrical property of resistive switching, RSDs can be divided into two categories: unipolar or bipolar. Programmable-metallization-cell (PMC) [11] is a promising bipolar switching technology. Its switching mechanism can be explained as forming or breaking the small metallic "nanowire" by moving the metal ions between two sold metal electrodes. Filament-based ReRAM is a typical example of unipolar switching [12] that has been widely investigated. The insulating material between two electrodes can be made conducting through a hopping or tunneling conduction path after the application of a sufficiently high voltage. The data storage could be achieved by breaking (RESET) or reconnecting (SET) the conducting path. Such switching mechanism can in fact be explained with the fourth circuit element, the **memristor** [13–15].

Memristor was predicted by Chua in 1971 [13], based on the completeness of circuit theory. Memristance (M) is a function of charge (q), which depends upon the historic behavior of the current (or voltage) profile [15, 16]. In 2008, the researchers at HP reported the first real device of a memristor in a solid-state thin film two-terminal device by moving the doping front along the device [14]. Afterwards, magnetic technology provides the other possible methods to build a memristive system [17, 18]. Due to its unique historic characteristic, memristor has very broad application including nonvolatile memory, signal processing, control and learning system etc [19].

Many companies are working on ReRAM technology and chip design, including Fujitsu, Sharp, HP lab, Unity Semiconductor Corp., Adesto Technology Inc. (a spin-off from AMD), etc. And in Europe, the research institute IMEC is doing independent research on ReRAMs with its partners Samsung Electronics Co. Ltd., Hynix

Table 1.1 Comparison of different memory technologies [21]

Features	SRAM	eDRAM	STT-RAM	PCRAM	ReRAM
Density	Low	High	High	Very high	Very high
Speed	Very Fast	Fast	Fast for read; slow for write	Slow for read; very slow for write	Slow for read/write
Dynamic Power	Low	Medium	Low for read; very high for write	Medium for read; high for write	Medium for read; high for write
Leakage Power	High	Medium	Low	Low	Low
Non-volatility	No	No	Yes	Yes	Yes

Semiconductor inc., Elpida Inc. and Micron Technology Inc. The main efforts on ReRAM research devote to material and devices [10]. Many circuit design issues have also been addressed, such as power-supply voltage and current monitoring. Recently, Sandisk and Toshiba demonstrated a 32 Gb ReRAM prototype in ISSCC 2013 [20].

Table 1.1 shows the comparison of these three emerging memory technologies against the conventional memory technologies used in traditional memory hierarchies.

1.3 Modeling

To help the architectural level and system-level design space exploration of the SRAM-based or DRAM-based cache and memory, various modeling tools have been developed during the last decade. For example, CACTI [22] and DRAMsim [23] have become widely used in the computer architecture community to estimate the speed, power, and area parameters of SRAM and DRAM caches and main memory.

Similarly, for computer architects to explore new design opportunities at architecture and system levels that the emerging memory technologies can provide, architectural level STT-RAM-based cache model [24, 25] and PCRAM-based cache/memory model [26] have been recently developed. Such architectural models provide the extraction of all important parameters, including access latency, dynamic access power, leakage power, die area, I/O bandwidth, etc., to facilitate architecture-level analysis, and to bridge the gap between the abundant research activities at process and device levels and the lack of a high-level cache and memory model for emerging NVMs.

The architectural modeling for cache and main memory built with emerging memory technologies (such as STT-RAM and PCRAM) raises many unique research issues and challenges.

- First, some circuitry modules in PCRAM/MRAM have different requirements from those originally designed for SRAM/DRAM. For example, the existing sense

amplifier model in CACTI [22] and DRAMsim [23] is based on voltage-mode sensing, while PCRAM data reading usually uses a current-mode sense amplifier.
- Second, due to the unique device mechanisms, the models of PCRAM/MRAM need specialized circuits to properly handle their operations. For example, in PCRAM, the specific pulse shapes are required to heat up GST material quickly and to cool it down gradually during the RESET and especially SET operations. Hence, a model of the slow quench pulse shaper need to be created.
- Finally, the memory cell structures between STT-RAM/PCRAM and SRAM/DRAM are different. PCRAM and STT-RAM typically use a simple "1T1R" (one-transistor-one-resistor) or "1D1R" (one-diode-one-resistor) structure, while SRAM and DRAM cell has a conventional "6T" structure and "1T1C" (one-transistor-one-capacitor) structure, respectively. The difference of cell structures directly leads to different cell sizes and array structures.

In addition, where to place these emerging memories in the traditional memory hierarchy also influences the modeling methodologies. For example, the emerging NVMs could be used as a replacement for on-chip cache or for off-chip DIMM (dual in-line memory module). Obviously, the performance/power of on-chip cache and off-chip DIMM would be quite different: When a NVM is integrated with logics on the same die, there is no off-chip pin limitation so that the interface between NVM and logic can be re-designed to provide a much higher bandwidth. Furthermore, off-chip memory is not affected by the thermal profile of the microprocessor core while the on-chip cache is affected by the heat dissipation from the hot cores. While higher on-chip temperature has a negative impact on SRAM/DRAM memory, it may have a positive influence on PCRAM because the heat can facilitate the write operations of PCRAM cell. The performance estimation of PCRAM becomes much more complicated in such a case.

Moreover, building an accurate PCRAM/MRAM simulator needs close collaborations with the industry to understand physics and circuit details, as well as architectural level requirements such as the interface/interconnect with the multi-core CPUs.

Chapter 2 of this book introduces a modeling tool called NVsim by Dong et al. This tool is widely used by research community as an open-source modeling tool for emerging memory technologies such as STT-RAM and PCRAM.

1.4 Leveraging Emerging Memory Technologies in Architecture Design

As the emerging memory technologies are getting mature, integrating such memory technologies into the memory hierarchies (as shown in Fig. 1.1) provides new opportunities for future memory architecture designs. Specifically, there are several characteristics of STT-RAM and PCRAM that make them promising as working class memories (i.e., on-chip caches and off-chip main memories), or as storage class memories: (1) Compared to SRAM/DRAM, these emerging memories usually have much higher density, with comparable fast access time; (2) Due to the non-volatility feature, they have zero standby power, and immune to radiation-induced soft errors; (3) Compared to NAND-Flash SSD, STT-RAM/PCRAM are byte-addressable. In addition, different hybrid compositions of memory hierarchy by using SRAM, DRAM, and PCRAM or MRAM can be motivated by different power and access behaviors of various memory technologies. For example, leakage power is dominant in SRAM and DRAM arrays; on the contrary, due to non-volatility, PCRAM or STT-RAM array consumes zero leakage power when idling but a much higher energy during write operations. Hence, the trade-off among using different memory technologies at various hierarchy levels becomes an interesting research topic. In addition, if these memory are used as on-chip cache or main memory rather than as storage, the data retention time for non-volatility is not that important since data are used and overwritten in a very short period of time. Consequently, data retention time can be traded for better performance and energy benefits (as demonstrated by Chap. 7).

In this book, Chaps. 3–9 covers different design options of using such emerging memory technologies at different level of memory hierarchies. Chapter 10 proposes a design space exploration framework for circuit-architecture co-optimization for NVM memory architecture design. Chapter 11 describes a prototyping effort that fabricated an NVM-based processor design.

1.4.1 Leveraging NVMs as On-Chip Cache

Replacing SRAM-based on-chip cache with STT-RAM/PCRAM can potentially improve performance and reduce power consumption. With larger on-chip cache capacity (due to its higher density), STT-RAM/PCRAM based on-chip cache can help reduce the cache miss rate, which helps improve the performance. Zero-standby leakage can also help reduce the power consumption. On the other hand, longer write-latency of such NVM-based cache may incur performance degradation and offset the benefits from the reduced cache miss rate. Although PCRAM is much denser than SRAM, the limited endurance makes it unaffordable to directly use PCRAM as on-chip caches, which have highly frequent accesses.

The performance/power benefits of STT-RAM for single-core processor were investigated by Dong et al. [24]. The research demonstrated that STT-RAM-based L2 cache can bring performance improvement and achieve more than 70 % power consumption reduction at the same time. The benefits of using STT-RAM shared L2 cache for multi-core processors were demonstrated by Sun et al. [25]. The simulation result shows that the optimized MRAM L2 cache improves performance by 4.91 % and reduces power by 73.5 % compared to the conventional SRAM L2 cache with a similar area. Wu et al. [21] studied a number of different hybrid-cache architectures (HCA) that are composed of SRAM/eDRAM/STT-RAM/PCRAM for IBM Power 7 cache architecture, and explored the potential of hardware support for intra-cache data movement and power consumption management within HCA caches. Under the same area constraint across a collection of 30 workloads, such aggressive hybrid-cache design provides 10–16 % performance improvement over the baseline design with a 3-level SRAM-only cache design, and achieves up to a 72 % reduction in power consumption.

In this book, Chaps. 6 and 7 give details on the evaluation of using NVM as on-chip cache, and the mitigation techniques to overcome some limitations such as performance/power overhead related to the write operations. device-architecture co-optimization can also be applied to achieve better performance/power benefits.

1.4.2 Leveraging NVMs as Main Memory

There are abundant recent investigations on using PCRAM as a replacement for the current DRAM-based main memory architecture. Lee et al. [27] demonstrated that a pure PCRAM-based main memory architecture implementation is about 1.6x slower and requires 2.2x energy than a DRAM-based main memory, mainly due to the overhead of write-operations. They proposed to re-design the PCM buffer organizations, with narrow buffers to mitigate high energy PCM writes. Also with multiple buffer rows, it can exploit locality to coalesce writes, hiding their latency and energy, such that the performance is only 1.2x slower with a similar energy consumption compared to the DRAM-based system. Qureshi et al. [28] proposed a main memory system consisting of PCM storage coupled with a small DRAM buffer, so that it can leverage the latency benefits of DRAM and the capacity benefits of PCM. Such memory architecture could reduce page faults by 5x and provide a speedup of 3x. A similar study conducted by Zhou et al. [29] demonstrated that the PCRAM-based main memory consumes only 65 % of the total energy of the DRAM main memory with the same capacity, and the energy-delay product is reduced by 60 %, with various techniques to mitigate the overhead of write-operations. All these work have demonstrated the feasibility of using PCRAM as main memory in the near future.

1.4.3 Leveraging NVM to Improve NAND-Flash SSD

NAND flash memory has been widely adopted by various applications such as laptops and mobile phones. In addition, because of its better performance compared to the traditional HDD, NAND flash memory has been proposed to be used as a cache in HDD, or even as the replacement of HDD in some applications. However, one well-known limitation of NAND flash memory is the "erase-before-write" requirement. It cannot update the data by directly overwriting it. Instead, a time-consuming erase operation must be performed before the overwriting. To make it even worse, the erase operation cannot be performed selectively on a particular data item or page but can only be done for a large block called the "erase unit". Since the size of an erase unit (typically 128 K or 256 K Bytes) is much larger than that of a page (typically 512 $\sim$ 8 K Bytes), even a small update to a single page requires all the pages within the erase unit to be erased and written again.

Compared to NAND flash memory, PCRAM/STT-MRAM has advantages of random access and direct in-place updating. Consequently, Chap. 3 gives details on how to use a hybrid storage architecture to combine the advantages of NAND flash memory and PCRAM/MRAM. In such hybrid storage architecture, PCRAM is used as the log region for NAND-flash. Such hybrid architecture has the following advantages: (1) the ability of "in-place updating" can significantly improve the usage efficiency of log region by eliminating the out-of-date log data; (2) the fine-granularity access of PCRAM can greatly reduce the read traffic from SSD to main memory; (3) the energy consumption of the storage system is reduced as the overhead of writing and reading log data is decreased with the PCRAM log region; and (4) the lifetime of the NAND flash memory in the hybrid storage could be increased because the number of erase operations is reduced.

1.4.4 Enabling Fault-Tolerant Exascale Computing

Due to the continuously reduced feature size, supply voltage, and increased on-chip density, computer systems are projected to be more susceptible to hard errors and transient errors. Compared to SRAM/DRAM memory, PCRAM/STT-RAM memory has unique features such as non-volatility and resilience to soft errors. The application of such unique features could enable novel architecture design for applications that can address the reliability challenges for future Exascale scale computing.

For example, checkpointing/rollback scheme, where the processor takes frequent checkpoints at a certain time interval and stores them to hard disk, is one of the most common approaches to ensure the fault-tolerance of a computing system. In the current peta-scale massive parallel processing (MPP) systems, such traditional checkpointing to hard disk could incur a large performance overhead and is not a scalable solution for future Exascale computing. For example, Dong et al. [30] proposed three variants of PCRAM-based hybrid checkpointing schemes, which reduce the checkpoint overhead and offer a smooth transition from the conventional pure HDD checkpoint to the ideal 3D PCRAM mechanism. With a 3D PCRAM approach, multiple layers of PCRAM memory are stacked on top of DRAM memory, integrated with the emerging 3D integration technology. With a massive memory bandwidth provided by the through-silicon-via (TSVs) enabled by 3D integration, fast and high-bandwidth local checkpointing can be realized. The proposed pure 3D PCRAM-based mechanism can ultimately take checkpoints with overhead less than 4 % on a projected Exascale system.

1.5 Mitigation Techniques for STT-RAM/PCRAM Memory

The previous section presents the benefits of using these emerging memory technologies in computer system design. However, such benefits can only be achieved with mitigation techniques that can help address the inherited disadvantages that related to the write operations: (1) Because of the non-volatility feature, it usually takes much longer and more energy for write operations compared to read operations; (2) Some emerging memory technologies such as PCRAM has the wear-out problem (lifetime reliability), which is one of the major concerns of using it as working memory rather than storage class memory. Consequently, introducing these emerging memory technologies into current memory hierarchy design gives rise to new opportunities but also presents new challenges that need to be addressed. In this section, we review mitigation techniques that help address the disadvantages of such emerging technologies.

1.5.1 Techniques to Mitigate Latency/Energy Overheads of Write Operations

In order to use the emerging NVMs as cache and memory, several design issues need to be solved. The most important one is the performance and energy overheads in write operations. A NVM has a more stable mechanism for data keeping, compared to a volatile memory such as SRAM and DRAM. Accordingly, it needs to take a longer time and consume more energy to over-write the existing data. This is the intrinsic characteristic of NVMs. PCRAM and MRAM are not exceptional. If we directly replace SRAM caches with PCRAM/MRAM ones, the long latency and high energy consumption in write operations could offset the performance and power benefits, and even result in degradation when the cache write intensity is high. Therefore, it is imperative to study techniques to mitigate the overheads of write operations in NVMs.

- *Hybrid Cache/Memory Architecture.* To leverage the benefits of both the traditional SRAM/DRAM (such as fast write-operations) and the emerging NVMs (such as high density, low leakage, and resilient to soft error), a hybrid cache/memory architecture can be used, such as STT-RAM/SRAM hybrid on-chip cache, which is described in details in Chap. 6, or PCRAM/DRAM hybrid main memory [28]. In such hybrid architecture, instead of building a pure STT-RAM-based cache or a pure PCRAM-based main memory, we could replace a portion of MRAM or PCRAM cells with SRAM or DRAM elements, respectively. The main purpose is to keep most of write intensive data within SRAM/DRAM part, and hence, to reduce write operations in NVM parts. Therefore, the dynamic power consumption can be reduced and performance can be further improved. The major challenges to this architecture are how to physically arrange two different types of memories and how to migrate data in between.

- *Novel Buffer Architecture.* The write buffer design in modern processors works well for SRAM-based caches, which have approximately equivalent read and write speeds. However, the traditional write buffer design may not be suitable for NVM-based caches, which feature a large variation between read and write latencies. Chapter 6 will give details on how to design a novel write buffer architecture to mitigate the write-latency overhead. For example, in the scenario where a write operation is followed by several read operations, the ongoing write operation may block the upcoming read ones and cause performance degradation. The cache write buffer can be improved to prevent the critical read operations from being blocked by long write operations. For example, a higher priority can be assigned to read operations when competition happens between read and write. In an extreme condition when write-retirements are always stalled by read operations, write buffer could become full, which can also degrade cache performance. Hence, how to properly deal with read/write sequence and whether this mechanism could be dynamically controlled based on applications also need to be investigated. A similar write cancellation and write pausing techniques are also proposed in Ref. [31]. In addition, Lee et al. [27] also proposed to redesign the PCRAM buffer, using narrow buffers to help mitigate high energy PCM writes. Multiple buffer rows can exploit locality to coalesce writes, hiding their latency and energy.
- *Eliminating Redundant Bit-Writes.* In a conventional memory access, a write updates an entire row of the memory cells. A large portion of such writes are redundant. A read-before-write operation can help identify such redundant bits and cancel those redundant write operations to save energy and reduce the impact on performance [32].
- *Data Inverting.* To further reduce the number of writes to PCRAM cells, a data inverting scheme [32, 33] can be adopted in the PCRAM write logic. When a new data is written to a cache block, we first read its old data value, and compute the Hamming distance (HD) between the two values. If the calculated HD is larger than the half of the cache block size, the new data value is inverted before the store operation. An extra status bit is set to 1 to denote that the stored value has been inverted.

1.5.2 Techniques to Improve Lifetime for NVMs

Write endurance is another severe challenge in PCRAM memory design. The state-of-the-art process technology has demonstrated that the write endurance for PCRAM is around 10^8–10^9 [29]. The problem is further aggravated by the fact that writes to caches and main memory can be extremely skewed. Consequently, those cells suffering from more frequent write operations will fail much sooner than the rest.

Techniques that proposed in the previous sub-section to reduce the number of write operations to STT-RAM/PCRAM can definitely help the lifetime of the memory, besides reducing the write energy overhead. In addition to those techniques, the following schemes can be used to further improve the lifetime of the memory.

- *Wear leveling.* Wear leveling technique, which has been widely implemented in NAND flash memory, attempts to work around the limitations of write endurance by arranging data access so that write operations can be distributed evenly across all the storage cells. Wear leveling technique can also be applied to PCRAM/MRAM-based cache and memory. a range of wear leveling techniques for PCRAM have been examined [27–29, 32, 34] recently. Such wear leveling techniques include: (1) Row Shifting. A simple shifting scheme can be applied to evenly distribute writes within a row. The scheme is implemented through an additional row shifter along with a shift offset register. On a read access, data is shifted back before being passed to the processor. (2) Word-line Remapping and Bit-line Shifting. Bit-line shifter and word-line remapper are used to spread the writes over the memory cells inside one cache block and among cache blocks, respectively. (3) Segment Swapping. Periodically, memory segments of high and low write accesses are swapped. The memory controller keeps track of the write counts of each segment, and a mapping table between the "virtual" and "true" segment number. Chapter 9 of this book will also cover more details about the wear-leveling techniques, including new considerations of intra-set and inter-set write variations when NVM is used as on-chip cache.
- *Graceful degradation.* In such scheme, the PCRAM allows continued operation through graceful degradation when hard faults occur [35]. The memory pages that contain hard faults are not discarded. Instead, they are dynamically formed pairs of complementary pages that act as a single page of storage, such that the total effective memory capacity is reduced but the the lifetime of PCRAM can be improved by up to 40× over conventional error-detection techniques.

1.6 Conclusion

This chapter reviews recent advance in memory architecture design with such emerging memory technology, discusses the benefits of using STT-RAM/PCRAM at various level of memory hierarchy, and also presents the mitigation techniques to overcome the challenges of applying such emerging memory technologies for future memory architecture design. These recent architectural-level studies have demonstrated that emerging memory technologies like STT-RAM/PCRAM/ReRAM have great potentials to improve future computer memory architecture design, and enable novel applications such as novel checkpointing techniques for future Exascale computing.

The rest of this book will give more details for different perspectives that are introduced in this chapter. With all these initial research efforts, we believe that the emerging of these new memory technologies will change the landscape of future memory architecture design.

References

1. International Technology Roadmap for Semiconductor, 2007.
2. Honjo, H., Saito, S., Ito, Y., Miura, S., Kato, Y., Mori, K., Ozaki, Y., Kobayashi, Y., Ohshima, N., Kinoshita, K., Suzuki, T., Nagahara, K., Ishiwata, N., Suemitsu, K., Fukami, S., Hada, H., Sugibayashi, T., Nebashi, R., Sakimura, n., & Kasai, N. (2009). A 90 nm 12 ns 32 Mb 2T1MTJ MRAM. *IEEE International Solid-State Circuits Conference (ISSCC)* (pp. 462–463).
3. Kawahara, T., Takemura, R., Miura, K., Hayakawa, J., Ikeda, S., Lee, Y. M., et al. (2008). 2 Mb SPRAM (SPin-transfer torque RAM) with bit-by-bit bi-directional current write and parallelizing-direction current read. *IEEE Journal of Solid-State Circuits*, *43*(1), 109–120.
4. Raoux, S., et al. (2008). Phase-change random access memory: A scalable technology. *IBM Journal of Research and Development*, *52*(4/5), 465–481.
5. Chen, Y.C., Rettner, C.T., Raoux, S., Burr, G.W., Chen, S.H., Shelby, R.M., Salinga, M., et al. (2006). Ultra-thin phase-change bridge memory device using gesb. *Proceedings of the IEEE International Electron Devices Meeting* (pp. 30.3.1–30.3.4).
6. Osada, K., Kotabe, A., Matsui, Y., Matsuzaki, N., Takaura, N., Moniwa, M., Kawahara, T., Hanzawa, S., & Kitai, N. (2007). A 512 kb embedded pram with 416 kbs write throughput at 100μa cell write current. *IEEE International Solid-State Circuits Conference (ISSCC)* (p. 26.2).
7. Cho, W.-Y., Kang, S., Choi, B.-G., Oh, H.-R., Lee, C.-S., Kim, H.-J., Park, J.-M., Wang, Q., Park, M.-H., Ro, Y.-H., Choi, J.-Y., Kim, K.-S., Kim, Y.-R., Chung, W.-R., Cho, H.-K., Lim, K.-W., Choi, C.-H., Shin, I.-C., Kim, D.-E., Yu, K.-S., Kwak, C.-K., Kim, C.-H., Lee, K.-J., & Cho, B. (2007) A 90 nm 1.8 V 512 Mb diode-switch pram with 266 Mb/s read throughput. *IEEE International Solid-State Circuits Conference (ISSCC)* (p. 26.1).
8. Pirola, A., Marmonier, L., Pasotti, M., Borghi, M., Mattavelli, P., Zuliani, P., Scotti, L., Mastracchio, G., Bedeschi, F., Gastaldi, R., Bez, R., De Sandre, G., & Bettini, L. (2010). A 90 nm 4 Mb embedded phase-change memory with 1.2 V 12 ns read access time and 1 Mb/s write throughput. *IEEE International Solid-State Circuits Conference (ISSCC)* (p. 14.7).
9. Barkley, G., Giduturi, H., Schippers, S., Vimercati, D., Villa, C., & Mills, D. (2010). A 45 nm 1 Gb 1.8 V phase-change memory. *IEEE International Solid-State Circuits Conference (ISSCC)* (p. 14.8).
10. Wong, P., Lee, H., Yu, S., et al. (1951). Metal-Oxide ReRAM. *Proceedings of the IEEE*, *100*(6), 2012.
11. Kozicki, M.N., Balakrishnan, M., Gopalan, C., Ratnakumar, C., & Mitkova, M. (2005) Programmable metallization cell memory based on ag-ge-s and cu-ge-s solid electrolytes. *Non-Volatile Memory Technology Symposium* (pp. 83–89).
12. Inoue, I.H., Yasuda, S., Akinaga, H., & Takagi, H. (2008). Nonpolar resistance switching of metal/binary-transition-metal oxides/metal sandwiches: Homogeneous/inhomogeneous transition of current distribution. *Physical Review B*, *77*, 035105.
13. Chua, L.O. (1971) Memristor—The missing circuit element. *IEEE Transactions on Circuit Theory*, *CT-18*, 507–519.
14. Tour, J. M., & He, T. (2008). The fourth element. *Nature*, *453*, 42–43.
15. Strukov, D.B., Snider, G.S., Stewart, D.R., & Williams, R.S. (2008) The missing memristor found. *Nature*, *453*, 80–83.
16. Chua, L. O. (1976). Memristive devices and systems. *Proceedings of IEEE*, *64*, 209–223.
17. Pershin, Y.V., & Di Ventra, M. (2008). Spin memristive systems: Spin memory effects in semiconductor spintronics. *Physical Review B: Condensed Matter*, *78*, 113309.
18. Wang, X., et al. (2009). Spin memristor through spin-torque-induced magnetization motion. *IEEE Electron Device Letters*, *30*, 294–297.
19. Chen, y., & Wang, X. (2009). Compact modeling and corner analysis of spintronic memristor. *IEEE/ACM International Symposium on Nanoscale Architectures 2009 (NANOARCH'09)* (pp. 7–12).

20. Liu, T., Yan, T., Scheuerlein, R., et al. (2013). A 130.7 mm^2 2-Layer 32 Gb ReRAM Memory Device in 24 nm Technology. *Proceedings of International Solid State Circuits Conference* (pp. 210–211).
21. Wu, X., Li, J., Zhang, L., Speight, E., Rajamony, R., & Xie, Y. (2009). Hybrid cache architecture with disparate memory technologies. *36th International Symposium on Computer Architecture, ISCA '09.*
22. Wilton, S. J. E., & Jouppi, N. P. (1996). CACTI: An enhanced cache access and cycle time model. *IEEE Journal of Solid-State Circuits, 31*, 677–688.
23. Wang, David, Ganesh, Brinda, Tuaycharoen, Nuengwong, Baynes, Kathleen, Jaleel, Aamer, & Jacob, Bruce. (2005). DRAMsim: a memory system simulator. *SIGARCH Computer Architecture News, 33*(4), 100–107.
24. Dong, X. , Wu, X. , Sun, G., Xie, Y., Li, H., & Chen, H. (2008). Circuit and microarchitecture evaluation of 3D stacking magnetic RAM (MRAM) as a universal memory replacement. *Proceedings of Conference on Design Automation* (pp. 554–559).
25. Sun, G., Dong, X., Xie, Y., Li, J., & Chen, Y. (2009). A novel architecture of the 3d stacked mram l2 cache for cmps. *IEEE 15th International Symposium on High Performance Computer Architecture* (pp. 239–249).
26. Dong, X., Jouppi, N., & Xie, Y. (2009) Pcramsim: System-level performance, energy, and area modeling for phase-change ram. *International Conference on Computer-Aided Design (ICCAD)* (pp. 269–275).
27. Lee, B.C., Ipek, E., Mutlu, D., & Burger, D. (2009). Architecting phase change memory as a scalable dram alternative. *Proceedings of ISCA* (pp. 2–13).
28. Qureshi, M.K., Srinivasan, V., & Rivers, J.A. (2009) Scalable high performance main memory system using phase-change memory technology. *ISCA '09: Proceedings of the 36th annual international symposium on Computer architecture*, New York, NY, USA, 2009. ACM (pp. 24–33).
29. Zhou, P., Zhao, B., Yang, J., & Zhang, Y. (2009) A durable and energy efficient main memory using phase change memory technology. *Proceedings of ISCA* (pp. 14–23).
30. Dong, X., Muralimanohar, N., Jouppi, N., Kaufmann, R., & Xie, Y. (2009) Leveraging 3d pcram technologies to reduce checkpoint overhead for future exascale systems. *International Conference on High Performance Computing, Networking, Storage and, Analysis (SC09).*
31. Qureshi, M., Franceschini, M., Lastras, M. (2010). Improving read performance of phase change memories via write cancellation and write pausing. *Proceedings of International Symposium on High Performance Computer Architecture (HPCA).*
32. Joo, Y., Niu, D., Dong, X., Sun, G., Chang, N., & Xie, Y. (2010). Energy- and endurance-aware design of phase change memory caches. *Proceedings of Design Automation and Test in Europe.*
33. Cho, S., & Lee, H. (2009). Flip-N-Write: A simple deterministic technique to improve PRAM write performance, energy and endurance. *Proceedings of International Symoposium on Microarchitecture (MICRO).*
34. Qureshi, M., Karidis, J., Franceschini, M., Srinivasan, V., Lastras, L., & Abali, B. (2009). Enhancing lifetime and security of phase change memories via start-gap wear leveling. *Proceedings of International Symoposium on Microarchitecture (MICRO).*
35. Ipek, E., Condit, J., Nightingale, E., Burger, D., & Moscibroda, T. (2010). Dynamically replicated memory: Building reliable systems from nanoscale resistive memories. *Proceedings of International Conference on Architecture Support for Programming Languages and Operating Systems.*

Chapter 2
NVSim: A Circuit-Level Performance, Energy, and Area Model for Emerging Non-volatile Memory

Xiangyu Dong, Cong Xu, Norm Jouppi and Yuan Xie

Abstract Various new non-volatile memory (NVM) technologies have emerged recently. Among all the investigated new NVM candidate technologies, spin-torque transfer memory (STT-RAM, or MRAM), phase change memory (PCRAM), and resistive memory (ReRAM) are regarded as the most promising candidates. As the ultimate goal of this NVM research is to deploy them into multiple levels in the memory hierarchy, it is necessary to explore the wide NVM design space and find the proper implementation at different memory hierarchy levels from highly latency-optimized caches to highly density-optimized secondary storage. While abundant tools are available as SRAM/DRAM design assistants, similar tools for NVM designs are currently missing. Thus, in this work, we develop *NVSim*, a circuit-level model for NVM performance, energy, and area estimation, which supports various NVM technologies including STT-RAM, PCRAM, ReRAM, and legacy NAND flash. *NVSim* is successfully validated against industrial NVM prototypes, and it is expected to help boost architecture-level NVM-related studies.

X. Dong · C. Xu · Y. Xie (✉)
Computer Science and Engineering Department,
Pennsylvania State University, IST Building, University Park, PA 16802, USA
e-mail: xiangyud@qualcomm.com

C. Xu
e-mail: congxu@cse.psu.edu

Y. Xie
e-mail: yuanxie@cse.psu.edu

N. Jouppi
Google, Inc., CA, USA
e-mail: jouppi@acm.org

Y. Xie (ed.), *Emerging Memory Technologies*,
DOI: 10.1007/978-1-4419-9551-3_2,

2.1 Introduction

Universal memory that provides fast random access, high storage density, and non-volatility within one memory technology becomes possible, thanks to the emergence of various new non-volatile memory (NVM) technologies, such as spin-torque transfer random access memory (STT-RAM, or MRAM), phase change random access memory (PCRAM), and resistive random access memory (ReRAM). As the ultimate goal of this NVM research is to devise a universal memory that could work across multiple layers of the memory hierarchy, each of these emerging NVM technologies has to supply a wide design space that covers a spectrum from highly latency-optimized microprocessor caches to highly density-optimized secondary storage. Therefore, specialized peripheral circuitry is required for each optimization target. However, since few of these NVM technologies are mature so far, only a limited number of prototype chips have been demonstrated and just cover a small portion of the entire design space. In order to facilitate the architecture-level NVM research by estimating the NVM performance, energy, and area values under different design specifications before fabricating a real chip, in this work, we build *NVSim*, a circuit-level model for NVM performance, energy, and area estimations, which supports various NVM technologies including STT-RAM, PCRAM, ReRAM, and legacy NAND flash.

The main goals of developing NVSim tool are as follows:

- Estimate the access time, access energy, and silicon area of NVM chips with a given organization and specific design options before the effort of actual fabrications;
- Explore the NVM chip design space to find the optimized chip organization and design options that achieve best performance, energy, or area;
- Find the optimal NVM chip organization and design options that are optimized for one design metric while keeping other metrics under constraints.

We build NVSim by using the same empirical modeling methodology as CACTI [39, 43] but starting from a new framework and adding specific features for NVM technologies. Compared to CACTI, the framework of NVSim includes the following new features:

- It allows to move sense amplifiers from inner memory subarrays to the outer bank level and factor them out to achieve overall area efficiency of the memory module;
- It provides more flexible array organizations and data activation modes by considering any combinations of memory data allocation and address distribution;
- It models various types of data sensing schemes instead of voltage sensing scheme only;
- It allows memory banks to be formed in a bus-like manner rather than the H-tree manner only;
- It provides multiple design options of buffers instead of latency-optimized option that uses logical effort;
- It models the cross-point memory cells rather than MOS-accessed memory cells only;

- It considers the subarray size limit by analyzing the current sneak path;
- It allows advanced target users to redefine memory cell properties by providing a customization interface.

NVSim is validated against several industry prototype chips within the error range of 30 %. In addition, we show how to use this model to facilitate the architecture-level performance, energy, and area analysis for applications that adopt the emerging NVM technologies.

2.2 Background of Non-volatile Memory

In this section, we first review the technology background of four types of NVMs modeled in NVSim, which are STT-RAM, PCRAM, ReRAM, and legacy NAND flash.

2.2.1 NVM Physical Mechanisms and Write Operations

Different NVM technologies have their particular storage mechanisms and corresponding write methods.

2.2.1.1 NAND Flash

The physical mechanism of the flash memory is to store bits in the floating gate and control the gate threshold voltage. The series bit cell string of NAND flash, as shown in Fig. 2.1a, eliminates contacts between the cells and approaches the minimum cell

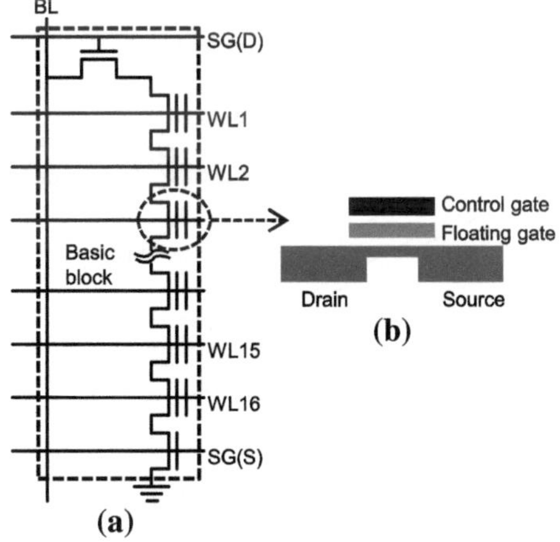

Fig. 2.1 The basic string block of NAND flash and the conceptual view of floating gate flash memory cell (*BL* bit line, *WL* word line, *SG* select gate)

size of $4F^2$ for low-cost manufacturing. The small cell size, low cost, and strong application demands make the NAND flash dominate the traditional non-volatile memory market. Figure 2.1b shows that a flash memory cell consists of a floating gate and a control gate aligned vertically. The flash memory cell modifies its threshold voltage V_T by adding electrons to or subtracting electrons from the isolated floating gate.

NAND flash usually charges or discharges the floating gate by using Fowler–Nordheim (FN) tunneling or hot-carrier injection (HCI). A program operation adds tunneling charges to the floating gate and the threshold voltage becomes negative, while an erase operation subtracts charges and the threshold voltage returns positive.

2.2.1.2 Spin-Torque Transfer RAM

Spin-torque transfer RAM (STT-RAM) uses magnetic tunnel junction (MTJ) as the memory storage and leverages the difference in magnetic directions to represent the memory bit. As shown in Fig. 2.2, MTJ contains two ferromagnetic layers. One ferromagnetic layer has fixed magnetization direction and it is called the reference layer, while the other layer has a free magnetization direction that can be changed by passing a write current and it is called the free layer. The relative magnetization direction of two ferromagnetic layers determines the resistance of MTJ. If two ferromagnetic layers have the same directions, the resistance of MTJ is low, indicating a "1" state; if two layers have different directions, the resistance of MTJ is high, indicating a "0" state.

As shown in Fig. 2.2, when writing "0" state into STT-RAM cells (RESET operation), positive voltage difference is established between SL and BL; when writing "1" state (SET operation), vice versa. The current amplitude required to reverse the direction of the free ferromagnetic layer is determined by the size and aspect ratio of MTJ and the write pulse duration.

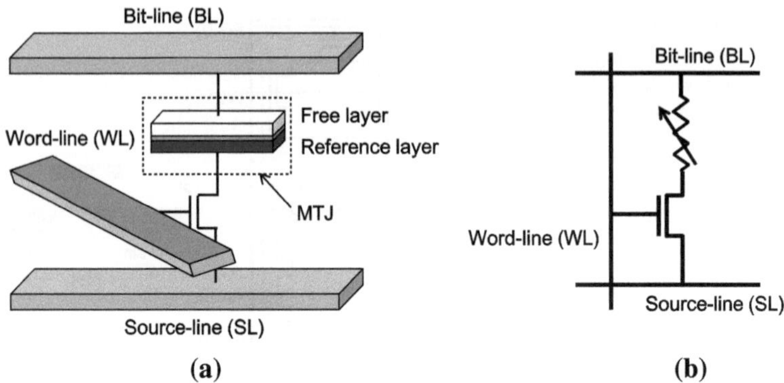

Fig. 2.2 Demonstration of a MRAM cell: **a** structural view; **b** schematic view (*BL* bit line, *WL* word line, *SL* source line)

2.2.1.3 Phase Change RAM

Phase change RAM (PCRAM) uses chalcogenide material (e.g., GST) to store information. The chalcogenide materials can be switched between a crystalline phase (SET state) and an amorphous phase (RESET state) with the application of heat. The crystalline phase shows low resistivity, while the amorphous phase is characterized by high resistivity. Figure 2.3 shows an example of a MOS-accessed PCRAM cell.

The SET operation crystallizes GST by heating it above its crystallization temperature, and the RESET operation melt-quenches GST to make the material amorphous as illustrated in Fig. 2.4. The temperature is controlled by passing a specific electrical current profile and generating the required Joule heat. High-power pulses are required for the RESET operation to heat the memory cell above the GST melting temperature. In contrast, moderate power but longer duration pulses are required for the SET operation to heat the cell above the GST crystallization temperature but below the melting temperature [33].

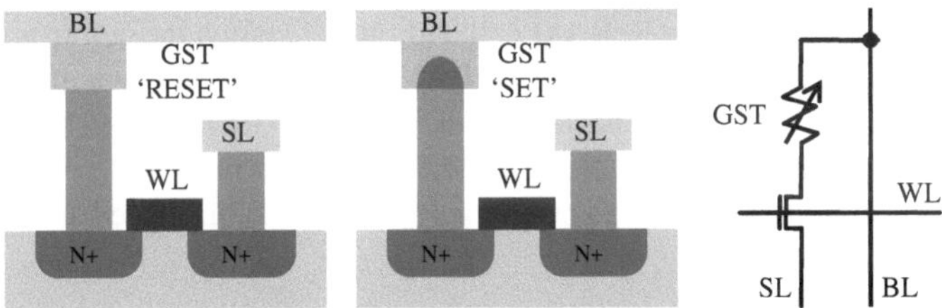

Fig. 2.3 The schematic view of a PCRAM cell with NMOS access transistor (*BL* bit line, *WL* word line, *SL* source line)

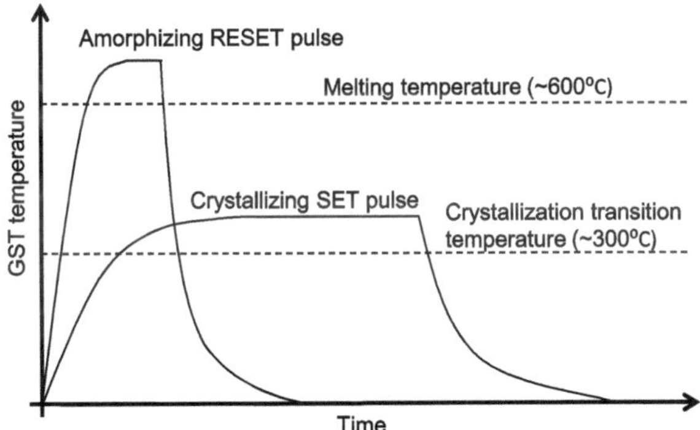

Fig. 2.4 The temperature–time relationship during SET and RESET operations

2.2.1.4 Resistive RAM

Although many non-volatile memory technologies (e.g., aforementioned STT-RAM and PCRAM) are based on electrically induced resistive switching effects, we define resistive RAM (ReRAM) as the one that involves electro- and thermochemical effects in the resistance change of a metal/oxide/metal system. In addition, we confine our definition to bipolar ReRAM. Figure 2.5 illustrates the general concept for the ReRAM working mechanism. An ReRAM cell consists of a metal oxide layer (e.g., Ti [45], Ta [42], and Hf [4]) sandwiched by two metal (e.g., Pt [45]) electrodes. The electronic behavior of metal/oxide interfaces depends on the oxygen vacancy concentration of the metal oxide layer. Typically, the metal/oxide interface shows Ohmic behavior in the case of very high doping and rectifying in the case of low doping [45]. In Fig. 2.5, the TiO_x region is semi-insulating indicating lower oxygen vacancy concentration, while the TiO_{2-x} is conductive indicating higher concentration.

The oxygen vacancy in metal oxide is n-type dopant, whose draft under the electric field can cause the change in doping profiles. Thus, applying electronic current can modulate the IV curve of the ReRAM cell and further switch the cell from one state to the other state. Usually, for bipolar ReRAM, the cell can be switched ON (SET operation) only by applying a negative bias and OFF (RESET operation) only by applying the opposite bias [45]. Several ReRAM prototypes [5, 22, 35] have been demonstrated and show promising properties on fast switching speed and low energy consumption.

2.2.2 *Read Operations*

The read operations of these NVM technologies are almost the same. Since the NVM memory cell has different resistance in ON and OFF states, the read operation can be accomplished either by applying a small voltage on the bit line and sensing the

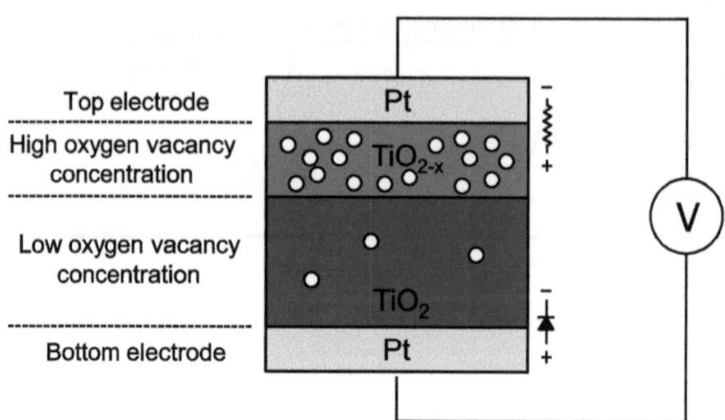

Fig. 2.5 The working mechanism of ReRAM cells

current that passes through the memory cell or by injecting a small current into the bit line and sensing the voltage across the memory cell. Instead of SRAM that generates complement read signals from each cell, NVM usually has a group of dummy cells to generate the reference current or reference voltage. The generated current (or voltage) from the to-be-read cell is then compared to the reference current (or voltage) by using sense amplifiers. Various types of sense amplifiers are modeled in NVSim as we discuss in Sect. 2.5.2.

2.2.3 Write Endurance Issue

Write endurance is the number of times that an NVM cell can be overwritten. Among all the NVM technologies modeled in NVSim, only STT-RAM does not suffer from the write endurance issue. NAND flash, PCRAM, and ReRAM all have limited write endurance, which is the number of times that a memory cell can be overwritten. NAND flash only has write endurance of 10^5–10^6. The PCRAM endurance is now in the range between 10^5 and 10^9 [1, 21, 32]. ReRAM research currently shows endurance numbers in the range between 10^5 and 10^{10} [20, 24]. A projected plan by ITRS for 2024 for emerging NVM, i.e., PCRAM and ReRAM, highlights endurance in the order of 10^{15} or more write cycles [14] . In NVSim, the write endurance limit is not modeled since NVSim is a circuit-level modeling tool.

2.2.4 Retention Time Issue

Retention time is the time that data can be retained in NVM cells. Typically, NVM technologies require retention time of higher than 10 years. However, in some cases, such a high retention time is not necessary. For example, Smullen et al. [36] relaxed the retention time requirement to improve the timing and energy profile of STT-RAMs. Since the trade-off between NVM retention time and other NVM parameters (e.g., the duration and amplitude of write pulses) is on the device level, as a circuit-level tool, NVSim does not model this trade-off directly but instead takes different sets of NVM parameters with various retention time as the device-level input.

2.2.5 MOS-Accessed Structure Versus Cross-Point Structure

Some NVM technologies (for example, PCRAM [18] and ReRAM [3, 18, 20]) have the capability of building cross-point memory arrays without access devices. Conventionally, in the MOS-accessed structure, memory cell arrays are isolated by MOS access devices and the cell size is dominated by the large MOS access device that is necessary to drive enough write current, even though the NVM cell itself is much smaller. However, taking advantage of the cell nonlinearity, an NVM array can be accessed without any extra access devices. The removal of MOS access devices leads to a memory cell size of only $4F^2$, where F is the process feature size. Unfortunately, the cross-point structure also brings extra peripheral circuitry design challenges and a trade-off between performance, energy, and area is always necessary as discussed in our previous work [44]. NVSim models both the MOS-accessed and the cross-point structures, and the modeling methodology is described in the following sections.

2.3 NVSim Framework

The framework of NVSim is modified from CACTI [38, 39]. We add several new features, such as more flexible data activation modes and alternative bank organizations.

2.3.1 Device Model

NVSim uses device data from ITRS report [14] and the MASTAR tool [15] to obtain the process parameters. NVSim covers the process nodes from 180, 120, 90, 65, 45, 32, to 22 nm and supports 3 transistor types, which are high performance (HP), low operating power (LOP), and low standby power (LSTP).

2.3.2 Array Organization

Figure 2.6 shows the array organization. There are 3 hierarchy levels in such organization, which are *bank*, *mat*, and *subarray*. Basically, the descriptions of these levels are as follows:

- *Bank* is the top-level structure modeled in NVSim. One non-volatile memory chip can have multiple banks. The bank is a fully functional memory unit, and it can be operated independently. In each bank, multiple mats are connected together in either H-tree or bus-like manner
- *Mat* is the building block of bank. Multiple mats in a bank operate simultaneously to fulfill a memory operation. Each mat consists of multiple subarrays and one predecoder block
- *Subarray* is the elementary structure modeled in NVSim. Every subarray contains peripheral circuitry including row decoders, column multiplexers, and output drivers.

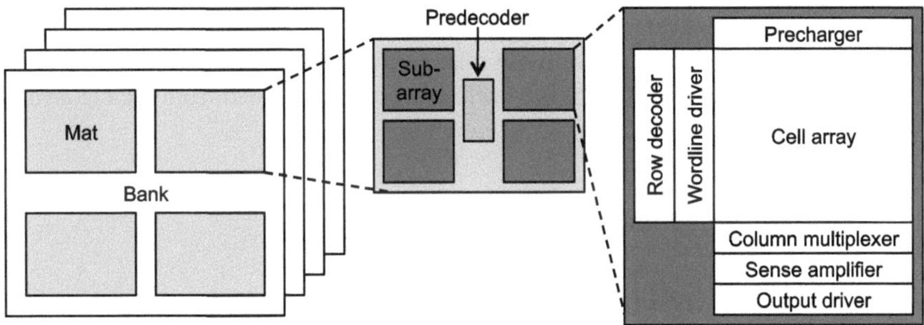

Fig. 2.6 The memory array organization modeled in NVSim: A hierarchical memory organization includes banks, mats, and subarrays with decoders, multiplexers, sense amplifiers, and output drivers

Conventionally, sense amplifiers are integrated on the subarray level as modeled in CACTI [38, 39]. However, in NVSim model, sense amplifiers can be placed either on the subarray level or on the mat level.

2.3.3 Memory Bank Type

For practical memory designs, memory cells are grouped together to form memory modules of different types. For instance,

- The main memory is a typical random access memory (RAM), which takes the address of data as input and returns the content of data;
- The set-associative cache contains two separate RAMs (data array and tag array) and can return the data if there is a cache hit by the given set address and tag;
- The fully associative cache usually contains a content-addressable memory (CAM).

To cover all the possible memory designs, we model 5 types of memory banks in NVSim: one for RAM, one for CAM, and three for set-associate caches with different access manners. The functionalities of these 5 types of memory banks are listed as follows:

1. *RAM*: Output the data content at the I/O interface given the data address
2. *CAM*: Output the data address at the I/O interface given the data content if there is a hit
3. *Cache with normal access*: Start to access the cache data array and tag array at the same time; the data content is temporarily buffered in each mat; if there is a hit, the cache hit signal generated from the tag array is routed to the proper mats, and the content of the desired cache line is output to the I/O interface
4. *Cache with sequential access*: Access the cache tag array at first; if there is a hit, then access the cache data array with the set address and the tag hit information and finally output the desired cache line to the I/O interface

5. *Cache with fast access*: Access the cache data array and tag array simultaneously; read the entire set content from the mats to the I/O interface; selectively output the desired cache line if there is a cache hit signal generated from the tag array.

2.3.4 Activation Mode

We model the array organization and the data activation modes using eight parameters, which are

- N_{MR}: number of rows of mat arrays in each bank;
- N_{MC}: number of columns of mat arrays in each bank;
- N_{AMR}: number of active rows of mat arrays during data accessing;
- N_{AMC}: number of active columns of mat arrays during data accessing;
- N_{SR}: number of rows of subarrays in each mat;
- N_{SC}: number of columns of subarrays in each mat;
- N_{ASR}: number of active rows of subarrays during data accessing;
- and N_{ASC}: number of active columns of subarrays during data accessing.

The values of these parameters are all constrained to be power of two. N_{MR} and N_{MC} define the number of mats in a bank, and N_{SR} and N_{SC} define the number of subarrays in a mat. N_{AMR}, N_{AMC}, N_{ASR}, and N_{ASC} define the activation patterns, and they can take any values smaller than N_{MR}, N_{MC}, N_{SR}, and N_{SC}, respectively. On the contrary, the limitation of array organization and data activation pattern in CACTI is caused by several constraints on these parameters such as $N_{\text{AMR}} = 1$, $N_{\text{AMC}} = N_{\text{MC}}$, and $N_{\text{SR}} = N_{\text{SC}} = N_{\text{ASR}} = N_{\text{ASC}} = 2$.

NVSim has these flexible activation patterns and is able to model sophisticated memory accessing techniques, such as single subarray activation [41].

2.3.5 Routing to Mats

In order to first route the data and address signals from the I/O port to the edge of memory mats and from mat to the edges of memory subarrays, we divided all the interconnect wires into three categories: *Address Wires*, *Broadcast Data Wires*, and *Distributed Data Wires*. Depending on the memory module types and the activation modes, the initial number of wires in each group is assigned according to the rules listed in Table 2.1. We use the terminology block to refer to the memory words in RAM and CAM designs and the cache lines in cache designs. In Table 2.1, N_{block} is the number of blocks, W_{block} is the block size, and A is the associativity in cache designs. The number of *Broadcast Data Wires* is always kept unchanged, the number of *Distributed Data Wires* is cut by half at each routing point where data are merged, and the number of *Address Wires* is subtracted by one at each routing point where data are multiplexed.

Table 2.1 The initial number of wires in each routing group

			N_{AW}	N_{BW}	N_{DW}
Random access memory (RAM)			$\log_2 N_{block}$	0	W_{block}
Content-addressable memory (CAM)				W_{block}	0
Cache	Normal access	Data array	$\log_2 (N_{block}/A)$	$\log_2 A$	W_{block}
		Tag array	$\log_2 (N_{block}/A)$	W_{block}	A
	Sequential access	Data array	$\log_2 N_{block}$	0	W_{block}
		Tag array	$\log_2 (N_{block}/A)$	W_{block}	A
	Fast access	Data array	$\log_2 (N_{block}/A)$	0	$W_{block} A$
		Tag array	$\log_2 (N_{block}/A)$	W_{block}	A

N_{AW} Number of address wires
N_{BW} Number of broadcast data wires
N_{DW} Number of distributed data wires

We use the case of the cache bank with normal access to demonstrate how the wires are routed from the I/O port to the edges of the mats. For simplicity, we suppose the data array and the tag array are two separate modules. While the data and tag arrays usually have different mat organizations in practice, we use the same 4×4 mat organization for the demonstration purpose as shown in Figs. 2.7 and 2.8. The total 16 mats are positioned in a 4×4 formation and connected by a 4-level H-tree. Therefore, N_{MR} and N_{MC} are 4. As an example, we use the activation mode in which two rows and two columns of the mat array are activated for each data access, and the activation groups are Mat $\{0, 2, 8, 10\}$, Mat $\{1, 3, 9, 11\}$, Mat $\{4, 6, 12, 14\}$, and Mat $\{5, 7, 13, 15\}$. Thereby, N_{AMR} and N_{AMC} are 2. In addition, we set the cache line size (block size) to 64 B, the cache associativity to $A = 8$, and the cache bank capacity to 1 MB, so that the number of cache lines (blocks) is $N_{block} = 8M/512 = 16{,}384$, the block size in the data array is $W_{block,data} = 512$, and the block size in the tag array is $W_{block,tag} = 16$ (assuming 32-bit addressing and labeling dirty block with one bit).

According to Table 2.1, the initial number of address wires (N_{AW}) is $\log_2 N_{block}/A = 11$ for both data and tag arrays. For data array, the initial number of broadcast data wires ($N_{BW,data}$) is $\log_2 A = 3$, which is used to transit the tag hit signals from the tag array to the corresponding mats in the data array; the initial number of distributed data wires ($N_{DW,data}$) is $W_{block,data} = 512$, which is used to output the desired cache line from the mats to the I/O port. For tag array, the broadcast data wire ($N_{BW,tag}$) is $W_{block,tag} = 16$, which is sent from the I/O port to each of the mat in the tag array; the initial number of distributed data wires ($N_{DW,tag}$) is $A = 8$, which is used to collect the tag hit signals from each mat to the I/O port and then send to the data array after a 8-to-3 encoding process.

From the I/O port to the edges of the mats, the numbers of wires in the three categories are changed as follows and demonstrated in Figs. 2.7 and 2.8, respectively.

1. At node A, the activated mats are distributed in both the upper and the bottom parts, so node A is a *merging* node. As per the routing rule, the address wires and broadcast data wires remain the same, but the distributed data wires are

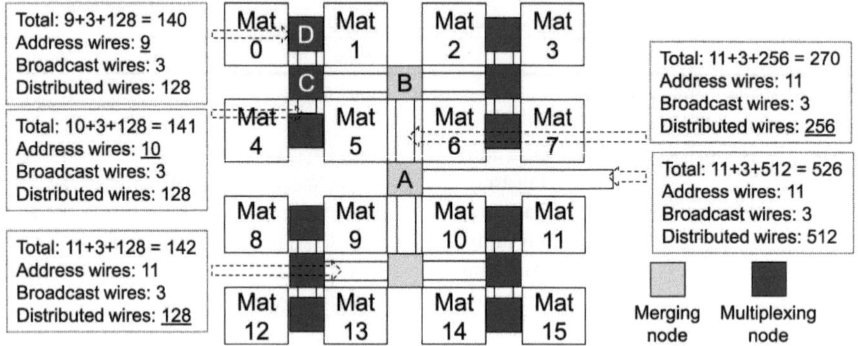

Fig. 2.7 The example of the wire routing in a 4 × 4 mat organization for the *data array* of a 8-way 1 MB cache with 64 B cache lines

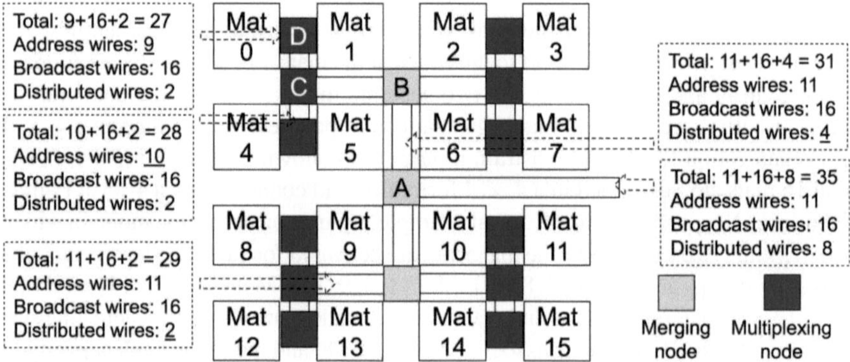

Fig. 2.8 The example of the wire routing in a 4 × 4 mat organization for the *tag array* of a 8-way 1 MB cache with 64 B cache lines

cut into half. Thus, the wire segment between node A and B have $N_{\text{AW}} = 11$, $N_{\text{BW,data}} = 3$, $N_{\text{DW,data}} = 256$, $N_{\text{BW,tag}} = 16$, and $N_{\text{DW,tag}} = 4$

2. Node B is again a *merging* node. Thus, the wire segment between node B and C have $N_{\text{AW}} = 11$, $N_{\text{BW,data}} = 3$, $N_{\text{DW,data}} = 128$, $N_{\text{BW,tag}} = 16$, and $N_{\text{DW,tag}} = 2$
3. At node C, the activated mats are allocated only in one side, either from Mat 0/1 or from Mat 4/5, so Node C is a *multiplexing* node. As per the routing rule, the distributed data wires and broadcast data wires remain the same, but the address wires are decremented by 1. Thus, the wire segment between node C and node D have $N_{\text{AW}} = 10$, $N_{\text{BW,data}} = 3$, $N_{\text{DW,data}} = 128$, $N_{\text{BW,tag}} = 16$, and $N_{\text{DW,tag}} = 2$
4. Finally, node D is another *multiplexing* node. Thus, the wire segments at the mat edges have $N_{\text{AW}} = 9$, $N_{\text{BW,data}} = 3$, $N_{\text{DW,data}} = 128$, $N_{\text{BW,tag}} = 16$, and $N_{\text{DW,tag}} = 2$.

Thereby, each mat in the data array takes the input of a 9-bit set address and a 3-bit tag hit signals (which can be treated as the block address in a 8-way associative set), and it generates the output of a 128-bit data. A group of 4 data mats provide the desired output of a 512-bit (64 B) cache line, and four such groups cover the entire 11-bit set address space. On the other hand, each mat in the tag array takes the input of a 9-bit set address and a 16-bit tag, and it generates a 2-bit hit signals (01 or 10 for hit and 00 for miss). A group of 4 tag mats concatenate their hit signals and provide the information whether a 16-bit tag hits in a 8-way associated cache with a 9-bit address space, and four such groups extend the address space from 9 bit to the desired 11 bit.

Other configurations in Table 2.1 can be explained in the similar manner.

2.3.6 Routing to Subarrays

The interconnect wires from mat to the edges of memory subarrays are routed using the same H-tree organization as shown in Fig. 2.9, and its routing strategy is the same wire partitioning rule described in Sect. 2.3.5. However, NVSim provides an option of building mat using a bus-like routing organization as illustrated in Fig. 2.10. The wire partitioning rule described in Sect. 2.3.5 can also be applied to the bus-like organization with a few extensions. For example, a *multiplexing* node with a fanout of N decrements the number of address wires by $\log_2 N$ instead of 1; a *merging* node with a fanout of N divides the number of distributed data wires by N instead of 2.

Furthermore, the default setting of including sense amplifiers in each subarray can cause a dominant portion of the total array area. As a result, for high-density memory

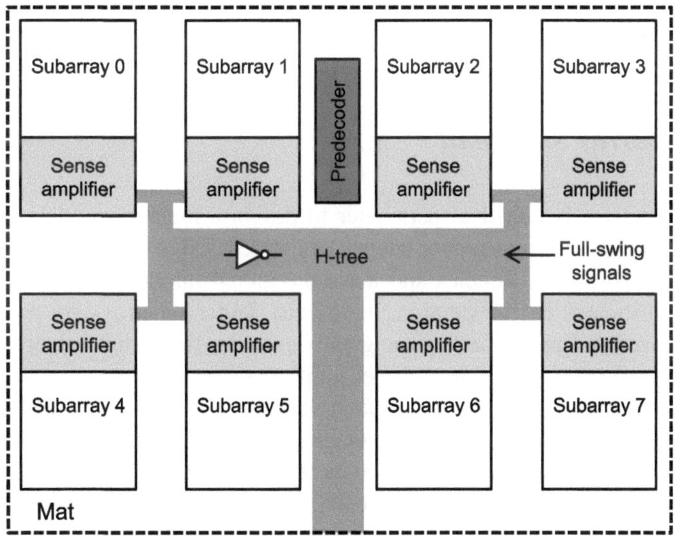

Fig. 2.9 An example of mat using internal sensing and H-tree routing

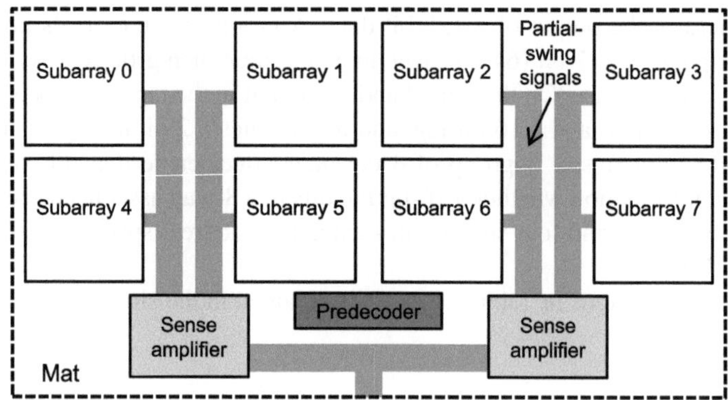

Fig. 2.10 An example of mat using external sensing and bus-like routing

module designs, NVSim provides an option of moving the sense amplifiers out of the subarray and using external sensing. In addition, a bus-like routing organization is designed to associate with the external sensing scheme.

Figure 2.9 shows a common mat using H-tree organization to connect all the sense amplifier-included subarrays together. In contrast, the new external sensing scheme is illustrated in Fig. 2.10. In this external sensing scheme, all the sense amplifiers are located at the mat level and the output signals from each sense amplifier-free subarray are partial swing. It is obvious that the external sensing scheme has much higher area efficiency compared to its internal sensing counterpart. However, as a penalty, sophisticated global interconnect technologies, such as repeater inserting, cannot be used in the external sensing scheme since all the global signals are partial swing before passing through the sense amplifiers.

2.3.7 Subarray Size Limit

The subarray size is a critical parameter to design a memory module. Basically, smaller subarrays are preferred for latency-optimized designs since they reduce local bit line and word line latencies and leave the global interconnects to be handled by the sophisticated H-tree solution. In contrast, larger subarrays are preferred for area-optimized designs since they can greatly amortize the peripheral circuitry area. However, the subarray size has its upper limit in practice.

For MOS-accessed subarray, the leakage current paths from unselected word lines are the main constraint to the bit line length. For cross-point subarray, the leakage current path issue is much more severe as there is no MOSFET in such subarray that can isolate selected and unselected cells [23]. The half-select cells in cross-point subarrays serve as current dividers in the selected row and columns, preventing the array size from growing unbounded since the available driving current is limited.

The minimum current that a column write driver should provide is determined by

$$I_{\text{driver}} = I_{\text{write}} + (N_r - 1) \times I(V_{\text{write}}/2) \tag{2.1}$$

where I_{write} and V_{write} are the current and voltage of either RESET or SET operation. Nonlinearity of memory cells is reflected by the fact that the current through cross-point memory cells is not directly proportional to the voltage applied on it, which means non-constant resistance of the memory cell. In NVSim, we define a nonlinearity coefficient, K_r, to quantify the current divider effect of the half-selected memory cells as follows:

$$K_r = \frac{R(V_{\text{write}}/2)}{R(V_{\text{write}})} \tag{2.2}$$

where $R(V_{\text{write}}/2)$ and $R(V_{\text{write}})$ are equivalent static resistance of cross-point memory cells biased at $V_{\text{write}}/2$ and V_{write}, respectively. Then, we derive the upper limit in a cross-point subarray size by

$$N_r = \left(\frac{I_{\text{driver}}}{I_{\text{write}}} - 1\right) \times 2 \times K_r + 1 \tag{2.3}$$

$$N_c = \left(\frac{I_{\text{driver}}}{I_{\text{write}}} - N_{\text{sc}}\right) \times 2 \times K_r + N_{\text{sc}} \tag{2.4}$$

where I_{driver} is the maximum driving current that the write driver attached to the selected row/column can provide and N_{sc} is the number of selected columns per row. Thus, N_r and N_c are the maximum numbers of rows and columns in a cross-point subarray.

As shown in Fig. 2.11, the maximum cross-point subarray size increases with larger current driving capability or larger nonlinearity coefficient.

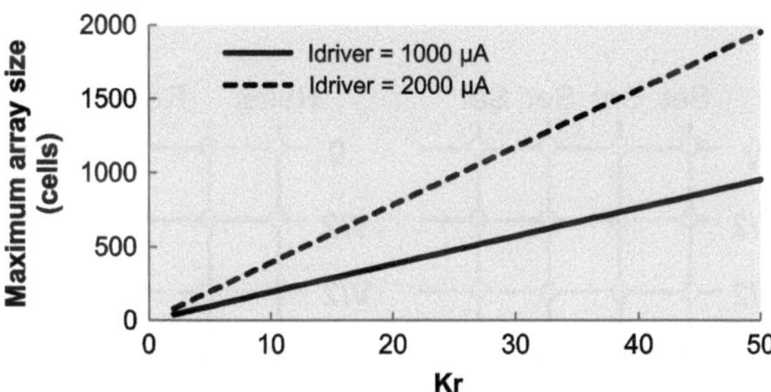

Fig. 2.11 Maximum subarray size versus nonlinearity and driving current

2.3.8 Two-Step Write in Cross-Point Subarrays

In cross-point structure, SET and RESET operations cannot be performed simultaneously. Thus, two steps of write operations are required in the cross-point structure when multiple cells are selected in a row.

In NVSim, we model two write methods for cross-point subarrays. The first one separates SET and RESET operations as Fig. 2.12 shows, and it is called SET-before-RESET. The second one erases all the cells in the selected row before the selective RESET operation as Fig. 2.13 shows, and it is called ERASE-before-RESET (EbR). Supposing the 4-bit word to write is "0101," we first write "x1x1" ("x" here means bias row and column of the corresponding cells at the same voltage to keep their original states) and then write "0x0x" in SET-before-RESET (SbR) method, or we first SET all the four cells and then write "0x0x" in ERASE-before-RESET method. The first method has smaller write latency since the erase operation can be performed before the arrival of the column selector signal, but it needs more write energy due to the redundant SET on the cells that are RESET back in the second step. Here, ERASE-before-RESET is chosen rather than ERASE-before-SET because SET operation usually consumes less energy than RESET operation does.

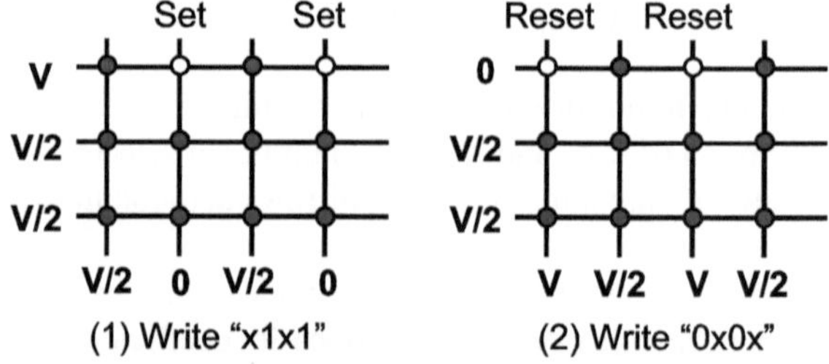

Fig. 2.12 Sequential write method: SET-before-RESET

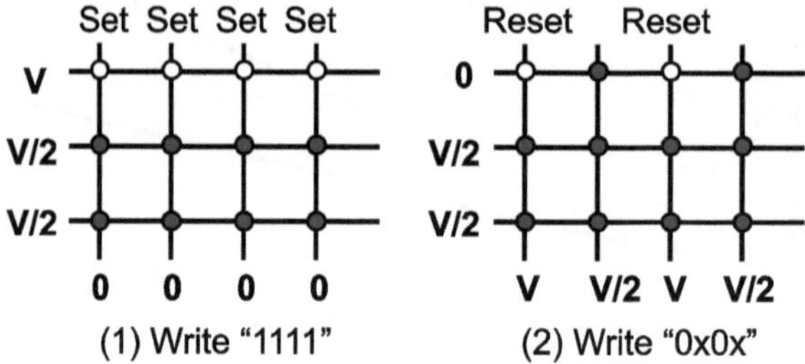

Fig. 2.13 Sequential write method: ERASE-before-RESET

2.4 Area Model

Since NVSim estimates the performance, energy, and area of non-volatile memory modules, the area model is an essential component of NVSim, especially given the facts that interconnect wires contribute a large portion of total access latency and access energy and the geometry of the module becomes highly important. In this section, we describe the NVSim area model from the memory cell level to the bank level in details.

2.4.1 Cell Area Estimation

Three types of memory cells are modeled in NVSim: MOS-accessed, cross-point, and NAND string.

2.4.1.1 MOS-Accessed Cell

MOS-accessed cell corresponds to the typical 1T1R (1-transistor-1-resistor) structure used by many NVM chips [1, 11, 13, 17, 19, 30, 40], in which an NMOS access device is connected in series with the non-volatile storage element (i.e., MTJ in STT-RAM, GST in PCRAM, and metal oxide in ReRAM) as shown in Fig. 2.14. Such an NMOS device turns on/off the access path to the storage element by tuning the voltage applied to its gate. The MOS-accessed cell usually has the best isolation among neighboring cells due to the property of MOSFET.

In MOS-accessed cells, the size of NMOS is bounded by the current needed by the write operation. The size of NMOS in each MOS-accessed cell needs to be sufficiently large so that the NMOS has the capability of driving enough write current.

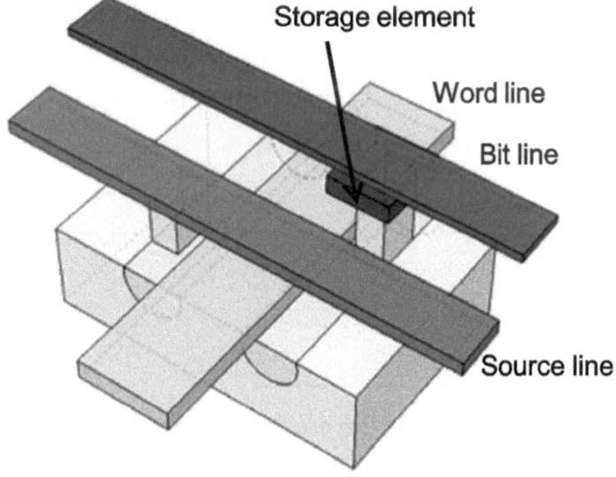

Fig. 2.14 Conceptual view of a MOS-accessed cell (1T1R) and its connected word line, bit line, and source line

The driving current of NMOS, I_{DS}, can be first-order estimated as follows[1],

$$I_{DS} = K\frac{W}{L}\left[(V_{GS} - V_{TH})\,V_{DS} - \frac{V_{DS}^2}{2}\right] \tag{2.5}$$

if NMOS is working at the linear region or calculated by

$$I_{DS} = \frac{K}{2}\frac{W}{L}(V_{GS} - V_{TH})^2\,(1 + \lambda V_{DS}) \tag{2.6}$$

if NMOS is working at the saturation region. Hence, no matter in which region NMOS is working, the current driving ability of NMOS is proportional to its width-to-length (W/L) ratio,[2] which determines the NMOS size. To achieve high cell density, we model the MOS-accessed cell area by referring to DRAM design rules [9]. As a result, the cell size of a MOS-accessed cell in NVSim is calculated as follows:

$$\text{Area}_{\text{cell,MOS-accessed}} = 3\,(W/L + 1)(F^2) \tag{2.7}$$

in which the width-to-length ratio (W/L) is determined by Eq. 2.5 or 2.6 and the required write current is configured as one of the input values of NVSim. In NVSim, we also allow advanced users to override this cell size calculation by directly importing the user-defined cell size.

2.4.1.2 Cross-Point Cell

Cross-point cell corresponds to the 1D1R (1-diode-1-resistor) [21, 22, 31, 46, 47] or 0T1R (0-transistor-1-resistor) [3, 18, 20] structures used by several high-density NVM chips recently. Figure 2.15 shows a cross-point array without diodes (i.e., 0T1R structure). For 1D1R structure, a diode is inserted between the word line and the storage element. Such cells either rely on the one-way connectivity of diode (i.e., 1D1R) or leverage materials' nonlinearity (i.e., 0T1R) to control the memory access path. As illustrated in Fig. 2.15, the widths of word lines and bit lines can be the minimal value of 1F and the spacing in each direction is also 1F; thus, the cell size of each cross-point cell is

$$\text{Area}_{\text{cell,cross-point}} = 4(F^2) \tag{2.8}$$

---

[1] Equations 2.5 and 2.6 are for long-channel drift/diffusion devices, and the equations are subjected to change depending on the technology, though the proportional relationship between the current and W/L still holds for very advanced technologies.

[2] Usually, the transitor length (L) is fixed as the minimal feature size, and the transistor width (W) is adjustable.

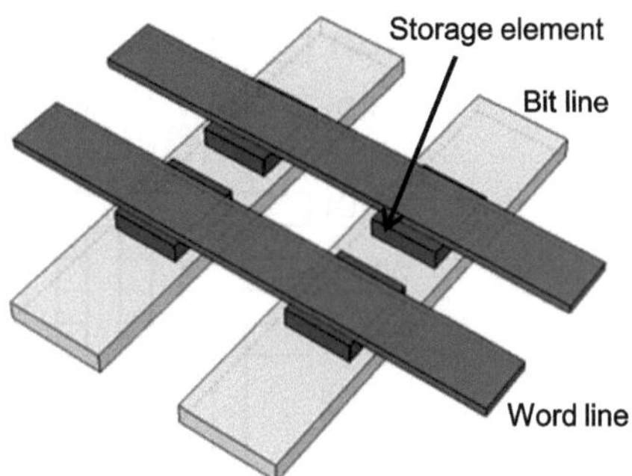

Fig. 2.15 Conceptual view of a cross-point cell array without diode (0T1R) and its connected word lines and bit lines

Compared to MOS-accessed cells, cross-point cells have worse cell isolation but provide a way of building high-density memory chip because they have much smaller cell sizes. In some cases, the cross-point cell size is constrained by the diode due to limited current density, and NVSim allows the user to override the default $4F^2$ setting.

2.4.1.3 NAND String Cell

NAND string cells are particularly modeled for NAND flash. In NAND string cells, a group of floating gates are connected in series and two ordinary gates with contacts are added at the string ends as shown in Fig. 2.16. Since the area of the floating gates can be minimized to 2 × 2F, the total area of a NAND string cell is

$$\text{Area}_{\text{cell,NANDstring}} = 2\,(2N + 5)\,(F^2) \tag{2.9}$$

where N is the number of floating gates in a string and we assume that the addition of two gates and two contacts causes 5F in the total string length.

2.4.2 Peripheral Circuitry Area Estimation

Besides the area occupied by memory cells, there is a large portion of memory chip area that is contributed to the peripheral circuitry. In NVSim, we have peripheral circuitry components such as row decoders, prechargers, and column multiplexers on the subarray level, predecoders on the mat level, and sense amplifiers and write drivers on either the subarray level or mat level, depending on whether internal or external data sensing scheme is used. In addition, on every level, interconnect wires might occupy extra silicon area if the wires are relayed using repeaters.

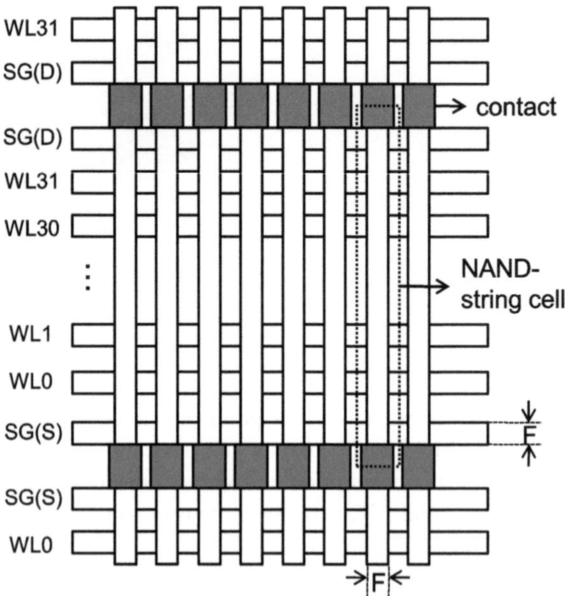

Fig. 2.16 The layout of the NAND string cell modeled in NVSim

In order to estimate the area of each peripheral circuitry component, we delve into the actual gate-level logic design as similar to CACTI [39]. However, in NVSim, we size transistors in a more generalized way than CACTI does.

The sizing philosophy of CACTI is to use logical effort [37] to size the circuits for minimum delay. NVSim's goal is to estimate the properties of NVM chips of a broad range, and these chips might be optimized for density or energy consumption instead of minimum delay; thus, we provide optional sizing methods rather than only applying logical effort. In addition, for some peripheral circuitry in NVM chips, the size of some transistors is determined by their required driving current instead of their capacitive load, and this violates the basic rules of using logical effort.

Therefore, we offer three transistor sizing choices in the area model of NVSim: one optimizing latency, one optimizing area, while another balancing latency and area. An example is illustrated in Fig. 2.17, demonstrating the different sizing methods when an output buffer with 4,096 times the capacitance of a minimum-sized inverter is to be designed. In a latency-optimized buffer design, the number of stages and all of the inverter sizing in the inverter chain are calculated by logical effort to achieve minimum delay (30 units) while paying a huge area penalty (1,365 units). In an area-optimized buffer design, there are only two stages of inverters, and the size of the last stage is determined by the minimum driving current requirement. This type of buffer has the minimum area (65 units), but is much slower than the latency-optimized buffer. The balanced option determines the size of last-stage inverter by its driving current requirement and calculates the size of the other inverters by logical effort. This results in a balanced delay and area metric.

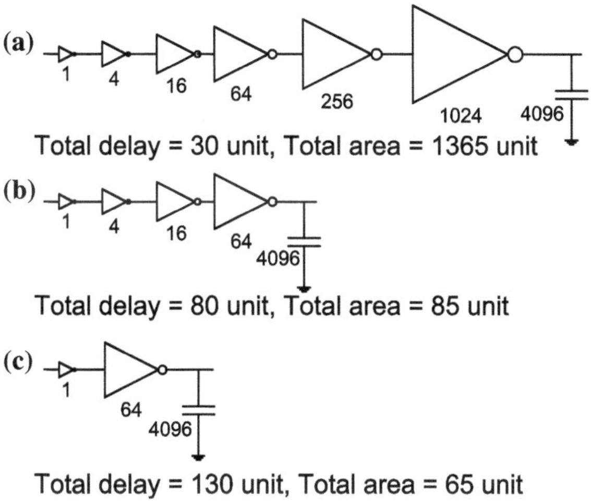

(a) 1 4 16 64 256 1024 4096

Total delay = 30 unit, Total area = 1365 unit

(b) 1 4 16 64 4096

Total delay = 80 unit, Total area = 85 unit

(c) 1 64 4096

Total delay = 130 unit, Total area = 65 unit

Fig. 2.17 Buffer designs with different transistor sizing: **a** latency-optimized; **b** balanced; **c** area-optimized

2.5 Timing and Power Models

As an analytical modeling tool, *NVSim* uses RC analysis for timing and power. In this section, we describe how resistances and capacitances are estimated in NVSim and how they are combined to calculate the delay and power consumption.

2.5.1 Generic Timing and Power Estimation

In NVSim, we consider the wire resistance and wire capacitance from interconnects, turn-on resistance, switching resistance, gate and drain capacitances from transistors, and equivalent resistance and capacitance from memory storage elements (e.g., MTJ in STT-RAM and GST in PCRAM). The methods of estimating wire and parasitic resistances and capacitances are modified from the previous versions of CACTI [39, 43] by several enhancements. The enhancements include updating the transistor models by latest ITRS report [14], considering the thermal impact on wire resistance calculation, adding drain-to-channel capacitance in the drain capacitance calculation, and so on. We build a lookup table to model the equivalent resistance and capacitance of memory storage elements since they are the properties of certain non-volatile memory technology. Considering NVSim is a system-level estimation tool, we only model the static behavior of the storage elements and record the equivalent resistances and capacitances of RESET and SET states (i.e., R_{RESET}, R_{SET}, C_{RESET}, C_{SET}).[3]

After calculating the resistances and capacitances of nodes, the delay of each logic component is calculated by using a simplified version of Horowitz's timing model [12] as follows:

$$\text{Delay} = \tau\sqrt{\left(\ln\frac{1}{2}\right)^2 + \alpha\beta} \qquad (2.10)$$

where α is the slope of the input, $\beta = 1/g_m R$ is the normalized input transconductance by the output resistance, and $\tau = \text{RC}$ is the RC time constant.

The dynamic energy and leakage power consumptions can be modeled as

$$\text{Energy}_{\text{dynamic}} = C V_{\text{DD}}^2 \qquad (2.11)$$

$$\text{Power}_{\text{leakage}} = V_{\text{DD}} I_{\text{leak}} \qquad (2.12)$$

where we model both gate leakage and subthreshold leakage currents in I_{leak}.

The overall memory access latency and energy consumption are estimated by combining all the timing and power values of circuit components together. NVSim follows the same methodology that CACTI [39] uses with minor modifications.

2.5.2 Data Sensing Models

Unlike other peripheral circuitries, the sense amplifier is an analog design instead of a logic design. Thus, in NVSim, we develop a separate timing model for the data sensing schemes. Different sensing schemes have their impacts on the trade-off between performance, energy, and area. In NVSim, we consider three types of sensing schemes: current sensing (CS), current-in voltage sensing (CIVS), and voltage divider sensing (VDS).

In the current sensing scheme as shown in Fig. 2.18, the state of memory cell (STT-RAM, PCRAM, or ReRAM) is read out by measuring the resulting current through the selected memory cell when a read voltage is applied: The current on the bit line is compared to the reference current generated by reference cells, the current difference is amplified by current-mode sense amplifiers, and they are eventually converted to voltage signals.

Figure 2.19 demonstrates an alternative sensing method by applying a current source on the selected memory cell and sensing the voltage via the voltage-mode sense amplifier.

The voltage divider sensing scheme is presented by introducing a resistor (R_x) in series with the memory cell as illustrated in Fig. 2.20. The resistance value is selected to achieve the maximum read sensing margin, and it is calculated as follows:

[3] One of the exceptions is that NVSim records the detailed IV curves for cross-point ReRAM cells without diode because we need to leverage the nonlinearity of the storage element.

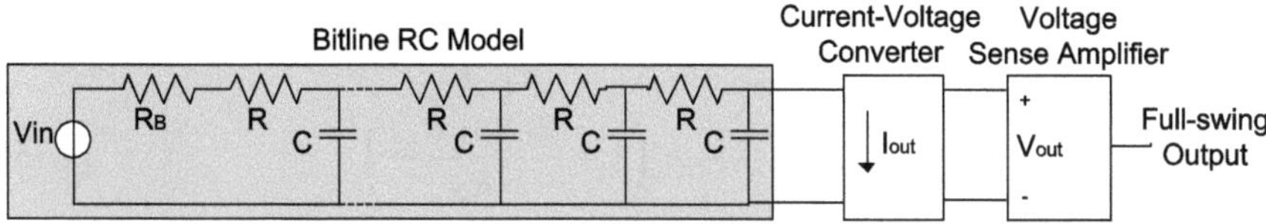

Fig. 2.18 Analysis model for current sensing scheme

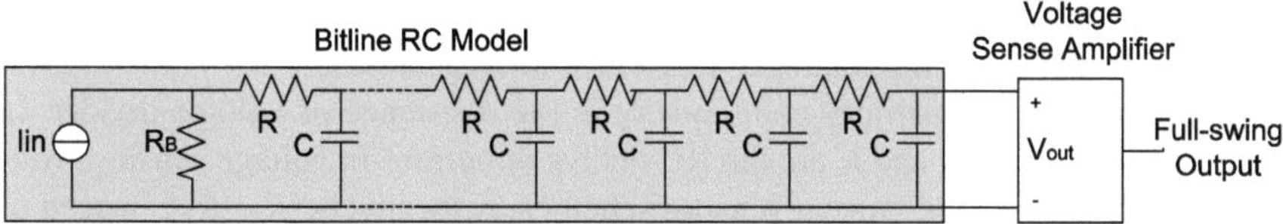

Fig. 2.19 Analysis model for current-in voltage sensing scheme

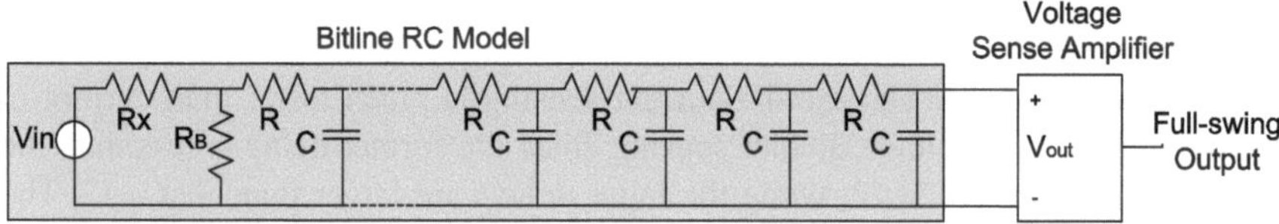

Fig. 2.20 Analysis model for voltage divider sensing scheme

$$R_x = \sqrt{R_{\mathrm{on}} \times R_{\mathrm{off}}} \tag{2.13}$$

where R_{on} and R_{off} are the equivalent resistance values of the memory cell in LRS and HRS, respectively.

2.5.2.1 Bitline RC Model

We model the bit line RC delay analytically for each sensing scheme. The most significant difference between the current-mode sensing and voltage-mode sensing is that the input resistance of ideal current-mode sensing is zero, while that of ideal voltage-mode sensing is infinite. And, the most significant difference between current-in voltage sensing and voltage divider sensing is that the internal resistance of an ideal current source is infinite, while the resistor R_x serving as a voltage divider can be treated as the internal resistance of a voltage source. Delays of current-in voltage sensing, voltage divider sensing, and current sensing are given by the following equations using Seevinck's delay expression [34]:

$$\delta t_v = \frac{R_T C_T}{2} \times \left(1 + \frac{2R_B}{R_T}\right) \tag{2.14}$$

$$\delta t_{vd} = \frac{R_T C_T}{2} \times \left(1 + \frac{2(R_B||R_x)}{R_T}\right) \tag{2.15}$$

$$\delta t_i = \frac{R_T C_T}{2} \times \left(\frac{R_B + \frac{R_T}{3}}{R_B + R_T}\right) \tag{2.16}$$

where R_T and C_T are the total line resistance and capacitance, R_B is the equivalent resistance of the memory cell, and R_x is the resistance of voltage divider. In these equations, t_v, t_{vd}, and t_i are the RC delays of current-in voltage sensing, voltage divider sensing, and current sensing schemes, respectively. $R_x||R_B$, instead of R_B, is used as the new effective pull-down resistance in Eq. 2.15 according to the transformation from a Thevenin equivalent to a Norton equivalent.

Equations 2.14 and 2.15 show that voltage divider sensing is faster than current-in voltage sensing with the extra cost of fabricating a large resistor. Comparing Eq. 2.16 with 2.14 and 2.15, we can see the current sensing is much faster than current-in voltage sensing and voltage divider sensing since the former delay is less than the intrinsic line delay $R_T C_T/2$, while the latter delays are larger than $R_T C_T/2$. The bit line delay analytical models are verified by comparing them with the HSPICE simulation results. As shown in Fig. 2.21, the RC delays derived by our analytical RC models are consistent with the HSPICE simulation results.

2.5.2.2 Current–Voltage Converter Model

As shown in Fig. 2.18, the current–voltage converter in our current-mode sensing scheme is actually the first-level sense amplifier, and the CACTI-modeled voltage sense amplifier is still kept in the bit line model as the final stage of the sensing scheme. The current–voltage converter senses the current difference $I_1 - I_2$, and then, it is converted into a voltage difference $V_1 - V_2$. The required voltage difference produced by current–voltage converter is set by default to 80 mV. Although this value is the minimum sensible voltage difference of the CACTI-modeled voltage sense amplifier, advanced user can override it for specific sense amplifier design. We refer to a previous current–voltage converter design [34], and the circuit schematic is shown in Fig. 2.22. This sensing scheme is similar to the hybrid I/O approach [28], which can achieve high-speed, robust sensing, and low power operation.

To avoid unnecessary calculation, the current–voltage converter is modeled by directly using the HSPICE-simulated values and building a look-up table of delay, dynamic energy, and leakage power (Table 2.2).

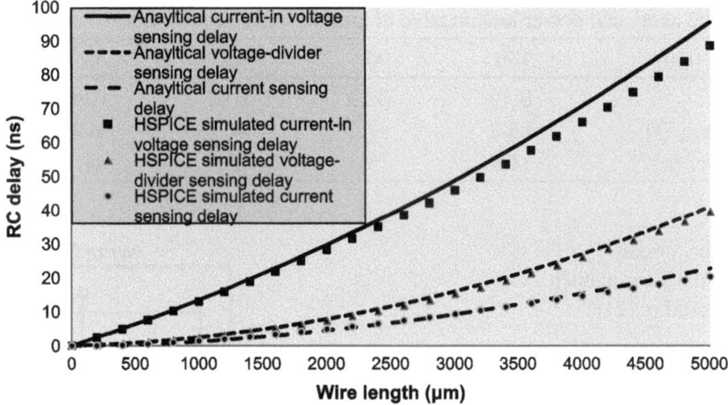

Fig. 2.21 Delay model verification of three sensing schemes comparing to HSPICE simulations

Fig. 2.22 The current–voltage converter modeled in NVSim

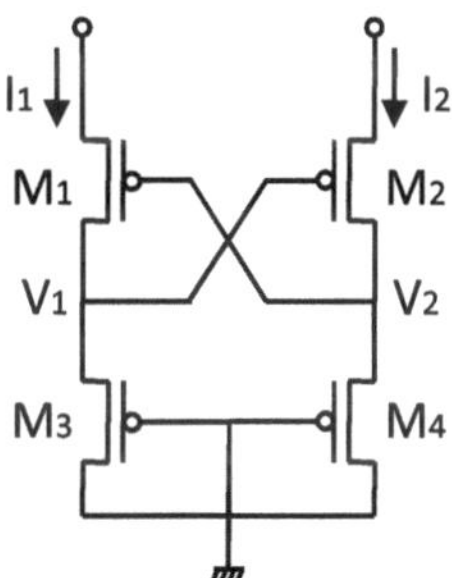

2.5.3 Cell Switching Model

Different NVM technologies have their specific switching mechanism. Usually, the switching phenomenon involves magnetoresistive, phase change, thermochemical, and electrochemical effects, and it cannot be estimated by RC analysis. Hence, the cell switching model in NVSim largely relies on the NVM cell definition. The predefined NVM cell switching properties include the SET/RESET pulse duration (i.e., t_{SET} and t_{RESET}) and SET/RESET current (i.e., I_{SET} and I_{RESET}) or voltage. NVSim does not model the dynamic behavior during the switching of the cell state, the switching latency (i.e., cell write latency) is directly the pulse duration and the switching energy (i.e., cell write energy) is estimated using Joule's first law that is

$$\begin{aligned}\text{Energy}_{\text{SET}} &= I_{\text{SET}}^2 R t_{\text{SET}} \\ \text{Energy}_{\text{RESET}} &= I_{\text{RESET}}^2 R t_{\text{RESET}}\end{aligned} \tag{2.17}$$

in which the resistance value R can be the equivalent resistance of the corresponding SET or RESET state (i.e., R_{SET} or R_{RESET}). However, for NVM technologies that have threshold switching phenomenon (e.g., PCRAM and ReRAM), the resistance

Table 2.2 The delay and power lookup table of current–voltage converter

Process node (nm)	130	90	65	45	32
Delay (ns)	0.49	0.53	0.62	0.80	1.07
Dynamic energy (fJ)	85.2	87.2	90.0	102.6	125.6
Leakage power (nW)	14.0	18.7	25.7	44.1	125.4

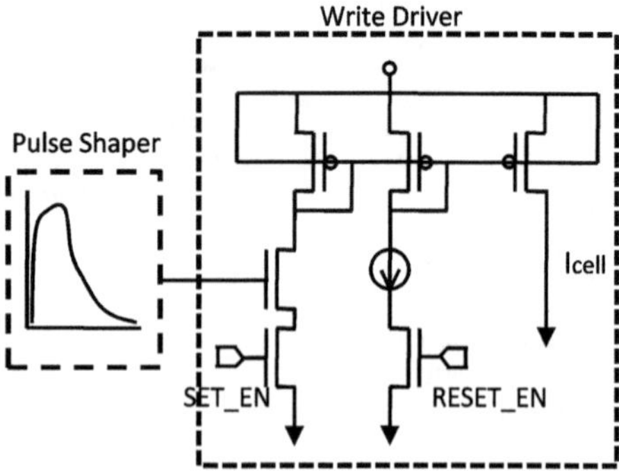

Fig. 2.23 The circuit schematic of the slow-quench pulse shaper used in [21]

value *R* always equals to the resistance of the low-resistance state. This is because when a voltage above a particular threshold is applied to these NVM cells in the high-resistance state, the resulting large electrical fields greatly increase the electrical conductivity [2].

2.6 Miscellaneous Circuitry

Some specialized circuitry is required for certain types of NVMs. For instance, some PCRAM chips need pulse shaper to reform accurate SET and RESET pulses, and NAND flash and some PCRAM chips need charge pump to generate the high-voltage power plane that is necessary for write operations.

2.6.1 Pulse Shaper

Some PCRAM needs specialized circuits to handle its RESET and SET operations. Specific pulse shapes are required to heat up the GST quickly and to cool it down gradually, especially for SET operations. This pulse shaping requirement is achieved by using a slow-quench pulse shaper. As shown in Fig. 2.23, the slow-quench pulse shaper is composed of an arbitrary slow-quench waveform generator and a write driver.

In NVSim, the delay impacts of the slow-quench shaper are neglected because they are already included in the RESET/SET calculation of the timing model. The energy impacts of the shaper are modeled by adding an energy efficiency during the RESET/SET operation, which we set the default value to 35 % [11], and it can be overridden by advanced user. The area of slow-quench shapers is modeled by measuring the die photographs [11, 21].

2.6.2 Charge Pump

The write operations of NAND flash and some PCRAM chips require voltage higher than the chip supply voltage. Therefore, a charge pump that uses capacitors as energy storage elements to create a higher voltage is necessary in a NAND flash chip design. In NVSim, we neglect the silicon area occupied by charge pump since the charge pump area can vary a lot depending on its underlying circuit design techniques, and the charge pump area is relatively small compared to the cell array area in a large-capacity NAND chip. However, we model the energy dissipated by charge pumps during the program and erase operations in NVSim because they contribute a considerable portion of the total energy consumption. The energy consumed by charge pumps is referred from an actual NAND flash chip design [16], which specifies that a conventional charge pump consumes 0.25 μJ at 1.8 V supply voltage. We use this value as the default in NVSim.

2.7 Validation Result

NVSim is built on the basis of generic assumptions of memory cell layouts, circuit design rules, and CMOS fabrication parameters, whereas the performance, energy, and area of a real non-volatile memory design depend on the specific choices of all these. However, as described in previous sections, we provide a set of knobs in NVSim to adjust the design parameters such as memory organization, wire type, transistor type, data sensing schemes, and others. Therefore, NVSim is capable of emulating a real memory chip, and comparing the NVSim estimation result to the actual memory chip parameters can show the accuracy of NVSim.

Hence, we validate NVSim against NAND flash chips and several industrial prototype designs of STT-RAM [40], PCRAM [1, 17], and ReRAM [35] in terms of area, latency, and energy. We first extract the information from real chip design specifications to set the input parameters required by NVSim, such as capacity, line size, technology node, and array organization. Then, we compare the performance, energy, and area estimation numbers generated from NVSim to the actual reported numbers in those chip designs. The validation results are listed in this section. Note that all the simulation results are for nominal cases since process variations are not supported in current version of NVSim.

Table 2.3 NVSim's NAND flash model validation with respect to a 50 nm 2 Gb NAND flash chip (B-SLC2) [10]

Metric	Actual	Projected	Error (%)
Area	23.85 mm^2	22.61 mm^2	−5.20
Read latency	21 μs	25.2 μs	+20.0
Program latency	200 μs	200.1 μs	+0.1
Erase latency	1.25 ms	1.25 ms	+0.0
Read energy	1.56 μJ	1.85 μJ	+18.6
Program energy	3.92 μJ	4.24 μJ	+8.2
Erase energy	34.5 μJ	36.0 μJ	+4.3

2.7.1 NAND Flash Validation

It is challenging to validate the NAND flash model in NVSim since the public manufacturer datasheets do not disclose sufficient data on the operation latency and power consumption for validation purpose. Instead, Grupp et al. [10] report both latency and power consumption measurements of several commercial NAND flash chips from different vendors. Grupp's report does not include the NAND flash silicon area; hence, we set the actual NAND flash chip area by assuming an area efficiency of 90 %. The comparison between the measurement [10] and the estimations given by NVSim is listed in Table 2.3. The estimation error is within 20 %.

2.7.2 STT-RAM Validation

We validate the STT-RAM model against a 65 nm prototype chip [40]. We let 1 bank = 32 × 8 mats and 1 mat = 1 subarray to simulate the memory array organization. We also exclude the chip area of I/O pads and duplicated cells to make the fair comparison. As the write latency is not disclosed, we assume the write pulse duration is 20 ns. The validation result is listed in Table 2.4. The result shows that the area and the latency estimation error are within 3 and 5 %, respectively.

2.7.3 PCRAM Validation

We first validate the PCRAM model against a 0.12 μm MOS-accessed prototype. The array organization is configured to have 2 banks, each has 8 × 8 mats. Every mat contains only one subarray. Table 2.5 lists the validation result, which shows a 10 % underestimation of area and 6 % underestimation of read latency. The projected write latency (SET latency as the worst case) is also consistent with the actual value.

Table 2.4 NVSim's STT-RAM model validation with respect to a 65 nm 64 Mb STT-RAM prototype chip [40]

Metric	Actual	Projected	Error
Area	39.1 mm^2	38.05 mm^2	−2.69 %
Read latency	11 ns	11.47 ns	+4.27 %
Write latency	$<$ 30 ns	27.50 ns	–
Write energy	N/A	0.26 nJ	–

Table 2.5 NVSim's PCRAM model validation with respect to a 0.12 μm 64 Mb MOS-accessed PCRAM prototype chip [1]

Metric	Actual	Projected	Error
Area	64 mm^2	57.44 mm^2	−10.25 %
Read latency	70.0 ns	65.93 ns	−5.81 %
Write latency	$>$ 180.0 ns	180.17 ns	–
Write energy	N/A	6.31 nJ	–

Another PCRAM validation is made against a 90 nm diode-accessed prototype [21].

2.7.4 ReRAM Validation

We validate the ReRAM model against a 180 nm 4 Mb HfO_2-based MOS-accessed ReRAM prototype [35]. According to the disclosed data, the subarray size is configured to 128 Kb. We further model a bank with 4 × 8 mats, and each mat contains a single subarray. The validation result is listed in Table 2.7. Note that the estimated chip area given by NVSim is much smaller than the actual value since the prototype chip has SLC/MLC dual modes, but the current version of NVSim does not model the MLC-related circuitry.

2.7.5 Comparison to CACTI

We also test the closeness between NVSim and CACTI by simulating identical SRAM caches and DRAM chips. The results show that NVSim models SRAM and DRAM more accurately than CACTI does since some false assumptions in CACTI are fixed in NVSim.

2.8 Case Studies by Using NVSim

In this section, we conduct two case studies to demonstrate how we can use NVSim in two ways: (1) use NVSim to optimize the NVM designs toward certain design metric

Table 2.6 NVSim's PCRAM model validation with respect to a 90 nm 512 Mb diode-selected PCRAM prototype chip [21]

Metric	Actual	Projected	Error (%)
Area	91.50 mm^2	93.04 mm^2	+1.68
Read latency	78 ns	59.76 ns	−23.40
Write latency	430 ns	438.55 ns	+1.99
Write energy	54 nJ	47.22 nJ	−12.56

Table 2.7 NVSim's ReRAM model validation with respect to a 0.18 μm 4 Mb MOSFET-selected ReRAM prototype chip [35]

Metric	Actual	Projected	Error
Area[a]	187.69 mm^2	33.42 mm^2	–
Read latency	7.2 ns	7.72 ns	+7.22 %
Write latency	0.3–7.2 ns	6.56 ns	–
Write energy	N/A	0.46 nJ	–

[a] A large portion of the chip area is contributed to the MLC control and test circuits, which are not modeled in NVSim

and (2) use NVSim to estimate the performance, energy, and area before fabricating a real prototype chip, especially when the emerging NVM device technology is still under development and there is no standard so far.

2.8.1 Use NVSim for Design Optimization

NAND flash is currently the widely used firmware storage or disk in embedded systems. However, codes stored in NAND must be copied to random-accessible memory like DRAM before execution since NAND's page-accessible structure causes poor random access performance. If emerging NVM technologies such as STT-RAM, PCRAM, and ReRAM can be adopted in such systems, and their byte-accessibility property can eliminate the need of DRAM modules in such systems. But, the issue of directly adopting emerging NVM technologies as the NAND flash substitute comes from the observation that the current prototype has a much slower read/write latency than DRAM. In this case study, we use PCRAM as an example without the loss of generality. The technology node used in this case study is 90 nm. Table 2.8 shows the latency difference between an NAND chip, a DRAM chip, and a PCRAM prototype chip with the same 512 MB capacity.

The comparison shows that the PCRAM prototype chip is much slower than its DRAM counterpart. To overcome this obstacle, it is necessary to optimize PCRAM chips for latency at the expense of area efficiency by aggressively cutting word lines and bit lines or inserting repeaters. Such area/performance trade-off is also available for DRAM designs. However, in this case study, we keep the DRAM chip parameters

Table 2.8 Using PCRAM as direct replacement of NAND

A typical 90 nm 512 Mb NAND (*source* K9F1208X0C datasheet)	
Access unit	Page
Read latency	15 μs
Write latency	200 μs
Erase latency	2 ms
A 90 nm 512 Mb PCRAM (*source* [21], Table 2.6 for more details)	
Access unit	Byte
Read latency	78 ns (59.76 ns, NVSim estimation)
Write latency	430 ns (438.55 ns, NVSim estimation)
A typical 90 nm 512 Mb DRAM (*source* K4T51043Q datasheet)	
Access unit	Byte
tRCD	15 ns
tRP	15 ns

unchanged since the current DRAM specification is already the sweet spot explored by DRAM industry for many years. But for PCRAM, such performance optimization is necessary.

Table 2.9 shows the comparison before and after NVSim optimization. The result shows that the PCRAM read latency can be reduced from 59.76 to 16.23 ns by only cutting subarrays into smaller size (from 1024 × 1024 to 512 × 32). Although the PCRAM write latency does not reduce too much due to the inherent SET/RESET pulse duration, write latency is typically not in the critical path and can be tolerated using write buffers. As a result, the optimized PCRAM chip projected by NVSim can properly replace the traditional NAND+DRAM solution in the embedded system. The latency optimization is at the expense of increasing chip area, which rises from 93.04 to 102.34 mm^2.

2.8.2 Use NVSim for Early-Stage Estimation

Considering the facts that the research of some emerging NVM technologies (e.g., ReRAM) is still in an early stage and there are only a limited number of NVM prototype chips available for high-level computer architects understanding the technologies, we expect NVSim would be helpful in providing performance, energy, and area estimations at an early design stage. In this case study, we demonstrate how NVSim can predict the full design spectrum of a projected ReRAM technology when such a device is fabricated as an 8 MB memory chip. Table 2.10 lists the projection.

Table 2.11 tabulates the full design spectrum of this 32 nm 8 MB ReRAM chip by listing the details of each design corner. As shown in the result, NVSim can optimize the same design toward different optimization targets by exploring the full design space, which means that NVSim automatically tunes all the design knobs such as

Table 2.9 New PCRAM parameters after NVSim latency optimization

Parameter	Before optimization	After optimization
Subarray size	1024 × 1024	512 × 32
Area	93.04 mm^2	102.34 mm^2
Read latency	59.76 ns	16.23 ns
Write latency	438.55 ns	416.23 ns

Table 2.10 The projection of a future ReRAM technology

	MOS-accessed	Cross-point
Cell size	$4F^2$	$20F^2$
Maximum NMOS driver size	$100F$	
RESET voltage and pulse duration	2.0 V, 100 ns	
SET voltage and pulse duration	−2.0 V, 100 ns	
READ input	0.4 V voltage source, or 2 μA current source	
LRS resistance	10 kΩ	
HRS resistance	500 kΩ	
Half-select resistance	–	100 kΩ

array structure, subarray size, sense amplifier design, write method, repeater design, and buffer design. If necessary, NVSim can also be explored to use different types of transistor or wire models to get the best result.

2.9 Related Work

Many modeling tools have been developed during the last decade to enable system-level design exploration for SRAM- or DRAM-based cache and memory. For example, CACTI [39, 43] is a tool that has been widely used in the computer architecture community to estimate the performance, energy, and area of SRAM and DRAM caches. Evans and Franzon [8] developed an energy model for SRAMs and used it to predict an optimum organization for caches. eCACTI [25] incorporated a leakage power model into CACTI. Muralimanohar et al. [29] modeled large-capacity caches through the use of an interconnect-centric organization composed of mats and request/reply H-tree networks.

In addition, CACTI has also been extended to evaluate the performance, energy, and area for STT-RAM [6], PCRAM [7, 26], cross-point ReRAM [44], and NAND flash [27]. However, as CACTI is originally designed to model an SRAM-based cache, some of its fundamental assumptions do not match the actual NVM circuit implementations, and thereby, the NVM array organization modeled in these CACTI-like estimation tools deviates from the NVM chips that have been fabricated.

Table 2.11 The predicted full design spectrum of a 32 nm 8 MB ReRAM chip

Optimization target	Area	Read latency	Write latency	Read energy	Write energy	Leakage power
Area (mm^2)	**0.664**	5.508	8.071	2.971	3.133	1.399
Read latency (ns)	107.1	**1.773**	1.917	5.711	6.182	426.8
Write latency (ns)	204.3	200.7	**100.6**	202.8	203.1	518.2
Read energy (nJ)	1.884	0.195	0.234	**0.012**	0.014	4.624
Write energy (nJ)	13.72	25.81	13.06	12.82	**12.81**	12.99
Leakage (mW)	1372	3872	7081	6819	7841	**26.64**
Array structure	Xpoint	Xpoint	1T1R	Xpoint	Xpoint	1T1R
Subarray size	512 × 512	128 × 128	1024 × 2048	512 × 512	256 × 256	2048 × 4096
Routing scheme	Non-H-tree	H-tree	H-tree	H-tree	H-tree	Non-H-tree
SA placement	External	Internal	Internal	Internal	Internal	External
SA type	CIVS	CS	CS	VDS	VDS	VDS
Write method	SbR	EbR	Normal	SbR	SbR	Normal
Interconnect	Normal	Repeated	Repeated	Low swing	Low swing	Normal
Output buffer optimized for	Area	Latency	Latency	Area	Area	Area

2.10 Conclusion

STT-RAM, PCRAM, and ReRAM are emerging memory technologies for future non-volatile memories. The versatility of these upcoming NVM technologies makes it possible to use these NVM modules at other levels in the memory hierarchy, such as execute in place (XIP) memory, main memory, or even on-chip cache. Such emerging NVM design options can vary for different applications by tuning circuit structure parameters such as the array organizations and the peripheral circuitry types or by using devices and interconnects with different properties. To enable the system-level design space exploration of these NVM technologies and facilitate computer architects leverage these emerging technologies, it is necessary to have a quick estimation tool. While abundant estimation tools are available as SRAM/DRAM design assistants, similar tools for NVM designs are currently missing. Therefore, in this work, we build *NVSim*, a circuit-level model for NVM performance, energy, and area estimation, which supports various NVM technologies including STT-RAM, PCRAM, ReRAM, and conventional NAND flash. This model is successfully validated against industrial NVM prototypes, and this new *NVSim* tool is expected to help boost NVM-related studies such as the next-generation memory hierarchy.

References

1. Ahn, S.J., et al. (2004). Highly manufacturable high density phase change memory of 64Mb and beyond. *Proceedings of the International Electron Devices Meeting* (pp. 907–910).
2. Burr, G. W., et al. (2010). Phase change memory technology. *Journal of Vacuum Science and Technology B*, *28*(2), 223–262.
3. Chen, Y.C., et al. (2003). An access-transistor-free (0T/1R) non-volatile resistance random access memory (RRAM) using a novel threshold switching, self-rectifying chalcogenide device. *Proceedings of the International Electron Devices Meeting* (pp. 750–753).
4. Chen, Y.S., et al. (2009). Highly scalable hafnium oxide memory with improvements of resistive distribution and read disturb immunity. *Proceedings of the International Electron Devices Meeting* (pp. 105–108).
5. Chien, W., et al. (2009). Multi-level operation of fully CMOS compatible WO_x resistive random access memory (RRAM). *Proceedings of the International Memory, Workshop* (pp. 228–229).
6. Dong, X., et al. (2008). Circuit and microarchitecture evaluation of 3D stacking magnetic RAM (MRAM) as a universal memory replacement. *Proceedings of the Design Automation Conference* (pp. 554–559).
7. Dong, X., et al. (2009). PCRAMsim: System-level performance, energy, and area modeling for phase-change RAM. *Proceedings of the International Conference on, Computer-Aided Design* (pp. 269–275).
8. Evans, R. J., & Franzon, P. D. (1995). Energy consumption modeling and optimization for SRAM's. *IEEE Journal of Solid-State Circuits*, *30*(5), 571–579.
9. Fishburn, F., et al. (2004). A 78nm 6F^2 DRAM technology for multigigabit densities. *Proceedings of the Symposium on VLSI Technology* (pp. 28–29).
10. Grupp, L.M., et al. (2009). Characterizing flash memory: Anomalies, observations, and applications. *Proceedings of the International Symposium on Microarchitecture* (pp. 24–33).
11. Hanzawa, S., et al. (2007). A 512kB embedded phase change memory with 416kB/s write throughput at 100μA cell write current. *Proceedings of the International Solid-State Circuits Conference* (pp. 474–616).

12. Horowitz, M. A. (1983). *Timing models for MOS circuits*. Tech. rep. California: Stanford University.
13. Hosomi, M., et al. (2005). A novel nonvolatile memory with spin torque transfer magnetization switching: Spin-RAM. *International Electron Devices Meeting* (pp. 459–462).
14. International Technology Roadmap for Semiconductors: Process Integration, Devices, and Structures 2010 Update. http://www.itrs.net/
15. International Technology Roadmap for Semiconductors: The Model for Assessment of CMOS Technologies And Roadmaps (MASTAR). http://www.itrs.net/models.html
16. Ishida, K., et al. (2009). A 1.8V 30nJ adaptive program-voltage (20V) generator for 3D-integrated NAND flash SSD. *Proceedings of the IEEE International Solid-State Circuits Conference* (pp. 238–239,239a).
17. Kang, S., et al. (2007). A 0.1 μm 1.8V 256Mb phase-change random access memory (PRAM) with 66MHz synchronous burst-read operation. *IEEE Journal of Solid-State Circuits*, *42*(1), 210–218.
18. Kau, D., et al. (2009). A stackable cross point phase change memory. *Proceedings of the IEEE International Electron Devices Meeting* (pp. 27.1.1-27.1.4).
19. Kawahara, T., et al. (2007). 2Mb spin-transfer torque RAM (SPRAM) with bit-by-bit bidirectional current write and parallelizing-direction current read. *IEEE International Solid-State Circuits Conference* (pp. 480–617).
20. Kim, K. H., et al. (2010). Nanoscale resistive memory with intrinsic diode characteristics and long endurance. *Applied Physics Letters*, *96*(5), 053,106.1-053,106.3.
21. Lee, K. J., et al. (2008). A 90nm 1.8V 512Mb diode-switch PRAM with 266MB/s read throughput. *IEEE Journal of Solid-State Circuits*, *43*(1), 150–162.
22. Lee, M.J., et al. (2007). 2-stack 1D–1R cross-point structure with oxide diodes as switch elements for high density resistance RAM applications. *Proceedings of the IEEE International Electron Devices Meeting* (pp. 771–774).
23. Liang, J., & Wong, H. S. P. (2010). Cross-point memory array without cell selectors: Device characteristics and data storage pattern dependencies. *IEEE Transactions on Electron Devices*, *57*(10), 2531–2538.
24. Lin, W., et al. (2010). Evidence and solution of over-RESET problem for HfO_x based resistive memory with sub-ns switching speed and high endurance. *Proceedings of the International Electron Devices Meeting* (pp. 19.7.1-19.7.4).
25. Mamidipaka, M., Dutt, N. (2004). eCACTI: An enhanced power estimation model for on-chip caches. Tech. Rep. TR04-28, Center for Embedded Computer Systems.
26. Mangalagiri, P., et al. (2008). A low-power phase change memory based hybrid cache architecture. *Proceedings of the Great Lakes Symposium on VLSI* (pp. 395–398).
27. Mohan, V., et al. (2010). FlashPower: A detailed power model for NAND flash memory. *Proceedings of Design, Automation and Test in, Europe* (pp. 502–507).
28. Moon, Y., et al. (2009). 1.2V 1.6Gb/s 56nm 6F^2 4Gb DDR3 SDRAM with hybrid-I/O sense amplifier and segmented sub-array architecture. *Proceedings of the International Solid-State Circuits Conference* (pp. 128–129).
29. Muralimanohar, N., et al. (2008). Architecting efficient interconnects for large caches with CACTI 6.0. *IEEE Micro*, *28*(1), 69–79.
30. Oh, H. R., et al. (2006). Enhanced write performance of a 64-Mb phase-change random access memory. *IEEE Journal of Solid-State Circuits*, *41*(1), 122–126.
31. Oh, J.H., et al. (2006). Full integration of highly manufacturable 512Mb PRAM based on 90nm technology. *Proceedings of the International Electron Devices Meeting* (pp. 49–52).
32. Pellizzer, F., et al. (2004). Novel μTrench phase-change memory cell for embedded and stand-alone non-volatile memory applications. *Proceedings of the International Symposium on VLSI Technology* (pp. 18–19).
33. Raoux, S., et al. (2008). Phase-change random access memory: A scalable technology. *IBM Journal of Research and Development*, *52*(4/5),
34. Seevinck, E., et al. (1991). Current-mode techniques for high-speed VLSI circuits with application to current sense amplifier for CMOS SRAM's. *IEEE Journal of Solid-State Circuits*, *26*(4), 525–536.

35. Sheu, S.S., et al. (2011). A 4Mb embedded SLC resistive-RAM macro with 7.2ns read-write random-access time and 160ns MLC-access capability. *Proceedings of the IEEE International Solid-State Circuits Conference* (pp. 200–201).
36. Smullen, C., et al. (2011). Relaxing non-volatility for fast and energy-efficient STT-RAM caches. *Proceedings of the International Symposium on High Performance Computer, Architecture* (pp. 50–61).
37. Sutherland, I. E., et al. (1999). *Logical effort: designing fast CMOS circuits*. Morgan Kaufmann.
38. Thoziyoor, S., et al. (2008). A comprehensive memory modeling tool and its application to the design and analysis of future memory hierarchies. *Proceedings of the International Symposium on Computer, Architecture* (pp. 51–62).
39. Thoziyoor, S., et al. (2008). CACTI 5.1 technical report. Tech. Rep. HPL-2008-20, HP Labs.
40. Tsuchida, K., et al. (2010). A 64Mb MRAM with clamped-reference and adequate-reference schemes. *Proceedings of the International Solid-State Circuits Conference* (pp. 268–269).
41. Udipi, A. N., et al. (2010). Rethinking DRAM design and organization for energy-constrained multi-cores. *ACM SIGARCH Computer Architecture News*, *38*(3), 175–186.
42. Wei, Z., et al. (2008). Highly reliable TaO_x ReRAM and direct evidence of redox reaction mechanism. *Proceedings of the International Electron Devices Meeting* (pp. 293–296).
43. Wilton, S. J. E., & Jouppi, N. P. (1996). CACTI: An enhanced cache access and cycle time model. *IEEE Journal of Solid-State Circuits*, *31*, 677–688.
44. Xu, C., et al. (2011). Design implications of memristor-based RRAM cross-point structures. *Proceedings of Design, Automation and Test in, Europe*, (pp. 1–6).
45. Yang, J. J., et al. (2008). Memristive switching mechanism for metal/oxide/metal nanodevices. *Nature Nanotechnology*, *3*(7), 429–433.
46. Yoshitaka, S., et al. (2009). Cross-point phase change memory with $4F^2$ cell size driven by low-contact-resistivity poly-Si diode. *Proceedings of the Symposium on VLSI Technology* (pp. 24–25).
47. Zhang, Y., et al. (2007). An integrated phase change memory cell with Ge nanowire diode for cross-point memory. *Proceedings of the IEEE Symposium on VLSI Technology* (pp. 98–99).

Chapter 3
A Hybrid Solid-State Storage Architecture for the Performance, Energy Consumption, and Lifetime Improvement

Guangyu Sun, Yongsoo Joo, Yibo Chen, Yiran Chen and Yuan Xie

Abstract In recent years, many systems have employed NAND flash memory as storage devices because of its advantages of high I/O performance, increasing capacity, and falling cost. On the other hand, the performance of NAND flash memory is limited by its "erase-before-write" requirement. Log-based structures have been used to alleviate this problem by writing updated data to the clean space. Log-based methods, however, cannot completely overcome the inherent limitation of NAND flash memory. It cannot avoid excessive erase operations when there are frequent updates, which quickly consume free pages, especially when some data are updated repeatedly. In this paper, we propose a hybrid architecture for the NAND flash memory storage, of which the log region is implemented using phase change random access memory (PCRAM). Compared to traditional log-based architectures, it has the following advantages: (1) the PCRAM log region allows in-place updating and byte-granularity access so that it significantly improves the usage efficiency of log pages by eliminating out-of-date log records; (2) it greatly reduces the traffic of reading from the NAND flash memory storage since the size of logs loaded for

G. Sun
Peking University, Beijing, China
e-mail: gsun@pku.edu.cn

Y. Joo
Ewha Womans University, Seoul, South Korea
e-mail: ysjoo@elpl.snu.ac.kr

Y. Chen
Google Inc., Mountain View, CA, USA
e-mail: cyb.thu@gmail.com

Y. Chen
University of Pittsburgh, Pittsburgh, USA
e-mail: yic52@pitt.edu

Y. Xie (✉)
Pennsylvania State University, Pennsylvania, USA
e-mail: yuanxie@cse.psu.edu

Y. Xie (ed.), *Emerging Memory Technologies*,
DOI: 10.1007/978-1-4419-9551-3_3,

the read operation is decreased; (3) the energy consumption of the storage system is reduced as the overhead of writing and reading log data is decreased with the PCRAM log region; (4) the lifetime of NAND flash memory is increased because the number of erase operations are reduced. To facilitate the PCRAM log region, we propose several management policies. The simulation results show that our proposed methods can substantially improve the performance, energy consumption, and lifetime of the NAND flash memory storage.

3.1 Introduction

As solid-state technologies advance, the past several years have been an exciting time for NAND flash memory. The cost of NAND flash memory has fallen dramatically as fabrication becomes more efficient and the market grows; the density has been improved with better process technologies. Consequently, NAND flash memory has been widely adopted by various applications such as laptops, PDA, mobile phones. In addition, because of its better I/O performance compared to traditional hard disk drives, NAND flash memory has also been proposed to be used as a cache for hard disk drives [1], or even as the replacement of hard disk drives in high-performance computing clusters and enterprise-scale data centers [2–6].

One well-known limitation of NAND flash memory is the "erase-before-write" requirement. It cannot update the data by directly overwriting it [7, 8]. In NAND flash memory, read and write operations are performed in the granularity of a page (typically 512 Bytes – 8 KB) [9]. A DRAM data buffer is normally employed to store a subset of pages of NAND flash memory for frequent accesses [8]. The modified pages are written back to flash memory when they are evicted from the DRAM data buffer. These pages, however, cannot be directly overwritten to the same place in flash memory. In another word, the NAND flash memory does not allow direct "in-place updating". Instead, a time-consuming erase operation must be performed before the overwriting. To make it even worse, the erase operation cannot be performed selectively on a particular data item or page but can only be done for a block of NAND flash memory called the "erase unit." Since the size of an erase unit (typically 128 KB or 256 KB) is much larger than that of a page, even a small update to a single page requires all pages in the same erase unit to be erased and written again.

In order to overcome the performance degradation and the energy overhead caused by the "erase-before-write," various techniques have been proposed [5, 10–13]. One popular idea is to use "out-of-place updating." It means that the updated data are written to other clean pages and the original data are invalidated. File systems employing this method are so-called log-based file system [10], and extensive research has been done in this field [10, 14]. Some research has pointed out that the performance of these file system is not good for the access pattern with frequent and small random updates, which is common for many applications including online transaction processing (OLTP) [5, 11]. Recently, Lee et al. proposed an in-page logging (IPL) approach, which can outperform the traditional log-based file systems for such an access pattern [5]. It caters for the characteristic of NAND flash memory by partitioning an erase unit into a data region and a log region. The data region stores normal data pages, and the log region stores updates to the data pages in the form of logs. When a modified data page is evicted from the data buffer, only the updated data are written back to NAND flash memory. Consequently, the data pages in an erase unit are not erased until clean space of the log region runs out. When the log region is full, a "merge operation" is triggered, through which data pages are written to a clean erase unit with all updates applied.

Although the IPL method helps in reducing the number of erase and write operations, it cannot completely overcome the inherent limitation of NAND flash memory because the log region itself is still implemented with NAND flash memory. For example, if data pages of an erase unit are frequently updated, these updates will quickly fill the log region and cause many merge operations; especially when a data item is repeatedly updated, many redundant log records are generated for the same data item, but only the latest one is valid. Hence, the utilization of log sectors becomes inefficient for this pattern of updates.

In recent years, various emerging nonvolatile memory (NVM) technologies are proposed, such as phase change random access memory (PCRAM), magnetic random access memory (MRAM), and resistive random access memory (RRAM). These technologies are considered as competitive candidates for future universal memory. As such emerging NVM technologies are getting mature, it has been a hot topic in computer architecture research to explore the usage of such emerging NVM technologies at different level of memory hierarchy to enable novel architecture design, such as NVM-based cache design [15–18], NVM-based memory architecture [15, 19–25], or NVM-based storage architecture [26, 27]. Compared to NAND flash memory, all these memory technologies have the important advantage of allowing direct in-place updating. The process technologies of these NVMs, however, are not as mature as that of NAND flash memory. It is still not feasible for them to directly replace NAND flash memory as massive storage because of their limitations of manufacture and high cost [17, 28]. Alternatively, we consider using a hybrid storage architecture to take advantages of NAND flash memory and other memory technologies.

In this work, we employ PCRAM as the log region of our hybrid storage for several reasons. First, PCRAM has the highest cell density among these emerging memory technologies [28]. Second, the capacity of the up-to-date PCRAM production is qualified for the log region of NAND flash memory that works as massive storage [29]. Third, prior work has shown that the feasibility of integrating PCRAM with NAND flash memory [26, 27]. Our contribution can be summarized as follows:

- We propose to use PCRAM as the log region of the NAND flash memory storage system.
- We develop a set of management policies for PCRAM log region to fully exploit its advantages of in-place updating ability, fine access granularity, long endurance, etc.

- We study the lifetime of the PCRAM log region. The technique is proposed to promise that the PCRAM log region will not wear out before the NAND flash memory.

Compared to the IPL method, the hybrid architecture has the following advantages: (1) the ability of "in-place updating" can significantly improve the usage efficiency of log region by efficiently eliminating the out-of-date log data; (2) the fine-granularity access of PCRAM can greatly reduce the read traffic from the NAND flash memory to the DRAM data buffer since the size of logs loaded for the read operation is decreased; (3) the energy consumption of the storage system is reduced as the overhead of writing and reading log data is decreased with the PCRAM log region; (4) the lifetime of the NAND flash memory could be increased because the number of erase operations are reduced, and the endurance of the PCRAM log region can be traded off to further improve the performance of the storage system. The simulation results show that, with proper PCRAM management policies, both performance and endurance of the hybrid storage are significantly improved.

3.2 Related Work

Recently, PCM has been extensively proposed as the alternative memory technology in both main memory and storage levels. In this section, we present a brief review of related work in different topics.

Most research work has proposed to use PCM as the replacement of DRAM for main memory design because PCM has advantages of better scalability, low standby power, etc., compared to DRAM. On the other hand, it has a well-known problem of "limited write cycle." For example, a PCM cell's write cycle is in the range of 10^6–10^8 with different technologies [23, 25]. NVM main memory design may wear out within several months without any optimization. The research on how to improve lifetime of NVM memory can be categorized into either write reduction or wear leveling techniques, which are described as follows.

Write Reduction

The purpose of this method is try to reduce the number of updated bits to NVM main memory in each write operation. Lee et al. proposed the method of partial write that only the cache lines written back from last-level cache (LLC) are updated in a memory line [30]. In DCW method [24], old data in NVN main memory are first read out and compared with new data bit by bit. Then, only the modified bits are updated in the write operation. The Flip-N-Write technique is proposed based on DCW method to further reduce the number of updated bits by calculating hamming distance between write data and old data [31]. These techniques can effectively reduce write intensity up to 85 % [24, 31].

Wear Leveling

Even with write reduction techniques, some cells in NVM memory may wear out faster than the others due to nonuniform write intensity [24]. Thus, the wear leveling techniques are also necessary to balance write intensity. Various techniques, such as table-based remapping, start-gap, security refresh, are proposed for wear leveling [23–25]. Although design details are different, the fundamental rule behind these techniques is the memory address remapping between *logic address* and *physical address*. In other words, memory addressing is periodically changed so that write intensity is uniformly distributed throughout the whole memory space.

Some prior works also leverage PCM for storage designs. For example, PCM has been proposed to store metadata of NAND flash [26, 27]. Other works argue that, in order to explore potential high performance of PCM, fast I/O interface like PCI-E is needed and the OS I/O stacked need to be modified [32, 33]. Moreover, the file system modification for PCM memory is also studied [34].

3.3 Background

In this section, we first illustrate the erase-before-write limitation for NAND flash memory. Then, we describe the existing IPL method and its limitation. Finally, we present a brief overview of PCRAM technology and the potential to solve the limitation of IPL method.

3.3.1 The Erase-Before-Write Limitation for NAND Flash

NAND flash memory shows asymmetry in how they read and write. While we can read any of the pages of a NAND flash memory, we must perform an erase operation before writing data to a page. The erase operation is performed in the granularity of an "erase unit" consisting multiple adjacent pages [5, 7]. Therefore, writing new data to the same page storing the old data, namely "in-place" update, cannot be performed directly. Instead, it is finished in several steps: (1) back up the other pages in the same erase unit; (2) perform an erase operation; (3) write both the new data and the backed up pages to the erase unit. Such a process is very time consuming.

Because of this erase-before-write limitation, log-based file systems are commonly used for flash memory devices. Figure 3.1 simply shows how a log-based file system works. In Fig. 3.1a, there is an update for a data item in page n. Because the data item cannot be overwritten to page n, the whole page holding the data item is written to another erased page, which is called the log page of page n. Then, the old copy of page n is invalidated. This process is called an "out-of-place" updating. Figure 3.1 also shows a merge operation mentioned in the previous section. The merge operations will happen when the device may run out clean pages, and some

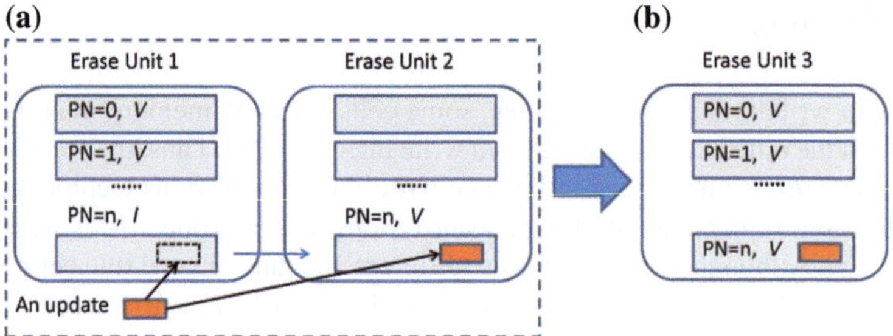

Fig. 3.1 **a** An update operation in NAND flash memory; **b** a merge operation in NAND flash memory. *'V'* valid, *'I'* invalid, *PN* page number

invalid pages need to be reclaimed. This process is also called *garbage collection*. During the garbage collection, valid pages in these two erase units need to be merged into a third erase unit, which is shown in Fig. 3.1b. This merge operation is very costly because all valid pages of the erase unit should be written to another erased unit.

3.3.2 IPL Method

It is easy to find that even if only a small portion of a page is updated, the whole page must be written to its log page in the log-based file systems, which can significantly degrade the write performance of the file system. Unfortunately, small-to-moderate-sized writes are quite a common access pattern for many cases including database applications such as OLTP [5, 11].

Recently, the IPL method [5] is proposed to overcome such a weakness of the log-based file systems for NAND flash memory. Figure 3.2 gives an illustration of the IPL method for NAND flash memory. Each erase unit consists of 60 data pages and 4 log pages. Each log page is divided into 16 log sectors. Unlike the traditional log-based file system, the log page in IPL method is used to record updates of data

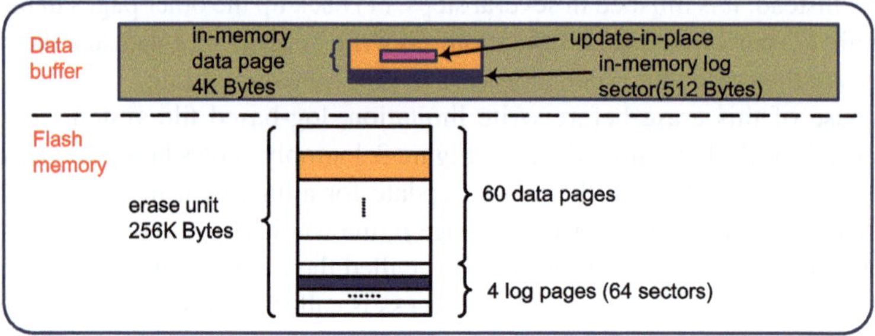

Fig. 3.2 An illustration of the IPL method

pages rather than copy a whole data page directly. Whenever an update is performed on a data page in the DRAM data buffer, the copy of data item in the buffer is updated in-place. At the same time, the IPL buffer manager adds a log record in the data buffer, which is allocated in the log sector assigned to the data page. A log sector in the data buffer can be allocated to the data page on demand, and it can contain more than one update record. A log sector in the data buffer can be released when its log records are written back to the log sector in flash memory. The log records are written to NAND flash memory when the log pages of an erase unit become full or when a dirty data page is evicted from the data buffer. When a dirty page is evicted, the data page itself does not need to be written back to flash memory, because all of its updates are stored in the form of log records. Consequently, only its log sectors are written back to NAND flash memory.

Since the number of the log sectors is fixed for each erase unit, the erase unit may eventually run out of empty log sectors after a certain number of updates are performed to the pages in the erase unit. For such a case, the IPL manager performs a merge operation as follows. First, it merges the old data pages with their log sectors to create up-to-date data pages. Second, it allocates a clean erase unit to write the generated up-to-date data pages. Finally, it invalidates all the pages in the old erase unit, so this erase unit can be reclaimed for future use.

In the IPL approach, only the log sectors that contain the updated data are written back to flash memory, and the usage of data pages is more efficient than that of traditional log-based flash memory. The size of log pages in an erase unit is limited, and these log sectors themselves do not allow in-place updating. This may cause significant performance degradation for some cases. Particularly, if there are frequent updates to the same erase unit, it would quickly run out its log sectors and cause merge operations frequently. Moreover, if there are multiple updates to the same data, only the latest updated one is valid. Hence, in such a case, the effective log sector capacity of an erase unit becomes much smaller, which would worsen the problem of the IPL approach.

3.3.3 PCRAM Process Technology

Different from the conventional RAM technologies (such as SRAM/DRAM), the information carrier of PCRAM is chalcogenide-based materials, such as $Ge_2Sb_2Te_5$ and $Ge_2Sb_2Te_4$ [35]. The crystalline and amorphous states of chalcogenide materials have a wide range of resistivity, about three orders of magnitude, and this forms the basis of data storage. The amorphous, high-resistance state is used to represent a bit '0', and the crystalline, low-resistance state represents a bit '1'.

Nearly, all prototype devices make use of a chalcogenide alloy of germanium, antimony, and tellurium (GeSbTe) called GST. When GST is heated to a high temperature (normally over 600 °C), it will get melted and its chalcogenide crystallinity is lost. Once cooled, it is frozen into an amorphous and its electrical resistance becomes high. This process is called RESET. One way to achieve the crystalline state is by applying a lower constant-amplitude current pulse for a time longer than the so-called RESET pulse. This is called SET process [36]. The time of phase transition is temperature dependent. Normally, it takes several 10 ns for the RESET and takes about 100 ns for the SET [36, 37].

As PCRAM technologies improve, PCRAM has shown more potential to replace NAND flash memory with advantages of allowing in-place updates and fast access speed. Table 3.1 compares the characteristics of PCRAM and NAND flash memory, which are estimated from prior work [38–40]. Note that the units of write and read operations for PCRAM and NAND flash memory are different. The PCRAM can be accessed in a fine granularity (byte based). We will show that this advantage makes the access to the log region much more flexible, compared to the traditional IPL method. Prior research has also shown that the PCRAM could achieve the same cell size as that of NAND flash memory with a vertically stacked memory element over the selection device [28, 37]. It means that it is feasible to replace NAND flash memory with PCRAM without inducing area overhead.

Currently, it is still not feasible to replace the whole NAND flash memory with PCRAM entirely due to its high cost and the limitation of manufacture [28, 29]. Consequently, we propose to use the PCRAM as the log region of NAND flash memory instead. A cell of NAND flash memory could be implemented to store multiple bits of data. The multi-level PCRAM is still under research [41]. In this work, the single-bit PCRAM is used.

3.4 Overview of the Hybrid Storage Architecture

Figure 3.3 shows the physical and structural views of NAND flash memory storage system with the PCRAM-based log region. Since the process technologies of PCRAM and NAND flash memory are different, it is difficult to physically place the PCRAM-based log region together with the flash-based data region. Unlike the IPL method, the PCRAM log region is separated from the flash-based data region in our work as shown in Fig. 3.3a. Based on log region management policies and run-time operations of applications, log sectors are dynamically assigned to each erase unit to store its own updates. The structural view of such relationship is shown in Fig. 3.3b. We need to maintain metadata for the relationship between data pages

Table 3.1 The comparison between PCRAM and NAND flash memory technologies

Tech.	Cell size	Write cycles	Access time			Access energy		
			Read	Write	Erase	Read	Write	Erase
Flash	$4F^2$	10^5	284 μs/ 4 KB	1833 μs/ 4 KB	>20 ms/ Unit	9.5 μJ/ 4 KB	76.1 μJ/ 4 KB	16.5 μJ/ 4 KB
PCRAM	$4F^2$	10^8	80 ns/ word	10 μs/ word	N/A	0.05 nJ/ word	0.094 nJ/ word	N/A

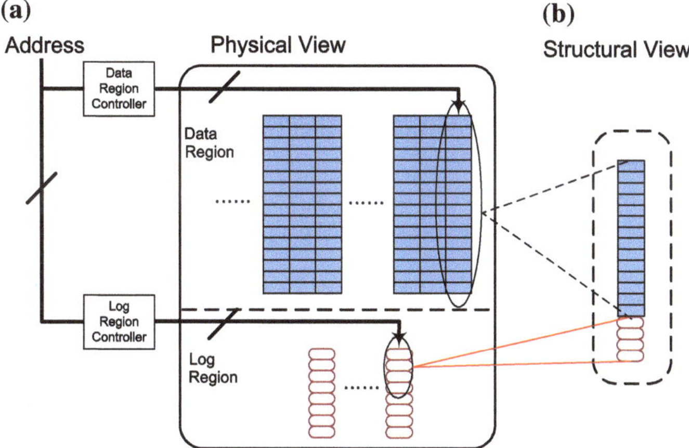

Fig. 3.3 Physical and structural views of the hybrid architecture

and their corresponding log sectors. The *log region controller* in Fig. 3.3 takes the responsibility of decoding addresses and managing the metadata for the log region.

The access process to the hybrid storage architecture is described as follows:

- For the read operation, the address of the accessed data is sent to both the data region and the log region. If there exist log sectors for the requested data page, they are loaded into the data buffer as well as the original data page to create the up-to-date data page.
- For the write operation, only updates are sent back to the PCRAM log region. Since PCRAM allows in-place updating, there are several scenarios we should consider:
  - Case 1: If there are no existing log records for the accessed data page, a new log sector is allocated to the accessed data page. Then, the current update is written to the log sector.
  - Case 2: If some log sectors have already been allocated to the accessed data page, the log records in these log sectors are compared with the current update. If there is a log record that has the same data address of the current update, the existing log record is overwritten by the current update, and no extra log record is generated.
  - Case 3: If the current update does not match any of existing log records, a new log record containing the current update is written to the log sectors of the data page. If log sectors assigned to the data page are fully occupied, a new log sector is requested for the current update as in Case 1.
- Merge operations are triggered when the merge conditions are satisfied. The merge conditions depend on the log region management policies, which will be introduced in following sections. During merge operations, updates in these log sectors are applied to corresponding data pages, and the data pages are written to another free

erase unit. The out-of-date data pages are invalidated, and the corresponding log sectors are released as clean ones for future use.

The data address of the current update needs to be compared to those of existing log records in order to achieve the in-place update. A relative address ($addr_{Relative}$) is used to represent the position of an update inside the accessed data page. It can be calculated by $addr_{Relative} = addr_{Update} - addr_{Page}$. The $addr_{Update}$ and $addr_{Page}$ represent the real addresses of the update and the accessed data page, respectively.

As we mentioned, an access to a PCRAM log region is managed with its own controller. The access to the log region is operated in parallel with that to the data region. Since the size of a log region is much smaller than that of a data region, the accessing delay on the peripheral circuitry of the log region is shorter than that to the data region. It means that using hybrid architecture will not induce extra delay. Instead, the total access latency could even be reduced in some scenarios. For example, in write operations, the latency is decided by the time of writing the update to the log region. Then, the performance could be improved with a shorter decoding time.

Besides the shorter access latency, the more important thing is that the hybrid architecture can take advantages of in-place updating capability of the PCRAM log region. Although it takes extra time to search whether the current update does not match existing log records, the overhead is trivial compared to the write latency because the read operation is much faster than the write one. If the in-place updating happens, much more benefits can be achieved. For example, since the update is written to the existing record, the time of allocating new log record and modifying metadata is saved. It is also beneficial in that the log region is used more efficiently: It would reduce the numbers of erase and write operations by increasing the effective capacity of the log region.

Furthermore, because the PCRAM log region can be accessed with byte granularity, the performance of read operations can also be improved. In the IPL method, since the log data are loaded in the page granularity, some log sectors that do not belong to the accessed data page are also loaded. If the log records of one data page are stored in more than one physical page of NAND flash memory, the overhead of loading the log records can be even higher than that of loading the data page. On the contrary, in our hybrid architecture, only the log records that belong to the accessed data page are loaded. It can greatly enhance the performance of read operations.

3.5 Management Policy of PCRAM Log Region

The ability of in-place updating and fine access granularity of PCRAM log region provide opportunities of optimizations. In this section, we propose a few assignment and corresponding management policies for the PCRAM log region.

3.5.1 Static Log Region Assignment

In the IPL method, each erase unit is uniformly assigned a fixed number of log pages. It is obvious that such a static assignment also works with the PCRAM log region. This method of assignment is named as the *basic static assignment*. Its illustration is shown in Fig. 3.4a. When the data pages in the erase unit are updated, only the log sectors that belong to the erase unit are assigned for these updates. Although the static log sector assignment is the same as that in IPL method, the performance can be improved using the in-place update capability of PCRAM. As in the IPL method, the merge operation is triggered when all log pages of an erase unit are fully occupied and no free log sector can be assigned to a new update. Note that the new update means that the incoming update is not matched to any existing log records. With the PCRAM log region, when all log pages of an erase unit are used, it is possible that the logs are written in-place without causing merge operations.

The main advantage of the static assignment is its simplicity in implementation. Since the log sectors assigned to each erase unit are predetermined and fixed, the overhead of keeping and searching the metadata of log sectors is negligible. Such uniform assignment, however, becomes inefficient when updates are not evenly distributed among erase units. In applications such as OLTP, the access/update intensity to each erase unit is typically not uniform. As shown in Fig. 3.5, the distribution of update numbers in respect of erase unit is highly skewed for a trace of TPC-C benchmark. Note that TPC-C is a popular benchmark that simulates a complete computing environment where a population of users executes transactions against a database [42]. The 3,000 most frequently updated erase units consume about 90 % of the total update numbers. The distribution of merge operations, in respect of the erase unit, is also shown in Fig. 3.5. The 3,000 most frequently updated erase units cause about 88 % of total merge operations with the static log region assignment. Thus, for such applications, it is inefficient to divide and assign the log region equally to each erase unit.

One way to mitigate the effect of uneven access distribution is to organize erase units into groups, which is named *group static assignment*. In this way, log pages in

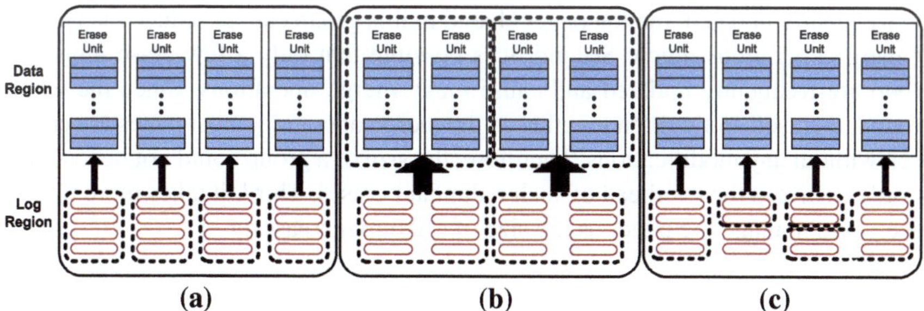

Fig. 3.4 **a** An example of the basic static log assignment. **b** An example of the group static log assignment. **c** An example of the dynamic static log assignment

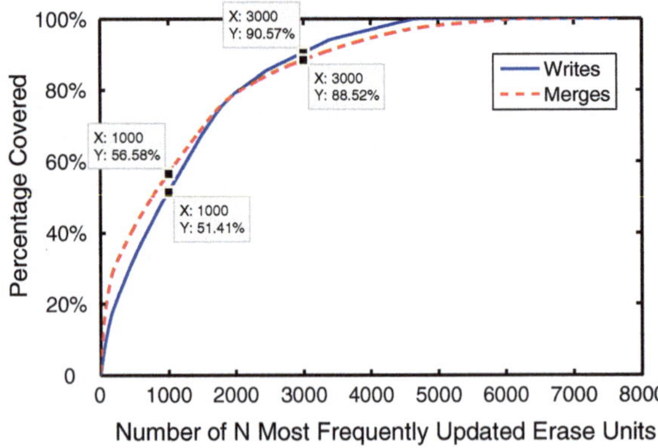

Fig. 3.5 Distributions of write and merge operations, in respect of most frequently updated erase units

a group can be shared among erase units, as shown in Fig. 3.4b. After sharing log pages in a group, the frequently updated erase units in a group could be assigned more log pages. Note that such a method of sharing log pages among erase units is not feasible for the IPL method using only NAND flash memory. It is because the log pages are placed together with data pages inside each erase unit of the IPL method. If we want to assign data pages with external log pages from other erase units, a pointer-like structure is needed to record locations of external log pages, which may increase the design complexity and timing overhead. Furthermore, since the log region is managed in page granularity, sharing log pages among erase units may generate some data pages, which have too many log pages. On the contrary, in our PCRAM log region, all log pages are placed together. There is no difference whether we manage log pages in the granularity of an erase unit or in a group with several erase units. Similar to the basic static assignment, the merge operation is also triggered when the log pages assigned to each group run out. Note that all erase units in the group need to be merged together.

Although the grouping method could help in mitigating the effect of unbalanced accesses, there are some limitations with it. First, if most of erase units inside a group are frequently updated, the log pages shared in this group cannot reduce the number of merge operations much. Second, it is possible that only a few erase units in a group are frequently updated and cause many merge operations. Since the whole group are erased and written together in the merge operation, most pages are erased even without any modifications. Therefore, the size of a group could not be too large in order to avoid the overhead of such redundant erases. Third, it takes more time to access and manage log pages as the size of a group increases.

3.5.2 Dynamic Log Region Assignment

In order to overcome the limitations of the static assignment, we propose a dynamic log page allocation method. The basic idea of the dynamic allocation method is that the number of log sectors of an erase unit could be assigned on demand based on the number of its updates. In other words, the log region is *shared* among all erase units. If an erase unit is frequently updated, the log region controller assigns more log sectors to it unless there is no remained free log sectors in the log region. Note that such a dynamic assignment is different from the group static assignment. The log sectors of each erase unit are managed individually, and the merge operation of one erase unit will not affect the others.

Figure 3.4c gives an illustration of the dynamic assignment. At the beginning, there are no updates to any data page, and all log sectors are free. During the access process, the number of log sectors assigned to each erase unit grows up based on its updates. The frequently updated erase unit is assigned more log sectors.

The dynamic assignment promises that frequently updated erase units can get more log pages so that the number of merge operations are significantly reduced for applications having unbalanced updates. Such a dynamic assignment is not feasible for the IPL method with the NAND flash memory log region, which does not allow in-place updating. It is obvious that the number of the log sectors is increased linearly in proportion to the size of updates when the in-place updating is not allowed. When the updates to erase units are unbalanced, the frequently updated erase unit may have too many log sectors, and the overhead of accessing such erase unit is not tolerable. On the contrary, in our PCRAM log region, the number of log sectors is not increased when some data are repeatedly updated. In the worst case, the size of logs is the same as that of the data. Therefore, accessing overhead for the dynamic assignment is moderate with the PCRAM log region.

Intuitively, the merge operations are triggered when the whole PCRAM log region is fully occupied. In that case, however, once the merge operation is triggered, all log sectors should be merged to the corresponding data pages at the same time, which makes the storage system not available for a long time. Hence, if the burst of updates arrives when the log region is nearly full, the whole system has to be stalled for a long time waiting for merge operations to be completed, which is not desirable. Instead, we set up a threshold of free log sectors, and the merge operations are triggered when the capacity of free log sectors are lower than it. Its advantage is that there are always some free log sectors reserved for the burst of updates. In addition, it is more likely to process merge operations during the idle time of the storage system.

With the dynamic assignment method, the whole log region is managed together. It is possible to reduce the number of merge operations with flexible managements. Because the PCRAM logs have much better endurance than the NAND flash memory data pages, the log sectors could be updated for many times before being merged. Therefore, log sectors could be merged selectively instead of being merged all together. Such policies are described as follows:

- If some erase units are just frequently updated, it is highly possible that these pages will be accessed repeatedly in the near future. Such erase units are named as *hot units* in this work. If we could prevent these hot pages from being merged, the number of merge operations could be reduced. A simple FIFO queue is employed to record most recently updated erase unit. When the merge operation is triggered, erase units in the queue are left untouched so that hot data will not be merged.
- As we know, the data of whole erase unit is copied to a clean space, and the timing overhead mainly depends on the erase time. Therefore, it is not worth merging the pages that have only a few log sectors. In another word, merging such erase units is not efficient because it will not releases many log sectors. Consequently, we could set up a threshold based on the number of log sectors of the erase unit, which is named as *merge threshold*. The erase unit is kept untouched during the merge process if the number of its log sectors are lower than the merge threshold.

The hot unit queue size and the threshold of log sector number have the important impact on the number of merge operations and the lifetime of the system. These issues are further discussed in the experimental sections.

It can be found that the size of metadata for dynamic assignment is larger than that for static assignment. Since we want to share the whole log region among data pages, we need more bits in metadata to store the location (address) of the log sectors. In this work, the metadata is located in the PCRAM because it is normally frequently accessed and modified [26]. It means that the available log region capacity is reduced by storing metadata. In addition, we need to keep the record of hot erase units with a FIFO queue. In this work, a link-based table structure is used for keeping the metadata of log sectors. Note that some data structures, such as the hash table, could be employed to reduce the space and timing complexity of searching and accessing the metadata. Such topics are out of the scope of this work and are not discussed. In the experimental section, we will discuss overhead of storing metadata in details, and we will show that the performance is still improved with the overhead.

3.6 Endurance of the Hybrid Storage Architecture

The lifetime endurance is one important issue of the NAND flash memory. There are various good approaches proposed for wear leveling of NAND flash memory [43–45]. When the in-place updating is enabled in the hybrid architecture, the endurance of the storage system can be improved. Since updates are written to the PCRAM-based log region, the write intensity to the data region is greatly reduced compared to the pure NAND flash memory. For the same applications, the lifetime of the data region is increased with the PCRAM-based log region. Because PCRAM has much better endurance than the NAND flash memory, the log region may still wear slower than the data region does. If we promise that the log region will not wear out before the data region, the endurance of the whole system is increased. As we mentioned in the previous section, we can use selective merge operations to reduce the number of merge operations. The lifetime of the data region is further improved at the same time, but the log region may wear faster. It is a trade-off between performance and the endurance of the log region. Consequently, the wear leveling is necessary for the PCRAM-based log region. In this section, we discuss the issues related to the lifetime of the log region and propose the technique for wear leveling.

3.6.1 Lifetime of the Log Region

In the IPL method, the log pages of an erase unit can only be written once before being merged with data pages. Therefore, log and data pages have the same level of wearing. On the contrary, in our PCRAM hybrid system, the merge operations should be controlled to promise that the log region will not wear out before the data region. Besides the basic merge conditions introduced in the previous section, several other merge conditions are enforced to prevent the log region from wearing too fast.

For the basic static log assignment, since the log sectors assigned to each erase unit are fixed, the lifetime of each erase unit could be controlled individually. Table 3.1 shows that the number of allowed writes of PCRAM is about 1,000 times larger than that of NAND flash memory. Theoretically, if there are 1,000 updates to the same log record, the erase unit should be merged. Then, the data and log regions have the same wear level. In this work, a threshold is set up for each log sector. If a log sector is updated up to 1,000 times, a merge operation is triggered for the erase unit. In the worst case, a single log record in the log sector will not be updated for more than 1,000 times before the erase unit is merged. Therefore, the log region will not wear out faster than the data region does. Such method also works with the group static log assignment. In fact, even for applications with high intensive write operations, such case of forced merge operations rarely happens. With the static assignment, the lifetime of the hybrid system is decided by the data region. Since the number of merge operations is reduced with the hybrid architecture, the endurance is also improved. Note that a 10-bit counter is needed for each log sector.

For the dynamic log assignment, the case is more complicated because the log region is shared among all data pages. It is possible that some log sectors are reused repeatedly by data pages with intensive updates. Therefore, using a threshold as in the static assignment may not efficiently prevent the log region wearing out before the data region does. In addition, the lifetime of the system is related to the management policies. First, the number of merge operations could be reduced if the policy prevents the hot pages being merged. The log sectors belonged to these hot pages may wear faster because they may be used for a long time without being merged. Second, the number of merge operations could be reduced if only data pages with large number of log sectors are merged. It is possible that some data pages are updated for a few times, and then, they are not accessed for a long time, which are called *cold pages*. Since these cold pages would not be merged for a long time, their log sectors will not also be used for a long time, which contributes to the unbalanced wear leveling. We introduce the techniques of wear leveling to deal with these problems and improve the lifetime of the log region.

3.6.2 Wear Leveling

With the dynamic log assignment, when a data page requires for a new log sector, the controller chooses one from the pool of free log sectors. The write intensities of log sectors may be unbalanced with random assignments of log sectors. Therefore, for each request, the log sector with the lowest write numbers should be assigned to even out the write intensities. Theoretically, we should keep tracing the total number of write operations for each log sector. As the PCRAM can normally endure 10^8 write cycles, it needs a 30-bit counter to record the total number of write operations. In order to reduce the area and timing overhead, we can use a small counter instead and reset its value periodically to track the approximate wear level of each log sector. As we introduce in Sect. 3.6.1, a 10-bit counter is needed for each log sector in the static log assignment. This counter can be employed in the dynamic log assignment for wear leveling.

A wear-aware dynamic log assignment is described as follows:

- The free log sector pool is organized as a link structure. Initially, all log sectors are randomly linked together.
- For each requirement, the head sector of the link is evicted and assigned.
- After each merge operation, the released log sectors are inserted into the link structure based on the number of write operations recorded in its counter. The link structure is kept in sorted order according to the number of writes of each log sector.
- When any of the counter reaches the maximum counting number, the write number of the log sectors at the tail of the link structure is subtracted from all counters.

With this structure, the free log sectors are sorted approximately based on their write cycles. The write intensities are balanced among log sectors.

3.7 Evaluation of Metadata and Management Overhead

As we mentioned before, the size of metadata depends on the log assignment and management policies of the PCRAM log region. In order to fully leverage the advantages of PCRAM, the metadata resides in the PCRAM. Therefore, the actual capacity of PCRAM log region is affected by the size of metadata. In this section, the maintenance and management overhead of the metadata is quantitatively analyzed for the wear-aware dynamic log assignment, which has the maximum size of metadata.

3.7.1 Space Overhead

The Record Entry. The metadata should record log sectors that belong to each data page. In this work, the information (address) of each log sector is stored individually

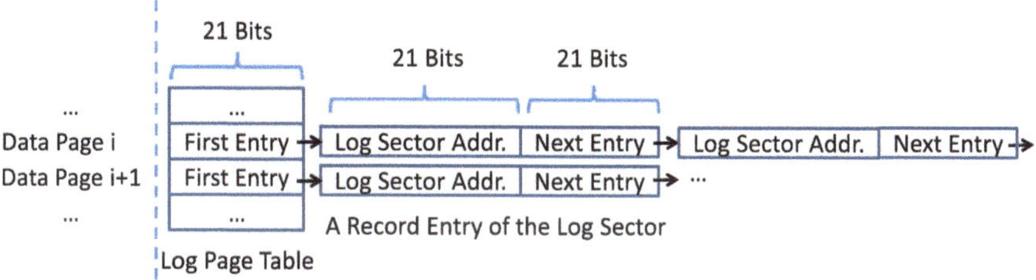

Fig. 3.6 An illustration of the metadata for the wear-aware dynamic log assignment

in the structure named *Record Entry*. Figure 3.6 illustrates a log page table and the link structure for each data page. The log page table points to the entry recording the first log sector that belongs to the data page. The rest entries of log sectors, which belong to the same data page, are stored in the link structure shown in Fig. 3.6. In each entry of the log sector, the address of the next entry is also recorded. Consequently, all log sectors of a data page can be accessed sequentially with this link structure. Since the number of log sectors are fixed, the address of the log sector stored in each entry can be fixed. It means that only the "next entry" part needs to be modified when log sectors are assigned to different data pages. Note that this is not the only structure of metadata. This structure, however, promises that the metadata will not wear out before the corresponding log sector. It is because the metadata space of a log sector is fixed in the PCRAM region, and the metadata is only modified after the log sector is updated.

Assume the size of NAND flash memory is 32 GB, and the total size of PCRAM log region is 1 GB. The size of a log sector is 512 Bytes. Then, the total number of log sectors is 32 GB/512 Bytes $= 2^{21}$. It means that we need 21 bits to represent the address of the log sector in each record entry. In addition, we need extra $14 + 7$ bits to store the address of the next entry. The 14 bits are used to point to the address of the log sector containing the next record entry; the 7 bits are used to represent the relative address inside this log sector. We will explain in the next paragraph why 21 bits are enough for the address of these log sectors.

Since the address of a record entry is 21 bits, the log page table will consume 21×2^{23} bits $= 21$ MB. As we mentioned in the previous subsection, at most 70 % of the log region is occupied. Therefore, at most, there are $0.7 \times 2^{21} \approx 1{,}468{,}007$ record entries of log sectors, which can exist at the same time. These entries will take $1{,}468{,}007 \times (21 + 21)$ bits ≈ 7.4 MB. Consequently, we need 7.4 MB/512 bytes $\approx 2^{14}$ log sectors to store all these record entries. Note that the space is always reserved for the metadata in the PCRAM. It means that the 7.4 MB will not be used for storing information other than metadata. Therefore, the index of these log sectors is separated from the rest log region. That is the reason why we only need 14 (2^14 log sectors in total) bits instead of 21 bits to represent the address of log sectors dedicated for metadata. Each log sector can store 256 KB/42 bits $< 2^7$ record entries. That is the reason we need 7 bits to represent the relative address of a record entry inside a log sector.

Metadata for Wear Leveling. In order to achieve the wear-aware log assignment, we maintain a self-sorted data structure to store the total number of write cycles. In this work, we use the data structure of a self-sorted binary tree. For each node, the write cycle numbers of its child nodes are larger than its own one. Therefore, the root of the tree always records the log sector with the least number of write cycles. Each node of the tree records the address (21 Bits) of the log sector and the total number of write cycles (10 Bits). In the worst case, all log sectors are free and there is one node in the tree for each log sector. In this case, the total size of the tree is $(21 + 10) \times 2^{21} \approx 8\,\text{MB}$. In fact, the experimental results show that the percentage of free log sectors is normally less than 50 %.

Considering all these contributions, we can find that the total size of metadata is normally less than 32 MB. Even in the worst case, the metadata consumes less than 4 % of the PCRAM, and the overhead of metadata is considered in the experiments. For other assignments, the size of metadata is smaller, and the details are not presented due to the page limitation.

3.7.2 Timing Overhead

The management of log assignments also incurs timing overhead. We also take the wear-aware dynamic log assignment as the example to introduce the timing overhead. For the read operation, the latency of accessing the metadata (record entries) varies for each page. With the in-place updating ability, the log size of a data page is no larger than that of one data page. Therefore, it takes less than 100 μs to access all log record entries for each data page.

For the write operation, the metadata need to be updated if there is new log sector assigned to a data page. In the worst case, the whole log sector is occupied with the new updates, and a new log sector is assigned to the data page. In this case, the size of updated PCRAM log region is the same as that in the NAND flash log region. Since the write latency to PCRAM is less than that to NAND flash memory, the latency of this part is reduced.

During the merge operation, it takes less time to release log pages in PCRAM log region than that in NAND flash one, because the average log size of a data page is reduced with the PCRAM log region. In order to achieve the wear-aware log assignment, it takes extra latency to update the sorted tree of free log sector. In the worst case, the timing complexity of searching the binary tree is $Log_2(N)$, where the N is the total number of nodes. Therefore, at most, we need to read 21 (we have 2^{21} nodes) times to search and update the binary tree. The time is less than 100 μs, which is trivial compared to that of a merge operation (>20 ms).

3.8 Experimental Results

In this section, we presents the simulation results of our hybrid architecture with different configurations. Then, the results are analyzed and compared with those of prior work.

3.8.1 Experimental Setup

We use TPC-C benchmarks [42] for the evaluations, which are generated using an open-source tool *Hammerora* [46]. The tool runs with a MySQL database server on a Linux platform under different configurations. In our experiments, the configuration of transactions in these benchmarks are described in the Table 3.2. Note that different sizes of data buffers are simulated.

The current operating systems do not support the management of data buffer as shown in Fig. 3.2. In order to obtain proper log-based accessing traces to NAND flash memory, we implemented a simulation tool of data buffer with the structure shown in Fig. 3.2. This simulation tool uses the processor's memory requests of the database as the input. Then, the accesses to NAND flash memory storage system are generated based on these memory requests with the log-based data buffer. In order to obtain quantitative evaluations of performance, power consumption, and lifetime, we implement the models of PCRAM and corresponding simulation tools in device, circuit, and system levels. For the NAND flash memory, we study recent work and products specs from industries [5, 39, 40, 47], and then, we adjust and integrate proper parameters into our simulation tools.

The sizes of a page, an erase unit, and a log sector are set to be 4 , 256 KB, and 512 Bytes, respectively. For the configuration of the IPL method using only NAND flash memory, there are four log pages in each erase unit. We have mentioned in Sect. 3.3 that if the multi-level storage is supported with the PCRAM log region, the area of log region is not increased if we replace NAND flash memory with PCRAM of the same capacity. In order to explore the advantages of using PCRAM, the size of PCRAM log region is reduced to be half of the flash log region. Therefore, the area of log region is still kept the same even if the multi-level storage is not employed for the PCRAM log region. We will show that performance is still improved. Thus, for the simple static assignment configuration of hybrid architecture, there are 8 KB of PCRAM logs in each erase unit. For the dynamic assignment, the merge operations

Table 3.2 The configuration of transactions in benchmarks

1G.(20-100)M.100u:	1 GByte database, 100 simulated users, the size of buffer pool varies from 20 to 100 MB

are triggered when the size of free log sectors are lower than 30 % of the total size of PCRAM. Then, we ensure that the system will not be stalled for a long time when burst write operations happen.

3.8.2 Write Performance Simulation

As we addressed before, the large overhead of write and erase operations is the obstacle of improving performance of NAND flash memory. Our main goal of using the PCRAM log region is to improve the write performance. Similar to prior work, we also consider the impact of the data buffer size on the write performance. As shown in Fig. 3.7a, the number of write operations is reduced as the size of data buffer increases. Note that the different buffer sizes are considered to show the efficiency of PCRAM log region under different environments. The actual buffer size may be decided in real cases for different applications.

The comparison of merge numbers with different methods is shown in Fig. 3.7b. The *IPL* represents the pure NAND flash memory using the IPL method. The *s-PCRAM* represents the hybrid architecture using the basic static log assignment. The *d-PCRAM* represents the hybrid architecture using the basic dynamic log assignment. It means that there are no hot unit queue or merge threshold for optimizations of merge operations.

The results show that the number of merge operations is reduced when the in-place updating is enabled in the PCRAM log region. For the static log assignment, the number of merge operations is reduced by 22.9 % on average, compared to that of the IPL method. We can find that the hybrid architecture works better when the size of data buffer is smaller. The reason is that, some write operations to the storage system may be filtered when the size of data buffer is increased. Then, the write intensities become more unbalanced. Therefore, the static log assignment method works less efficient (as discussed in Sect. 3.4). We can also find that using the dynamic assignment of log pages further reduces the number of merge operations significantly. On average, the number of merge operations is reduced by 58.5 %, compared to the IPL method. This observation is consistent with the previous discussion. All log

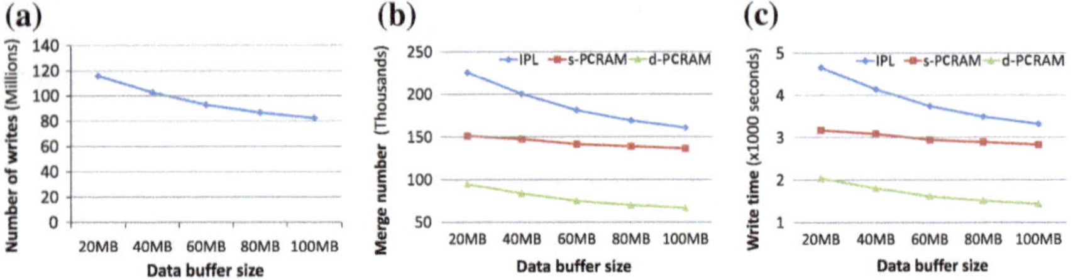

Fig. 3.7 **a** The impact of data buffer size on the write operations; **b** the comparison of merge numbers; **c** the comparison of write time

sectors are assigned on demand with the dynamic assignment. Consequently, we get more benefits from the in-place updating with the dynamic assignment of log pages. The more important thing is that the size of data buffer has little impact on the working efficiency of the dynamic assignment method. As we discussed before, the dynamic assignment method works well even when the write intensities are highly unbalanced.

The write time is also simulated and compared in Fig. 3.7c. The similar conclusion can be drawn for the write time because the results have the same trend as that of merge operations. It is reasonable because the time consumed by merge operations is dominating in total time of write operations.

In Fig. 3.8a, the numbers of merge operations are shown for the hybrid architecture using the group static log assignment. The results include the numbers of merge operations with different group sizes. The number of merge operations decreases as the group size increases, but the improvement is not significant. And we have mentioned that the overhead of management is increased with the size of group. The results show that we can get more benefits from the group method when the size of data buffer is larger. It is because the group method is used to mitigate the effect of unbalanced write intensities, and the write intensities become more unbalanced with a larger size of data buffer.

In Fig. 3.8b, the results of using different sizes of hot unit queues are compared. At the beginning, the number of merge operations decreases as the size of hot queue increases. It means that we can get more benefits from in-place updating of hot erase units, if they are not merged. The number of merge operations is increased when the queue size is too large. It is because too many erase units are kept untouched during the process of merge operations. The capacity of free log sectors is reduced greatly. Furthermore, many erase units in the queue are not hot ones when the queue size is too large. The results show that the method works best with the queue size of 16. The number of merge operations is reduced by 6.9 % on average.

Figure 3.8c shows the results after we set up the threshold number of log sectors for merge operations. The number of merge operations could be reduced significantly after we set up the threshold because the log sectors are released more efficiently for each merge operation. Similarly, the threshold cannot be too large. Otherwise,

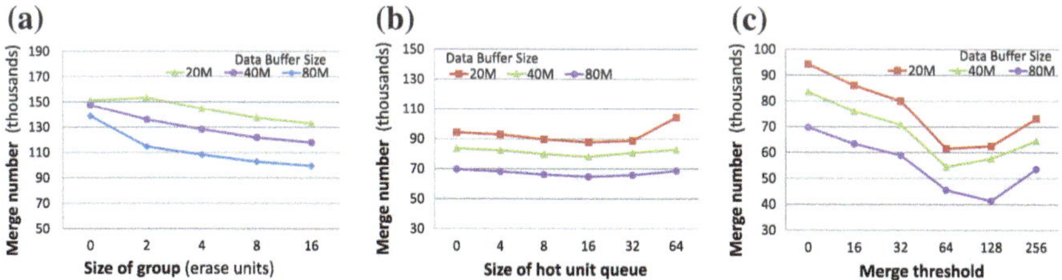

Fig. 3.8 **a** The impact of the group size on the group static assignment. **b** The impact of the hot unit queue size on the dynamic assignment. **c** The impact of the merge threshold on the dynamic assignment

the total number of free log sectors are reduced greatly, and the number of merge operations is increased. The results show that the method works best with the merge threshold of 128. The number of merge operations is reduced by 35.2 % on average.

3.8.3 Read Performance Simulation

For different methods, the sizes of data and log that are read from the NAND flash memory are listed in Table 3.3. The *g-PCRAM-4* represents the hybrid architecture using the group static log assignment, and there are four erase units in each group. The *d-PCRAM-16-128* indicates the hybrid architecture, which uses the dynamic log assignment with the hot unit queue and merge threshold. The size of hot unit queue is set to 16, and the merge threshold is set to 128. The *Average Overhead* shows the average log data's percentage in total data per read operation. The results of static and dynamic assignments are compared to the IPL method, The *Reduction* represents the decrease in overhead for reading log data in each read operation, when results are compared to the IPL ones.

The results show that the average time of a read operation is also reduced with the PCRAM log region because of two reasons. First, for each read operation, the number of loaded log records is reduced with the in-place updating. In addition, the read speed of PCRAM is faster than that of NAND flash memory. The effect of the data buffer is also considered for the read operation. The three sets of results show that we can get similar benefit from using the PCRAM log pages with different sizes of data buffers. One interesting observation is that more log pages are read if we use the dynamic assignment method. It is because the frequently accessed erase units are assigned more log pages. The frequent updates also generate more log pages for the same data page. And the numbers of read operations to these erase units are

Table 3.3 Read performance evaluations for PCRAM log pages

Methods	Data pages read	Log records read	Avg. overhead (%)	Reduction (%)
1 GB *database*, 20 MB *data buffer*				
IPL	103220	6178 pages	5.99	–
g-PCRAM-4	103220	23680 KB	2.87	52.1
d-PCRAM-16-128	103220	33320 KB	4.04	32.6
1 GB *database*, 40 MB *data buffer*				
IPL	87600	5715 pages	6.52	–
g-PCRAM-4	87600	21040 KB	3.00	54.0
d-PCRAM-16-128	87600	29768 KB	4.25	34.9
1 GB *database*, 80 MB *data buffer*				
IPL	68240	4742 pages	6.95	–
g-PCRAM-4	68240	18080 KB	3.31	52.3
d-PCRAM-16-128	68240	24304 KB	4.45	35.9

much larger than those of other ones because of intensive accesses. Consequently, the total number of read operations are increased. Nevertheless, it is worth using the dynamic allocation method because the write performance is increased greatly. With write and read evaluation results, the estimated total execution time is compared for managements policies, which is shown in Fig. 3.10a.

3.8.4 Energy Consumption Evaluation

The energy consumption for different methods is compared in Fig. 3.9. The write operations energy is shown in Fig. 3.9a. The dynamic log assignment consumes the least energy for the same benchmark because the number of merge operations is greatly decreased with this method. Furthermore, the total size of log written to the log region is reduced with the ability of in-place updating. For the read operation, using PCRAM log region can also help to reduce the energy for two reasons. First, the average size of log data loaded with one read operation is reduced. Second, it takes less energy to access the PCRAM than that to access the same size of NAND flash memory. Note that the dynamic assignment method consumes a little more read operation energy than that of the static method. It is also because the frequently accessed erase units are assigned more log pages. The total energy consumption is also compared in Fig. 3.9c. Since the energy of write and merge operations dominates, the dynamic log assignment still consumes the least energy consumption when both read and write operations are considered.

3.8.5 Lifetime Evaluation

In this work, we assume the write cycles of NAND flash memory are 10^5, and the write cycles of PCRAM are 10^8. In order to evaluate the lifetime of the data region, the benchmark traces are kept feeding into the simulation tool. And we keep tracking the number of write operations till one cell of data region wears out. The lifetime

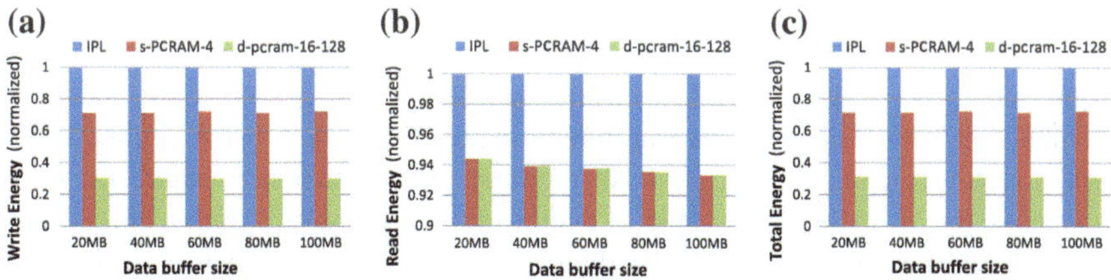

Fig. 3.9 **a** The comparison of write operation energy; **b** the comparison of read operation energy; **c** the comparison of total energy consumption

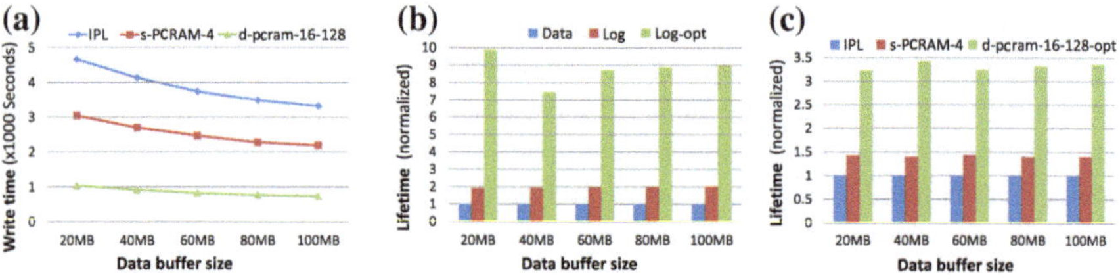

Fig. 3.10 **a** The comparison of the total execution time; **b** the comparison of lifetime between the data region and the log region; **c** the comparison of the whole storage system lifetime with different configurations

of the log region is simulated in the same way. Then, based on the traces, we can estimate and compare the lifetime.

The lifetime of data and log regions with the best dynamic assignment (d-PCRAM-128-6) are compared in Fig. 3.10b. The first column is the lifetime of the data region. The second column is the lifetime of the log region without using any wear leveling technique. The third column is the optimized lifetime of the log region with the wear leveling technique. We can find that, without wear leveling, the lifetime of the log region is just a little longer than that of the data region for benchmarks in this work. It is possible that the log region wears out first for benchmarks with a little higher write intensities. After we use the wear leveling technique, the lifetime of the log region is increased to about 10 times of the data region's lifetime. It promises that the log region will not wear out before the data region does, even for benchmarks with much higher write intensities. Then, the lifetime of the whole storage system is decided by that of the data region. The lifetime of the whole storage system are compared in Fig. 3.10c with different configurations. The results show that the lifetime of the whole storage system is improved after using the PCRAM log region. It is because the number of erase operations is decreased with the same write intensity. Consequently, the hybrid storage system with the dynamic log assignment has the longest lifetime.

3.9 Conclusion

The performance of NAND flash memory is limited because of its erase-before-write requirement, which does not allow in-place updating. The optimizations and managements on NAND flash memory architecture are also limited by this problem. The hybrid architecture using NAND flash memory and another nonvolatile memory, such as PCRAM, makes it possible to exploit the advantages of both technologies. Our PCRAM-based log region method has shown that the performance of the flash memory could be improved significantly, if the log region allows the in-place updating even without the support of multi-level storage. We also show that our method can decrease the overhead of read operations and increase the lifetime of the storage system. Furthermore, the energy consumption of both read and write operations is reduced. More important, the in-place updating capability of PCRAM provides more flexible management policies that enable further optimizations of performance and lifetime. In the future, we plan to study the hybrid nonvolatile memory architecture that combines other emerging memory technologies (such as MRAM and RRAM) with the NAND flash memory to further explore the efficiency of using nonvolatile memories.

References

1. Kgil, T., Roberts, D., & Mudge, T. (2008). Improving NAND flash based disk caches. *In Proceedings of International Symposium on Computer Architecture* (pp. 327–338). Doi: http://dx.doi.org/10.1109/ISCA.2008.32
2. Tiwari, D., Vazhkudai, S.S., Kim, Y., Ma, X., Boboila, S., Desnoyers, P.J. (2012). Reducing data movement costs using energy efficient, active computation on ssd. In *Proceedings of the 2012 USENIX conference on Power-Aware Computing and Systems, HotPower'12* (pp. 4–4). USENIX Association, Berkeley, CA, USA. http://dl.acm.org/citation.cfm?id=2387869.2387873
3. Reinsel, D., & Janukowicz, J. (2008). White paper: Datacenter SSDs: Solid footing for growth. http://www.samsung.com/global/business/semiconductor/ products/flash/FlashApplicationNote.html
4. Lee, S., Moon, B., Park, C., Kim, J., & Kim, S. (2008). A case for flash memory SSD in enterprise database applications. In *Proceedings of ACM International Conference on Management of Data*, (pp. 1075–1086). Doi: http://doi.acm.org/10.1145/1376616.1376723
5. Lee, S., & Moon, B. (2007). *Design of flash-based DBMS: An in-page logging approach.* In *Proceedings of ACM International Conference on Management of Data.*
6. Lee, K., Kim, H., Woo, K., Chung, Y., & Kim, M. (2009). Design and implementation of MLC NAND flash-based DBMS for mobile devices. *Journal of Systems and Software* (March, 2009). Doi: http://doi.acm.org/10.1145/1289881.1289911
7. Toshiba America Electronic Components, Inc.: NAND flash applications design guide (2004).
8. Leventhal, A. (2008). Flash storage memory. *Journal of Communications of the ACM, 51*(7), 47–51. Doi: http://doi.acm.org/10.1145/1364782.1364796
9. Chiang, M., & Chang, R. (1999). Cleaning policies in mobile computers using flash memory. *Journal of Systems and Software, 48*(3), 213–231. Doi: http://dx.doi.org/10.1016/S0164-1212(99)00059-X
10. Rosenblum, M., & Ousterhout, J. (1992). The design and implementation of a log-structured file system. *ACM Transaction on Computer Systems, 10*(1), 26–52. Doi: http://doi.acm.org/10.1145/146941.146943
11. Birrel, A., Isard, M., Thacker, C., & Wobber, T. (2005). A design for high-performance flash disks. *Technical Report MSR-TR-2005-176*, Microsoft Research.
12. Caulfield, A.M., Grupp, L.M., & Swanson, S. (2009). Gordon: using flash memory to build fast, power-efficient clusters for data-intensive applications. In *Proceeding of the 14th International Conference on Architectural Support for Programming Languages and Operating Systems* (pp. 217–228). Doi: http://doi.acm.org/10.1145/1508244.1508270
13. Gupta, A., Kim, Y., & Urgaonkar, B. (2009). DFTL: a flash translation layer employing demand-based selective caching of page-level address mappings. In *Proceeding of the 14th International Conference on Architectural Support for Programming Languages and Operating Systems*, (pp. 229–240). Doi: http://doi.acm.org/10.1145/1508244.1508271

14. Kim, S., & Jung, S. (2006). A log-based flash translation layer for large NAND flash memory. *In Proceedings of International Conference on Advanced Communication Technology, 3*, 1641–1644. doi:10.1109/ICACT.2006.206302.
15. Sun, G., Dong, X., Xie, Y., Li, J., & Chen, Y. (2009). A novel architecture of the 3d stacked mram l2 cache for cmps. In *High Performance Computer Architecture*, 2009. HPCA 2009. 15th International Symposium on IEEE (pp. 239–249). doi: 10.1109/HPCA.2009.4798259.
16. Wu, X., Li, J., Zhang, L., Speight, E., Rajamony, R., & Xie, Y. (2009). Hybrid cache architecture with disparate memory technologies. In *Proceedings of the 36th Annual International Symposium on Computer Architecture, ISCA '09* (pp. 34–45). doi: 10.1145/1555754.1555761.
17. Dong, X., Wu, X., & Sun, G., et al. (2008). Circuit and Microarchitecture Evaluation of 3D Stacking Magnetic RAM (MRAM) as a Universal Memory Replacement. In *DAC '08: Proceedings of the 45th Annual Conference on Design Automation* (pp. 554–559).
18. Zhang, W., & Li, T. (2009). *Exploring phase change memory and 3D die-stacking for power/thermal friendly, fast and durable memory architectures*. In International Conference on Parallel Architectures and Compilation Techniques.
19. Zhou, P., Zhao, B., Yang, J., & Zhang, Y. (2009). A durable and energy efficient main memory using phase change memory technology. In *Proceedings of ISCA* (pp. 14–23).
20. Lee, B.C., Ipek, E., Mutlu, O., & Burger, D. (2009). Architecting phase change memory as a scalable DRAM alternative. In *Proceedings of ISCA* (pp. 2–13).
21. Qureshi, M.K., Srinivasan, V., & Rivers, J.A. (2009). Scalable high performance main memory system using phase-change memory technology. In *Proceedings of ISCA* (pp. 24–33).
22. Qureshi, M., Franceschini, M., & Lastras-Montano, L., (2010). Improving read performance of phase change memories via write cancellation and write pausing. In High Performance Computer Architecture (HPCA), (2010). *IEEE 16th International. Symposium*, (pp. 1–11). doi:10.1109/HPCA.2010.5416645.
23. Qureshi, M.K., Karidis, J., Franceschini, M., Srinivasan, V., Lastras, L., & Abali, B. (2009). Enhancing lifetime and security of pcm-based main memory with start-gap wear leveling. In *Proceedings of the 42nd Annual IEEE/ACM International Symposium on Microarchitecture, MICRO 42* (pp. 14–23). ACM, New York, NY, USA. doi:10.1145/1669112.1669117.
24. Zhou, P., Zhao, B., Yang, J., & Zhang, Y. (2009). A durable and energy efficient main memory using phase change memory technology. In *Proceedings of the 36th annual international symposium on Computer architecture, ISCA '09* (pp. 14–23). doi:10.1145/1555754.1555759.
25. Seong, N.H., Woo, D.H., & Lee, H.H.S. (2010). Security refresh: prevent malicious wear-out and increase durability for phase-change memory with dynamically randomized address mapping. In *Proceedings of the 37th annual international symposium on Computer architecture, ISCA '10* (pp. 383–394). ACM, New York, NY, USA. doi:10.1145/1815961.1816014.
26. Park, Y., Lim, S., Lee, C., & Park, K. (2008). PFFS: a scalable flash memory file system for the hybrid architecture of phase-change RAM and NAND flash. In *Proceedings of ACM Symposium on Applied Computing*.
27. Kim, J., Lee, H., Choi, S., & Bahng, K. (2008). A PCRAM and NAND flash hybrid architecture for high-performance embedded storage subsystems. In *Proceedings of ACM International conference on Embedded software* (pp. 31–40). Doi: http://doi.acm.org/10.1145/1450058.1450064
28. Lam, C. (2008). Cell design considerations for phase change memory as a universal memory. *In Proceedings of International Symposium on VLSI Technology, Systems and Applications* (pp. 132–133). doi:10.1109/VTSA.2008.4530832.
29. Lee, K., et al. (2007). A 90nm 1.8V 512Mb diode-switch PCRAM with 266MB/s read throughput. In *Proceedings of IEEE International Solid-State Circuits Conference* (pp. 472–616). doi:10.1109/ISSCC.2007.373499.
30. Lee, B.C., Ipek, E., Mutlu, O., & Burger, D. (2009). Architecting phase change memory as a scalable dram alternative. *SIGARCH Computer Architecture News, 37*(3), 2–13. doi:10.1145/1555815.1555758.
31. Cho, S., & Lee, H. (2009). Flip-n-write: a simple deterministic technique to improve pram write performance, energy and endurance. In *Proceedings of the 42nd Annual IEEE/ACM International Symposium on Microarchitecture, MICRO 42* (pp. 347–357). ACM, New York, NY, USA. doi:10.1145/1669112.1669157.

32. Caulfield, A.M., De, A., Coburn, J., Mollow, T.I., Gupta, R.K., & Swanson, S. (2010). Moneta: A high-performance storage array architecture for next-generation, non-volatile memories. In *Proceedings of the 2010 43rd Annual IEEE/ACM International Symposium on Microarchitecture, MICRO '43* (pp. 385–395). IEEE Computer Society, Washington, DC, USA. doi:10.1109/MICRO.2010.33.
33. Akel, A., Caulfield, A.M., Mollov, T.I., Gupta, R.K., & Swanson, S. (2011). Onyx: a protoype phase change memory storage array. In *Proceedings of the 3rd USENIX conference on Hot topics in storage and file systems, HotStorage'11* (pp. 2–2). USENIX Association, Berkeley, CA, USA http://dl.acm.org/citation.cfm?id=2002218.2002220
34. Venkataraman, S., Tolia, N., Ranganathan, P., & Campbell, R.H. (2011). Consistent and durable data structures for non-volatile byte-addressable memory. In *Proceedings of the 9th USENIX conference on File and storage technologies, FAST'11* (pp. 5–5). USENIX Association, Berkeley, CA, USA. http://dl.acm.org/citation.cfm?id=1960475.1960480
35. Kang, D., Ahn, D., Kim, K., Webb, J., & Yi, K. (2003). One-dimensional heat conduction model for an electrical phase change random access memory device with an $8f^2$ memory cell (f=0.15 μm). *Journal of Applied Physics, 94*, 3536–3542. doi:10.1063/1.1598272.
36. Hudgens, S. (2006). OUM nonvolatile semiconductor memory technology overview. In *Proceedings of Materials Research Society Symposium.*
37. Zhang, Y., et al. (2007). An integrated phase change memory cell with Ge nanowire diode for cross-point memory. *In Proceedings of IEEE Symposium on VLSI Technology* (pp. 98–99). doi:10.1109/VLSIT.2007.4339742.
38. Park, S., Jung, D., Kang, J., Kim, J., & Lee, J. (2006). CFLRU: a replacement algorithm for flash memory. Architecture and Synthesis for Embedded Systems: In *Proceedings of International Conference on Compilers* (pp. 234–241).
39. Samsung Electronics: datasheet K9G8G08UOM (2006).
40. Samsung Electronics: datasheet KPS1215EZM (2006).
41. Nirschl, T., et al. (2007). Write strategies for 2 and 4-bit multi-level phase-change memory. *In Proceedings of IEEE International Electron Devices Meeting* (pp. 461–464). doi:10.1109/IEDM.2007.4418973.
42. http://www.tpc.org
43. Increasing flash solid state disk reliability. Technical Report, SiliconSystems (2005).
44. Chang, Y., Hsieh, J., & Kuo, T. (2007). Endurance enhancement of flash-memory storage systems: An efficient static wear leveling design. In *Proceedings of Design Automation Conference* (pp. 212–217). Doi: http://doi.acm.org/10.1145/1278480.1278533
45. Jung, D., Chae, Y., Jo, H., Kim, J., & Lee, J. (2007). A group-based wear-leveling algorithm for large-capacity flash memory storage systems. In *Proceedings of International Conference on Compilers, Architecture and Synthesis for Embedded Systems* (pp. 160–164). doi: http://doi.acm.org/10.1145/1289881.1289911
46. http://hammerora.sourceforge.net/
47. Shibata, N., et al. (2007). A 70 nm 16GB 16-Level-Cell NAND flash memory. *Proceedings of IEEE Symposium on VLSI Circuits, 43*(4), 929–937. doi:10.1109/JSSC.2008.917559.

2. Caulfield, A.M., [illegible] Coburn, J., Mollov, T.I., Gupta, R.K., [illegible] Swanson, S.: [illegible] high-performance, energy-efficient [illegible] for emerging [illegible] memories. In: [illegible] Annual IEEE/ACM International Symposium on Microarchitecture, [illegible] IEEE Computer Society, [illegible] (2010) [illegible]
3. [illegible]
4. [illegible]
5. [illegible]
6. [illegible]
7. [illegible]
8. [illegible] In: Proceedings of [illegible] Computer [illegible] (2008)
9. Samsung electronics datasheet K9L8G08U0M (2006)
10. Samsung electronics datasheet K9WAG08U1A (2006)
11. [illegible] et al. (2008) [illegible] 2 and 3 bits/cell level phase change memory [illegible]
12. http://www.[illegible]
13. [illegible] Technical Report, [illegible]
14. Chang, Y., [illegible] Kuo, T.: [illegible] In: Proceedings of Design Automation Conference, pp. 212–217. Doi: http://doi.acm.org/10.1145/[illegible]
15. Jung, D., Chae, Y., Jo, H., Kim, J.S., Lee, J. (2007). A group-based wear-leveling algorithm for large-capacity flash memory storage systems. In: Proceedings of International Conference on Compilers, Architecture and Synthesis for Embedded Systems, pp. 160–164. doi: http://doi.acm.org/10.1145/1289881.1289911
16. http://[illegible]
17. [illegible], N., et al. (2007). A 70 nm 16 Gb 16-level-cell NAND flash memory. Proceedings of [illegible] doi:10.1109/JSSC.2007.917[illegible]

Chapter 4
Energy Efficient Systems Using Resistive Memory Devices

Meng-Fan Chang and Pi-Feng Chiu

Abstract Energy efficient systems use volatile memory (VM) for fast and low-voltage access and use nonvolatile memory (NVM) for power-off-mode data storage. However, memory is becoming a bottleneck for electronic systems attempting to achieve low energy consumption. The resistive memory (memristor) is a promising NVM for energy efficient computing because of its fast write speed and low-power operations. This chapter provides an overview of the circuit design technologies and applications of resistive memory devices for energy efficient systems, including resistive RAM (ReRAM), nonvolatile logic, and nonvolatile SRAM.

4.1 Introduction

Energy efficient computing is critical to mobile or battery-powered electronic systems. Battery-powered chips currently face the challenges of high computing performance, low-power operation, a small form factor, and reliable data backup in the event of a sudden power failure. The thermal problem is also a key factor preventing the incorporation of many functional blocks on a 2D or 3D chip, resulting in problems with reliability, speed degradation, incremental power leakage, and thermal runaway effects.

Figure 4.1 shows a comparison of the performance of volatile memory (VM), nonvolatile memory (NVM), and emerging nonvolatile memory (ENVM). SRAM achieves the highest speeds and lowest operational voltage, but is unable to store data when the power supply is turned off. Logic ROMs [1–5] have nonvolatile feature and fast read speed, but lack flexibility in on-field data updating. Embedded Flash [2, 6–8]

M.-F. Chang (✉) · P.-F. Chiu
Department of Electrical Engineering, National Tsing Hua University (NTHU), 30013 Hsin Chu, Taiwan, Republic of China
e-mail: mfchang@ee.nthu.edu.tw

Y. Xie (ed.), *Emerging Memory Technologies*,
DOI: 10.1007/978-1-4419-9551-3_4,

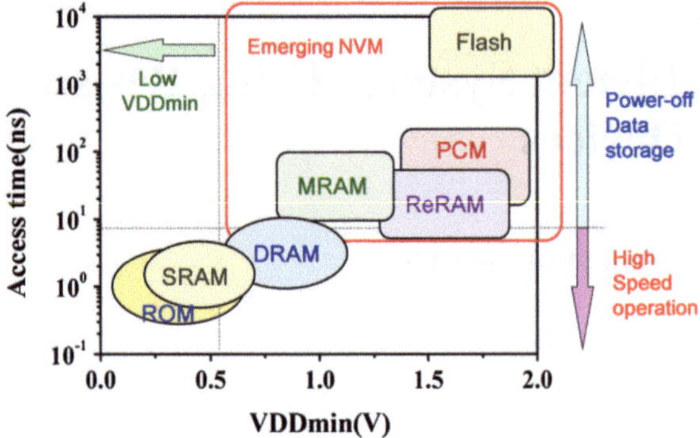

Fig. 4.1 Speed versus VDDmin for various memory types

allows on-field programming and has nonvolatile features, but requires high-voltage operation and suffers from a slow write speed.

The intelligent use of the power-off mode can help suppress the power consumption of chips, particularly nanometer chips with large standby power, as long as the data backup/restore operation (power-on/off) does not consume too much power. However, critical data must be first flushed to NVM to prevent data loss, as shown in Fig. 4.2. If NVM are capable of write operations at lower power levels, energy consumption can be reduced even further.

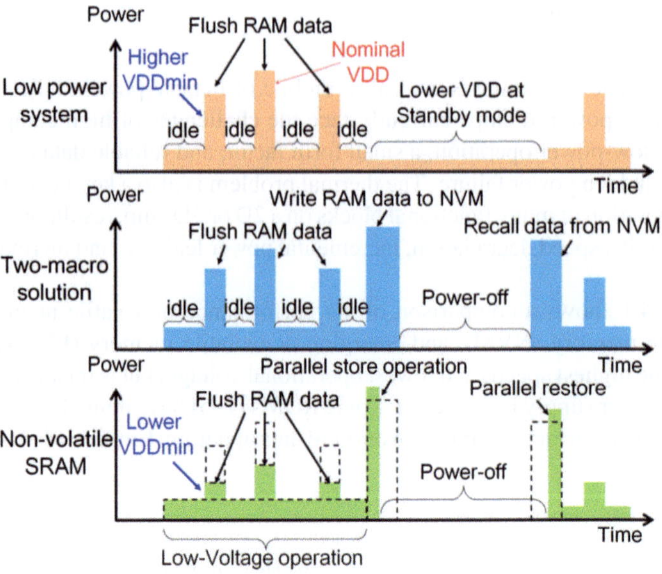

Fig. 4.2 Power consumption of a energy efficient system during active and standby modes

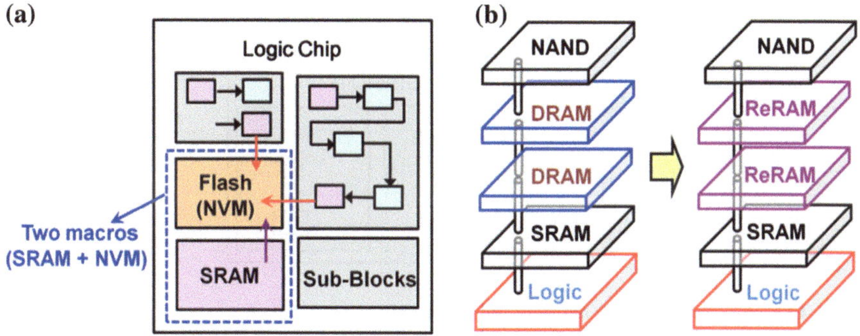

Fig. 4.3 Conventional two-macro scheme **a** 2D and **b** 3D structures

To address the gaps in speed and minimum operation voltage (VDDmin) between SRAM and NVM, many low-power chips use SRAM/DRAM for fast or low-VDDmin access [2, 6–10] and flash memory for data storage in power-off mode, instead of using a single NVM macro for both computing and power-off storage. This two-macro (VM + NVM) scheme (Fig. 4.3a) also reduces the number of NVM accesses and relaxes the endurance requirements for NVM. If nonvolatile random access memory (nv-RAM) can achieve fast random read/write speeds and high endurance, it can provide alternative memory architectures for 3D-ICs without the need for DRAM, as shown in Fig. 4.3b. Thus, nv-RAM can reduce power consumption by eliminating the DRAM self-refresh power required for additional DRAM-NVM data transfer.

As discussed in [7, 8, 11, 12], the two-macro scheme consumes a considerable amount of energy during power-off data backup operations and requires long store/restore times because of word-by-word (serial) SRAM read/write and long NVM write/read procedures. This is mainly because of the slow and power-consuming NVM access. This store/restore behavior results in extended power-on/power-off time. The long store time makes the two-macro scheme vulnerable to data loss in the event of sudden power failure. The large peak current consumed by flash memory also causes significant power integrity degradation for a system, particularly for 3D-ICs.

In conventional power-off operations, a chip only stores a limited amount of data and global operating states from SRAM/DRAM in NVM. This is because of the constraints of NVM capacity and the store time required for power-off operation. This causes the chip to lose detailed information (or local states) regarding sub-blocks, which are stored in local registers or flip-flops, following power-off operation. Without restoring the local states back to sub-blocks, the chip requires additional time and power to repeat computations in returning to its prior-power-off state, following a power-on procedure. In the worst-case scenario, and particularly after a sudden power failure, the system may fail to recover its prior-power-off state. Nonvolatile logic (nvLogic) [7, 8, 11–14] enables the store/restore of local states during power-off/power-on operations without the risk of data loss or the consumption of additional

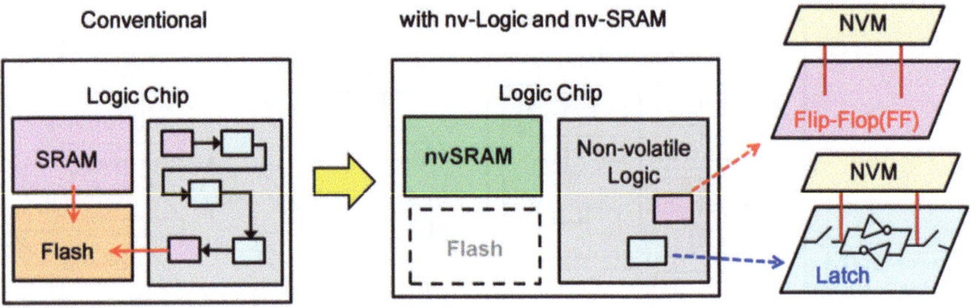

Fig. 4.4 A chip structure with nvLogic and nvSRAM

power/time in re-computing the pre-power-off state. Section 3.3 presents a discussion of the nvLogic.

Researchers have proposed concept called nonvolatile SRAM (nvSRAM) to deliver faster store speeds than those offered by conventional two-macro schemes. The nvSRAM integrates SRAM cells and NVM devices within a single cell, forming a direct bit-to-bit connection in a 2D/3D arrangement to achieve rapid parallel data transfer and fast power-on/power-off speed. Sections 3.3 and 3.4 discuss the nvSRAM.

By using both nvLogic and nvSRAM, a chip may have an alternative structure to achieve low-power operation and reliable data backup while a small area overhead, as shown in Fig. 4.4.

To ensure low power consumption, and thus a long battery lifetime, mobile systems require a low dynamic power consumption and a low active-mode standby current. Many portable chips employ dynamic voltage scaling (DVS) schemes [15, 16] or use a low supply voltage (VDD) to reduce power consumption. These methods reduce dynamic power consumption and active-mode leakage current, especially in ultra-low-power systems and health/biomedical applications [17–19]. Figure 4.5 shows an example of dynamic power reduction using a lower VDD for a 90 nm SRAM macro.

In both 2D and 3D chips, the thermal effect is one of the bottlenecks in reducing system standby power. Figure 4.6 displays the measured leakage current of a 512 Kb SRAM macro at three operating temperatures. Increasing the operating temperature from its lowest value to its highest increases the SRAM leakage current by a factor of more than 150x. To suppress the thermal effect and power consumption, the VDD must be reduced.

Currently emerging resistive memory (ReRAM/memristor) [20–32] shows considerable promise because it offers the advantages of rapid write time, low write voltage, large R-ratio, multilevel capability, and relatively low write power consumption compared to Flash and other NVMs. These features make memristor technology a good candidate for implementation in nvLogic and nvSRAM devices requiring low store power and fast store time. However, current memristor devices suffer from limited endurance, which influences the design of the circuit structure for memristor-based nvLogic and nvSRAM. This section reviews the various circuit structures of

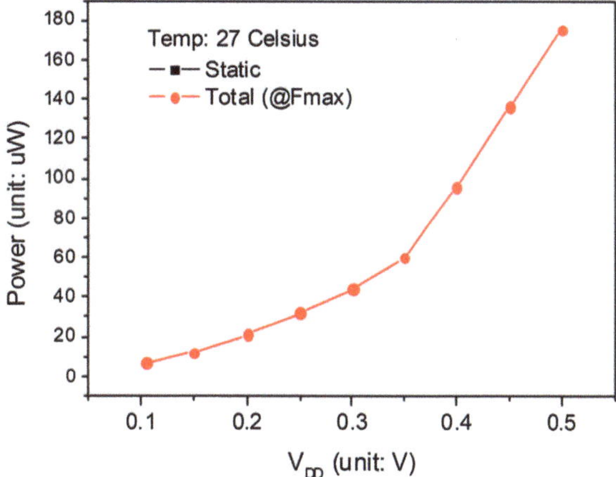

Fig. 4.5 Dynamic power of an on-chip SRAM macro

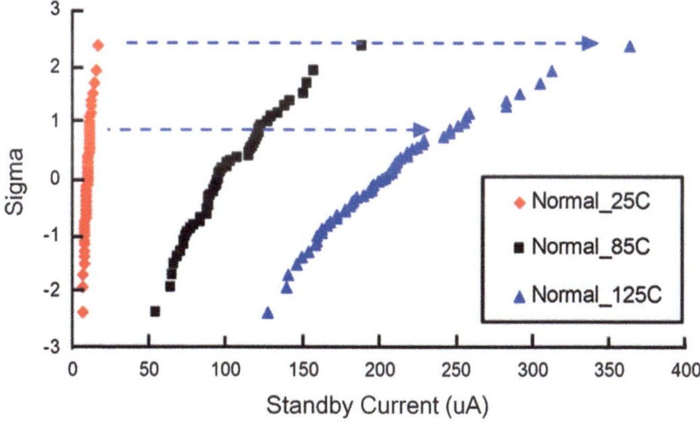

Fig. 4.6 Leakage current of an on-chip SRAM macro

endurance-aware memristor-based nvLogic and nvSRAM, especially for low-voltage applications.

4.2 Fundamental of ReRAM (Memristor) Technology

4.2.1 An Example of ReRAM Devices: HfO_2-Based Bipolar ReRAM

Figure 4.7 shows a resistive memory device consisting of a TiN/TiOx/HfOx/TiN resistive memory device [33–36] and a NMOS switch transistor. This resistive memory cell uses the back-end-of-line (BEOL) process for the resistive device and

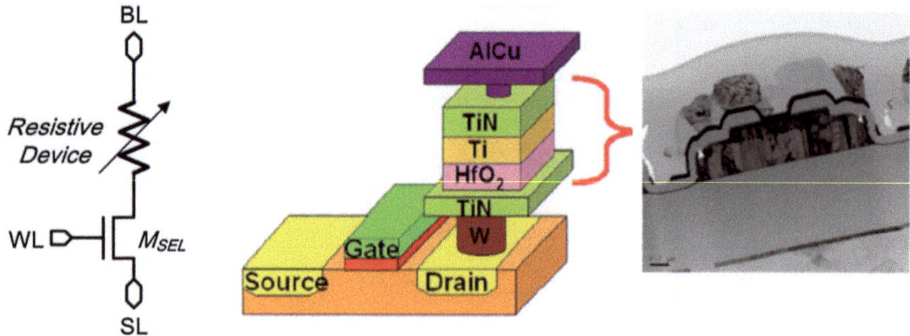

Fig. 4.7 Structure of employed HfO_2-based bipolar ReRAM cell

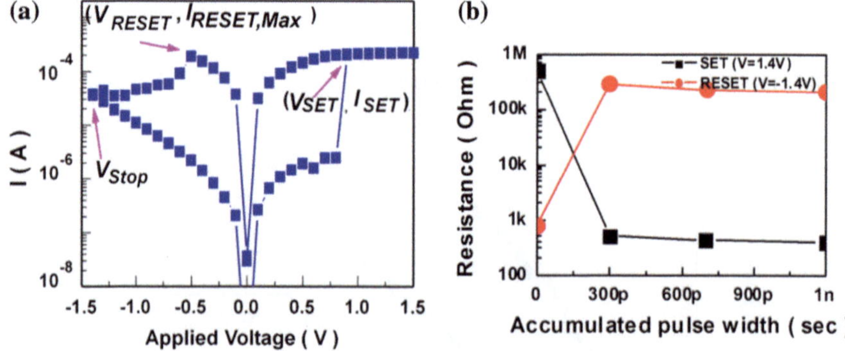

Fig. 4.8 **a** IV curve and **b** switching behavior of the HfOx-based ReRAM device

Table 4.1 SLC operation

	0 (SET)	1 (RESET)	Read
WL	V_{G_SET}	VDD	VDD
BL	V_{SET}	0	V_{BL-R}
SL	0	V_{RESET}	0
State	$LRS\ (R_L)$	$HRS\ (R_H)$	'1'/'0'

standard CMOS logic technology for the NMOS switch. This structure makes the resistive memory cell appropriate for combination with CMOS logic technology.

This resistive memory cell is capable of two direct overwrite operations during SLC operation: SET and RESET. Figure 4.8a shows the IV curve of this HfO_2-based bipolar ReRAM device. The bipolar resistive switching behavior of this device defines write '0' and write '1' operations according to the polarity of the bias voltage. Data-0 is represented by a low-resistance state (LRS), and data-1 is represented by a high-resistance state (HRS). Table 4.1 shows the operating conditions of this resistive memory cell. The SET operation changes the resistive device from HRS to LRS by applying a SET voltage (V_{SET}) to the bit line (BL) and connecting the source line (SL) to the ground. A RESET operation applies a RESET voltage (V_{RESET}) to the SL and connects the BL to the ground. Similar to the SLC operation, this resistive memory

Table 4.2 MLC operation

	00	01	10	11	Read
WL	V_{G_SET1}	V_{G_SET2}	V_{G_SET3}	VDD	VDD
BL	V_{SET}	V_{SET}	V_{SET}	0	V_{BL-R}
SL	0	0	0	V_{RESET}	0
State	LRS_1 (R_{00})	LRS_2 (R_{01})	LRS_3 (R_{10})	HRS (R_{11})	00~11

cell supports MLC operation by applying different RESET operating conditions, as shown in Table 4.2.

As shown in Fig. 4.8b [36], this resistive device has fast switching behavior and small write current (<25 μA), characteristics, but requires a V_{BL-R} below 0.3 V to prevent read disturbance.

4.2.2 Speed, Power, and Endurance of Resistive Memory (Memristor) Devices

Many resistive memory (memristor) devices have been reported in recent years. Figure 4.9 presents a comparison of write current and switching time for various NVM devices. ReRAM/memristor devices clearly achieve faster write times with a smaller write current than other NVM devices. Figure 4.10 shows the write

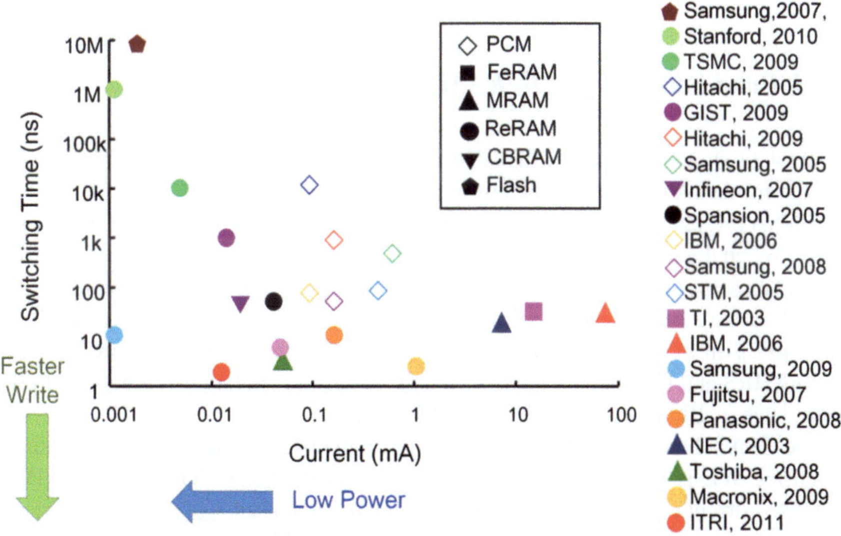

Fig. 4.9 Write current and switching time of NVM devices

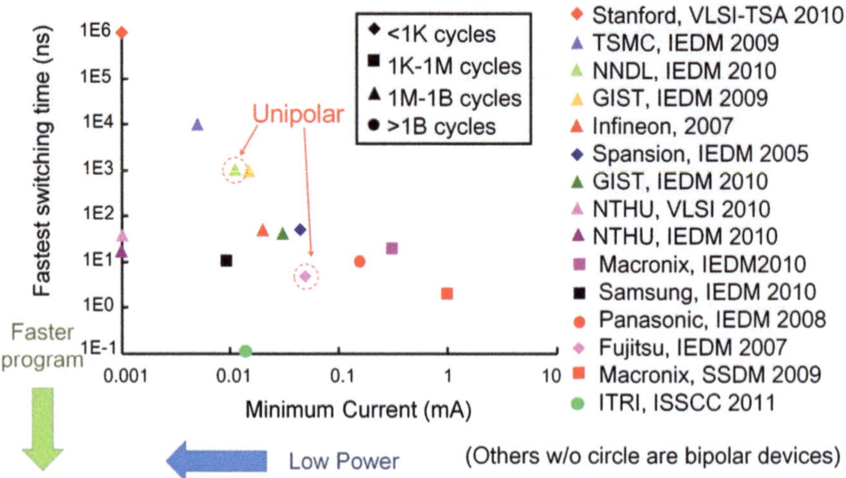

Fig. 4.10 Write current and switching time of recent ReRAM devices

performance of various ReRAM/memristor devices, including bipolar and unipolar devices. Several resistive memory devices consume only a small amount of current and provide rapid switching time. Thus, ReRAM is one of the most promising candidates for use as nvLogic and nvSRAM among all forms of NVM devices.

However, current ReRAM devices have limited endurance (Fig. 4.10). This endurance issue has influenced the design of nvLogic and nvSRAM circuit structures.

4.2.3 Read Schemes for ReRAM Arrays and Macros

Figure 4.11 shows the waveforms of two commonly used sense amplifiers (SA): the voltage-mode SA (VSA) and the current-mode SA (CSA).

A VSA pre-charges the selected BLs to a targeted voltage (V_{PRE}) in the pre-charge phase. When the word line (WL) is on, the BL is discharged by I_{CELL} to read an LRS cell and remains at V_{PRE} to read an HRS cell. If the BL loading is large and the I_{CELL} is small, a long BL developing time is required for high-yield sensing. Adopting a low V_{BL-R} in a VSA limits the BL voltage difference (sensing margin) between reading HRS and LRS cells. This in turn makes the read operations vulnerable to BL noise, such as BL cross talk and WL-to-BL coupling.

The CSA scheme uses a fixed bias voltage on the BL to induce I_{CELL} for reading. When the WL is on, the LRS cell generates a larger I_{CELL} than that of an HRS cell. The current comparator then compares the to-be-sensed I_{CELL} with a reference current (I_{REF}) to determine the sensing result. Figure 4.12 shows the read access time of the VSA and CSA schemes for various BL lengths in ReRAM. When reading a

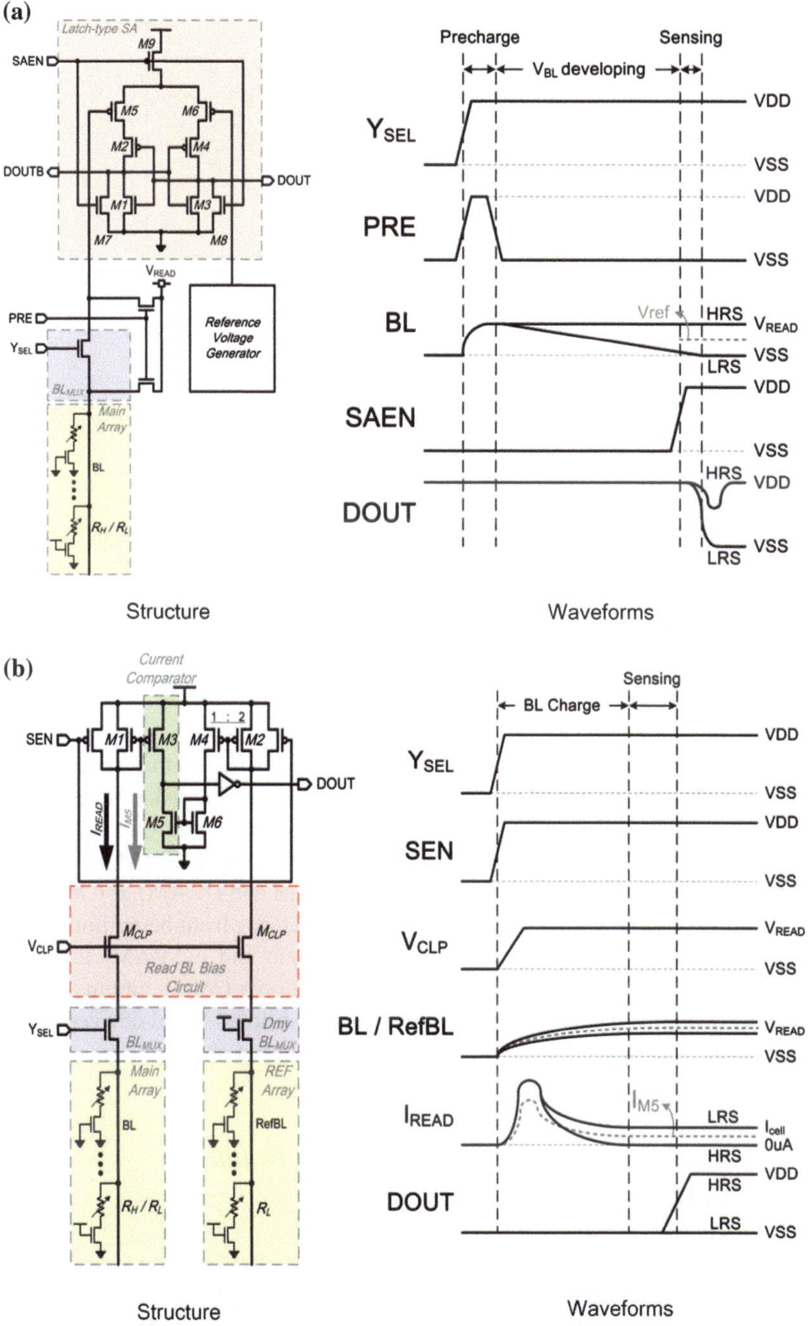

Fig. 4.11 Sensing schemes in NVM **a** voltage-mode sensing; **b** current-mode sensing

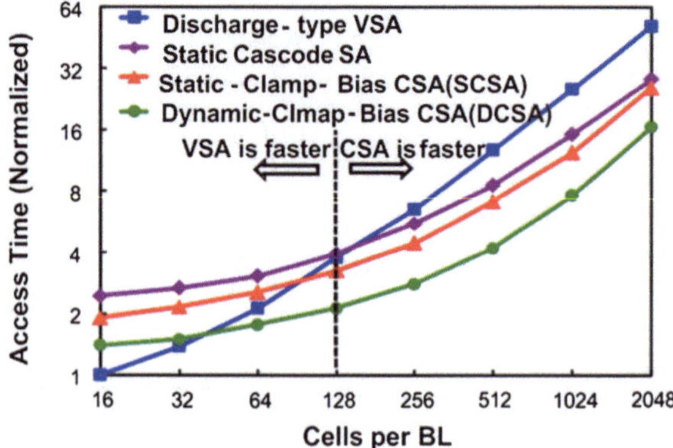

Fig. 4.12 Random read and write access times across various BL lengths for a HfO_2-based ReRAM device

0.18 um ReRAM LRS cell with $I_{CELL} = 20$ uA, a CSA achieves a faster read speed than a VSA when the BL length exceeds 128 rows. Thus, choosing the appropriate sensing scheme for ReRAM depends on the resistance and cell array structures.

4.3 Memristor-Based nvLogic and nvSRAM

As discussed in Sect. 1.1, nonvolatile logic (nvLogic) enables the store/restore of local states during power-off/power-on operations without the risk of losing data or consuming additional power/time in re-computing the pre-power-off state.

Various types of NVM devices, including Flash, MRAM, PCRAM, and ReRAM (memristor), have far less endurance than VM (SRAM and DRAM) or CMOS logic (latch, flip-flop, logic gates). This prevents NVM devices from being implemented in regular computing or as access units for nvLogic and nvSRAM. Endurance-aware nvLogic and nvSRAM must employ conventional CMOS circuits for computing/accessing. When used in conjunction with additional NVM devices, they act as power-on/power-off data storage units, as shown in Fig. 4.13.

4.3.1 Nonvolatile-Latch (nvLatch) and Nonvolatile SRAM Cells

Figure 4.14 shows a conventional latch or 6T SRAM cell. Figure 4.15 shows several reported nonvolatile-latch (nvLatch) cells or nvSRAM cells derived from conventional 6T SRAM cells. These nvSRAM cells are equipped with additional transistors to provide NVM devices with storage and isolation functions. These nonvolatile

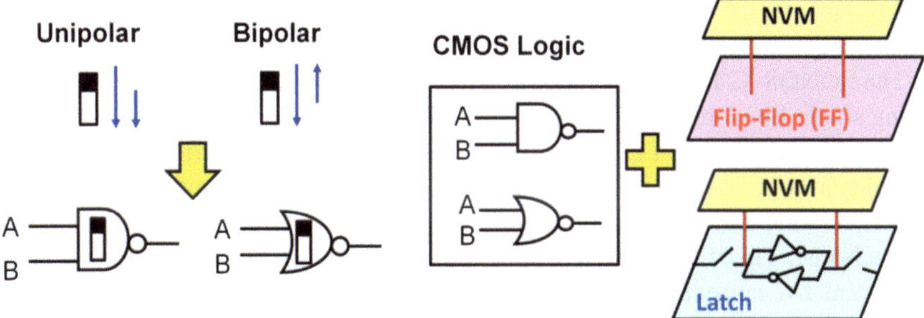

Fig. 4.13 NvLogic using nvLatch or nvFF

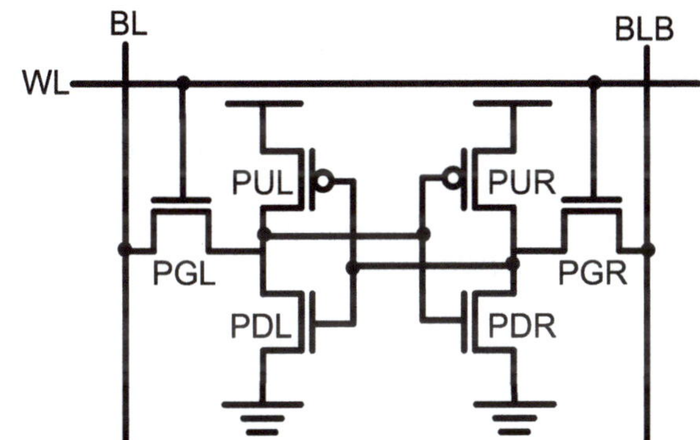

Fig. 4.14 Conventional 6T latch/SRAM cell

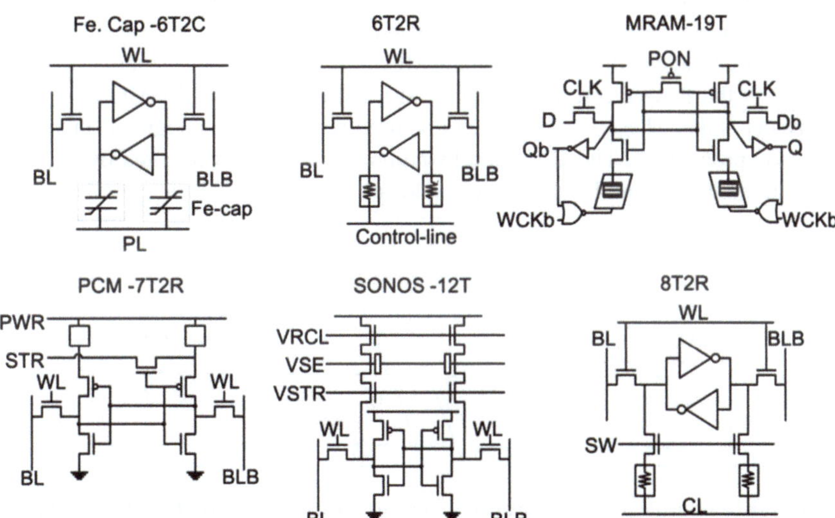

Fig. 4.15 Reported nonvolatile-latch (nvLatch) cells and nvSRAM cells

SRAM designs achieve faster store/restore speeds than the two-macro approach for mobile chips because of their bit-to-bit parallel data storage.

The SONOS-12T cell [37] requires a large area and a high store voltage and has a long single-bit store time (T_{STORE}) because of the slow program time of SONOS devices. The MRAM-19T2R latch [38] consumes a large store current (I_{STORE}) and occupies a large area. The Fe-capacitor-6T2C cell [39] has a much smaller area than 12T-19T nvLatch cells, but requires an extra 1/2VDD bias to read SRAM and access NVM. This 1/2VDD bias requires an extra voltage regulator and consumes significant DC current. The PCM-7T2R [40] cell has a competitive cell area, but requires a large I_{STORE} and a moderate T_{STORE} because of the write behavior of the PCM device. The PCM-7T2R cell has degraded cell VDD (CVDD) integrity. The ReRAM-6T2R cell [41] has a compact area, but exhibits degraded SRAM cell stability and large cell leakage because of the elimination of NVM switches.

4.3.2 Memristor-Based Nonvolatile Flip-Flop for nvLogic

Speed, power, and area are all important parameters for logic circuits. The access/isolation functions of NVM in nvLogic circuits require additional transistors to provide store/restore nonvolatile functionality. However, the added transistors and parasitic load resulting from the nonvolatile function incur a large area overhead and decrease the speed and power of logic gates. Thus, it is impractical to add NVM devices to each logic gate.

Alternatively, applying nonvolatile functionality only to critical local states, which are stored at latches or flip-flops, can optimize the design of nvLogic circuits. Figure 4.16 shows the circuit structure of two nonvolatile flip-flops (nvFF) [42] for logic libraries: nvFF-M and nvFF-S. The nvFF-M employs the Rnv8T [43] cell as the master stage and a conventional CMOS circuit as the slave stage. The nvFF-S employs the Rnv8T cell as the slave stage and a conventional CMOS circuit as the master stage. These two nvFFs can be used separately or integrated in various power-on procedures. These two nvFFs require only 16 % of the area overhead of conventional FFs or nvFFs, yet achieve a lower VDD.

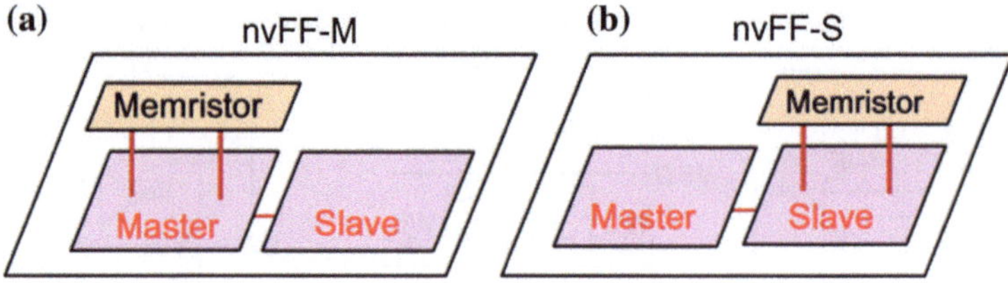

Fig. 4.16 **a** Nonvolatile flip-flop with memristor in master stage (nvFF-M); **b** nonvolatile flip-flop with memristor in slave stage (nvFF-S)

4.4 Low-Voltage nvSRAM

Figure 4.17 shows the application and cell structures of the nvSRAM cell. This nvSRAM design integrates SRAM cells and NVM devices within a single cell, forming a direct bit-to-bit connection in a 2D or vertical arrangement to achieve fast parallel data transfer and fast power-on/power-off speed in mobile systems.

As with conventional 6T SRAM cells, previous 6T-SRAM-based nvSRAM cells are unable to achieve a low VDDmin because of read/write failures, particularly in nanometer processes. To prevent nvSRAMs from limiting the power reduction provided by DVS schemes, low-VDDmin nvSRAM must offer a low-VDD power-on operation. Therefore, new cell structures with read-/write-assist features are needed for nvSRAM to achieve a low VDDmin.

4.4.1 Low-Voltage SRAM Review

Current DVS and low-voltage devices require low-VDDmin memory macros. Unfortunately, many embedded memories cannot achieve a low VDDmin because of read or write failure. Consequently, memory has become a bottleneck to reducing the VDDmin of low-power mobile systems.

Figure 4.18 plots the threshold voltage (V_{TH}) variation of a 90 nm process versus transistor width. Clearly, a larger transistor is associated with lower V_{TH} variation. However, most memory cells have a small area and cannot use large transistors. Nanometer memories are highly sensitive to V_{TH} variation.

Many SRAM designs use column-multiplexing and read–write column circuitry to achieve a small macro area and reduce the soft-error problem. Figure 4.19 displays an SRAM miniarray. When a WL is activated during a read or write operation, the selected SRAM cell and the other unselected SRAM cells on the same WL are accessed. The unselected cells (on unselected columns) are in the dummy-read mode, also called the half-selected (HS) mode. The data-0 storage node Q suffers

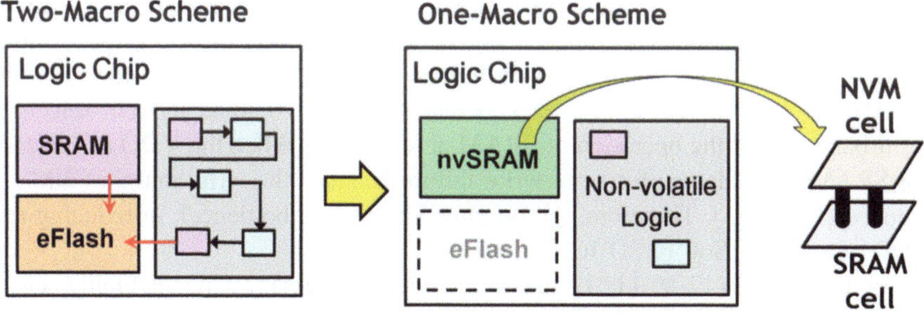

Fig. 4.17 Concept of application and cell structures of nvSRAM cell

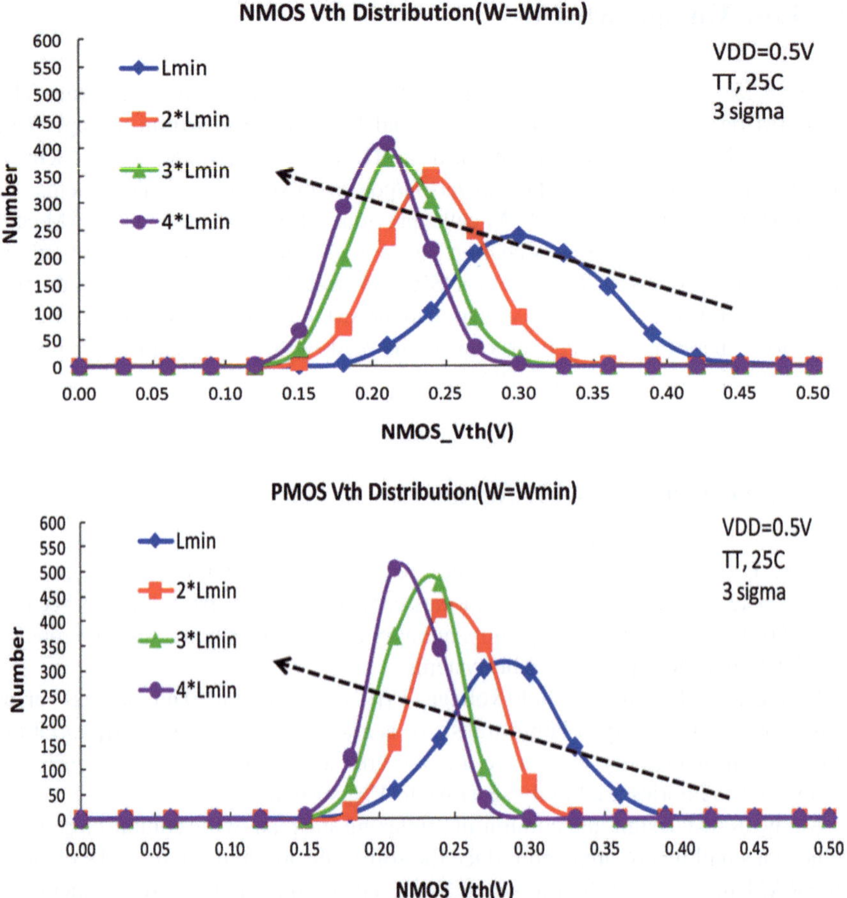

Fig. 4.18 Threshold voltage variation versus transistor width in a nanometer process

from a voltage bump when a cell is accessed. The voltage bump at Q, associated with V_{TH} variation, may cause data flipping in an SRAM cell during a read or dummy-read operation. This flipping affects the read stability or half-select stability of SRAM. Unfortunately, the cell stability, and specifically the static noise margin (SNM) [44–46], worsens as the VDD decreases (Fig. 4.20). Read and half-selection disturbances limit the operation of SRAM at low VDD. When the VDD is reduced, the SRAM cells also suffer from write failure because the write margin (WM) is reduced (Fig. 4.21). Therefore, the on-chip SRAM is a bottleneck in reducing the VDD of a mobile SoC or 3D-IC.

As discussed in [7–9, 11, 12, 44–46], SRAM cell performance is highly sensitive to variations in threshold voltage (V_{TH}). A decrease in VDD compromises the cell's stability (static noise margin (SNM)) [44–46], read and half-selection (HS)

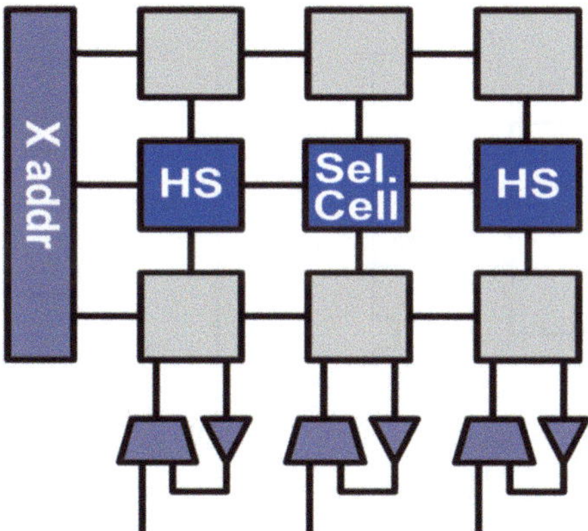

Fig. 4.19 Column-multiplexing SRAM cell array

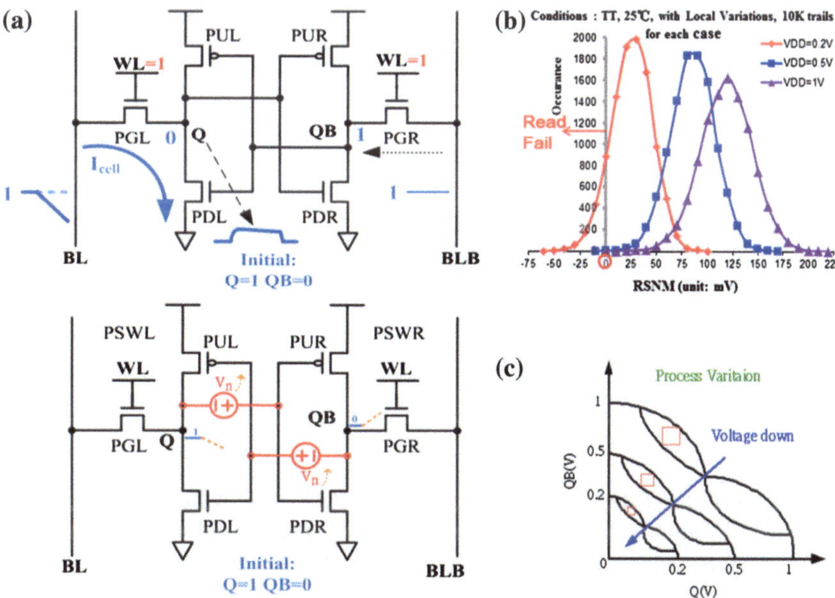

Fig. 4.20 a Read disturbance behavior, b RSNM of 6T SRAM cell, and c butterfly plot

disturbance, and the WM. Thus, SRAM creates a bottleneck in any attempt to reduce VDD in a chip.

Researchers have proposed various read- or write-assist schemes to improve the VDDmin of 6T SRAM cells [47–66]. However, the VDDmin of conventional 6T SRAM cells still exceeds 0.6 V.

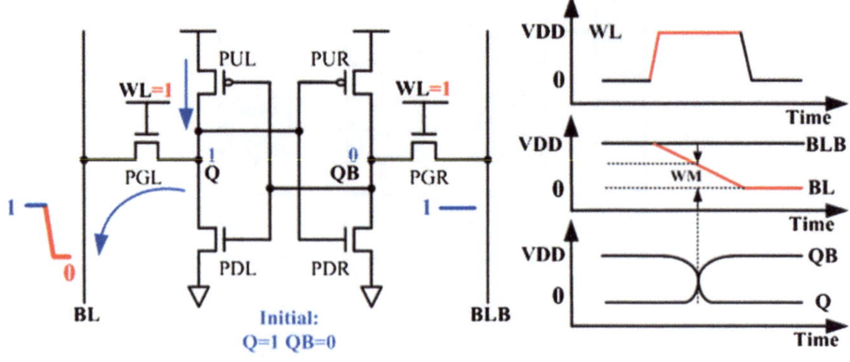

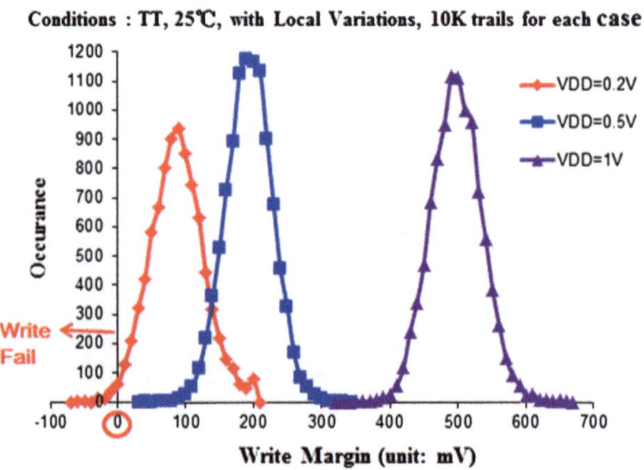

Fig. 4.21 Write margin versus VDD

Researchers have also proposed many new SRAM cells with more than 6T to lower the VDDmin. A 7T cell [67] is read disturb free, but suffers from degraded WM. A read-decoupled (RD) 8T cell [58–60], [68–73] has superior read stability, but fails to increase the WM and still suffers from write-mode half-select stability issues (Fig. 4.22). Several read-decoupled 9Ts [74, 75] or 10T SRAM cells [60, 76–80] have achieved ultra-low-VDD operation with good read stability, but they require a large area and have a limited WM. Asymmetrical 6T SRAM cells [81] can improve the WM at the expense of doubling the area. Single-ended 6T SRAM cells [18] improve the read-mode on–off ratio, but require a large area and suffer from WM degradation. Figure 4.23 presents a summary of the trade-off between area and VDDmin in SRAM cells. Clearly, using the above cells for nvSRAM still poses challenges in terms of degraded WM and large area overhead.

As mentioned earlier, DVS systems use a low to suppress active-mode power consumption. However, conventional SRAM suffers read disturb and write failure at

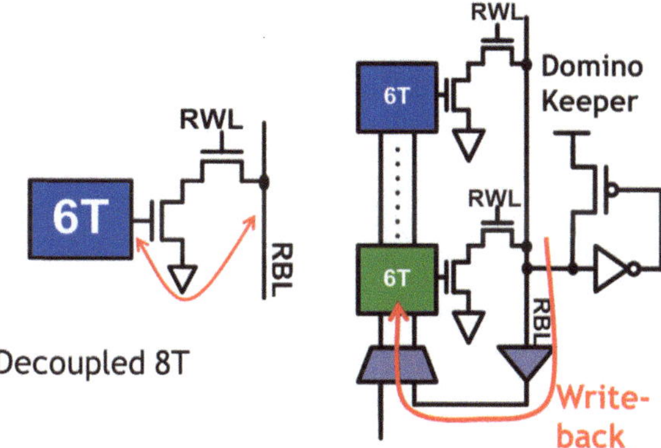

Fig. 4.22 Read-disturb-free 8T SRAM cell

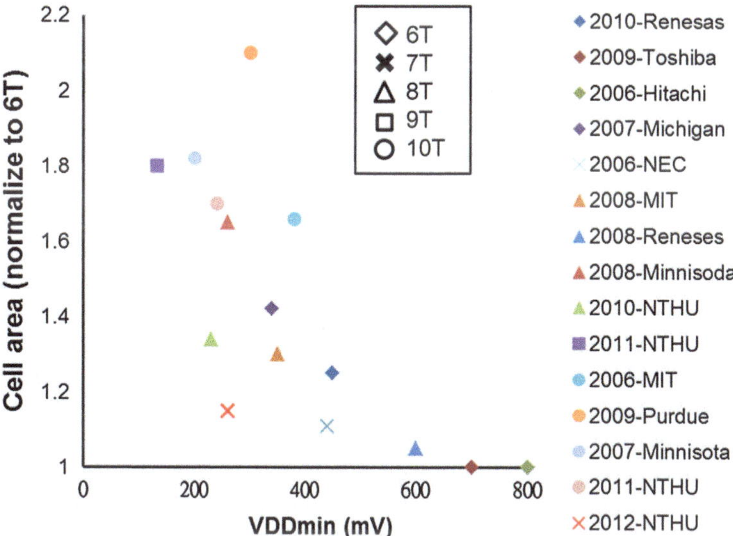

Fig. 4.23 VDDmin and cell area of the reported SRAM cells

a low VDD because of process variations. The addition of NVM devices at SRAM storage nodes yields an nvSRAM with a VDDmin that is worse than a typical SRAM. Many nvSRAM cells suffer degraded cell stability because of leakage through non-isolated NVM devices [82] and a loss of VDD/VSS integrity [38, 40]. Because of the enlarged parasitic capacitance at storage nodes caused by the connection of NVM devices/switches, nvSRAM cells equipped with NVM isolation switches suffer decreased write speeds, particularly at a low VDD. Accordingly, DVS or low-VDD chips also require a low VDDmin type of nvSRAM.

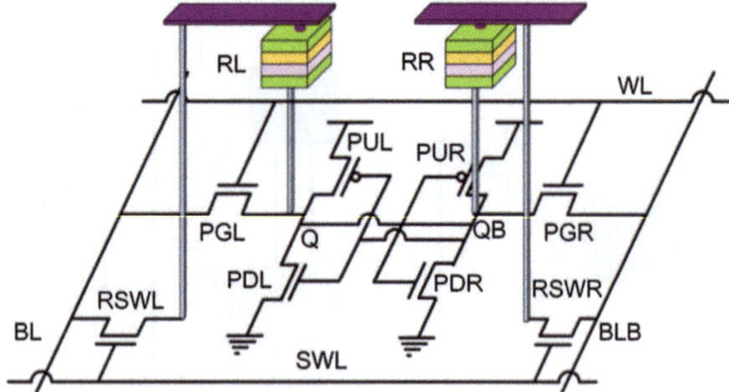

Fig. 4.24 Structure of the Rnv8T cell

4.4.2 Low-Voltage nvSRAM Example

4.4.2.1 Rnv8T Cell Structure

Previous study proposes a resistive nonvolatile 8T2R (Rnv8T) SRAM cell (or a memristor latch) [12] using 3D-stacked resistive memory (memristor or RRAM), as shown in Fig. 4.24. The Rnv8T cell enables fast-store and low-store power operations. This study also proposes a bit line–control line (BL-CL) sharing scheme to provide simple control of the memristors and a write-assist function with relaxed write constraints to enable read-favored sizing (RFS) against read/write failure at a reduced VDD.

The circuit scheme of the proposed resistive memory-based nonvolatile 8T2R (Rnv8T) memristor latch and SRAM cell consists of a 6T SRAM cell, 2T RRAM-switch (RSW), and two resistive devices (Fig. 4.24). The Rnv8T cell inherits all of the advantages of 6T SRAM, including its fast read/write and low-voltage operation. This design connects the memristors directly to SRAM storage nodes (*Q* and QB) to enable the storage of complementary backup data while maintaining nonvolatile characteristics. Unlike other nvSRAM cells, which require an additional control line (CL) to perform store operations, the Rnv8T cell shares BLs with the NVM CL to reduce the area overhead. This BL-CL sharing scheme enables RSWs to provide SRAM-mode write-assist functions exceeding their original storage capability. To reduce area overhead, the two memristor devices are vertically stacked above the 8T SRAM cell.

Figure 4.25 shows the operation flow of the Rnv8T cell. In normal mode (or SRAM mode), the Rnv8T cell performs SRAM functions and achieves fast read/write operations. When the system changes to standby mode, the Rnv8T macro proceeds to the store operation by flushing the SRAM data into memristor devices. After successful data backup, macro power can be completely shutdown to eliminate standby leakage. When the system wakes up, a restore operation recalls data from the memristor devices back to the SRAM storage nodes.

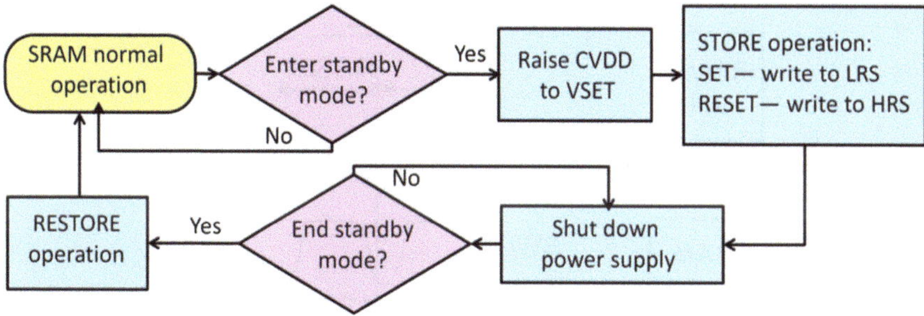

Fig. 4.25 Operation flow of the Rnv8T cell

4.4.2.2 Read/Write Operation

During read operations, the switch line (SWL) is kept low to turn off the RSWs (RSWL and RSWR) to avoid disrupting the stability of the SRAM cell. The Rnv8T cell performs a differential read in the same manner as a conventional 6T SRAM.

Write operations pull both the WL and the SWL high. The data to be written on the BL/BLB are written into the SRAM cell through the pass gates and RSWs. The RSWs provide a parallel write path and decrease the pull-up-PMOS to pass-gate-NMOS ratio, also called the pull-up ratio (PR). This method provides a larger WM and a faster write speed than that provided by a 6T SRAM cell. The WM used in this work is the write-trip point (WTP), which is the voltage differential between the BL and VSS when cell data flipping occurs. The transistor size of both the 6T and Rnv8T (before read-favored sizing) cells are the same as a commercial 6T SRAM cell under different technology nodes. To observe the worst effect of process variation, Fig. 4.26 shows the smallest values of the WM in the Monte Carlo simulation. Compared to 6T SRAM, the Rnv8T cell increases the WM by 4.13 and 5.78x at 0.5 V (0.18 μm) and 0.35 V (65 nm), respectively (Fig. 4.26).

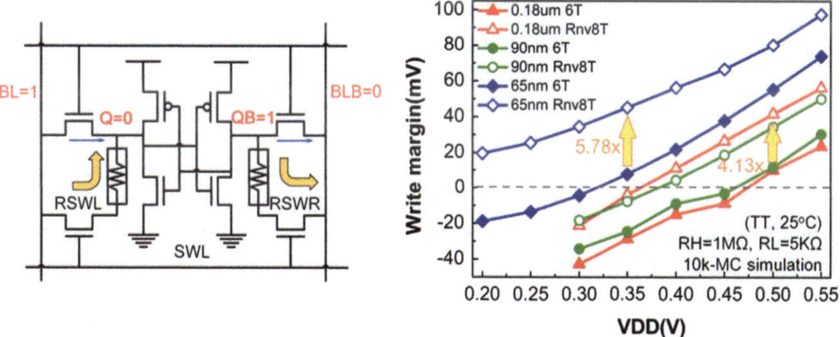

Fig. 4.26 Write margin versus VDD

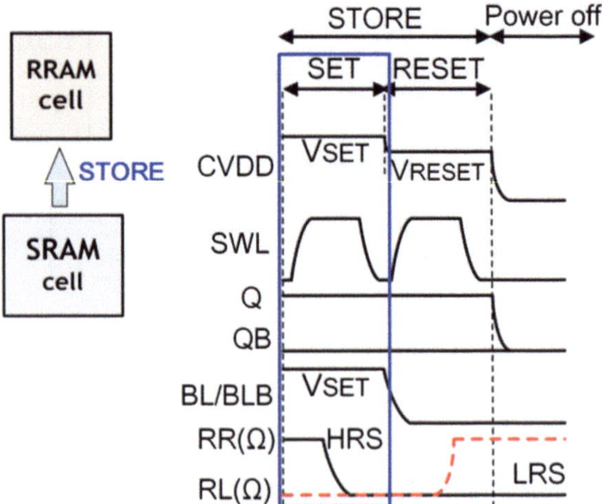

Fig. 4.27 Store operation of the Rnv8T cell

4.4.2.3 Store/Restore Operation

In store mode, two sub-operations (SET and RESET) are required to flush SRAM data to the memristor devices (Fig. 4.27). In store mode, the SWL is kept high for both SET and RESET sub-operations. SET operations raise the BL and BLB to a high voltage (V_{SET}) to provide the RL (if $Q = 0$) with positive voltage bias, switching it to a low-resistance state (LRS). The subsequent RESET operation pulls the BL and BLB to ground and set the CVDD to V_{RESET}. This provides RR (if $\mathrm{QB} = 1$) with a negative voltage bias and switches it to a high-resistance state (HRS).

The restore mode recalls data from the two memristor devices (RL and RR) to the SRAM storage nodes (Q and QB) during the system power-on period, as shown in Fig. 4.28. First, SWL is raised to turn on RSWs (RSWL and RSWR) and both BL and BLB are pulled to ground. This clears residual charges at the storage nodes (Q/QB) and ensures that the two nodes are in an equal state to allow a fair comparison at a later stage. As the CVDD increases, Q/QB is charged by PMOS, while RR/RL provides paths to ground. The difference in resistance between RL and RR creates different discharge currents and a voltage differential ($V_{Q-\mathrm{QB}}$) at Q and QB during power on. The $\mathrm{R_L}$ provides a large discharge current, resulting in a lower storage node voltage, whereas the $\mathrm{R_H}$ provides a small discharge current, resulting in a higher storage node voltage. Finally, the cross-coupled latches amplify $V_{Q-\mathrm{QB}}$ and restore the digital data to the SRAM storage nodes.

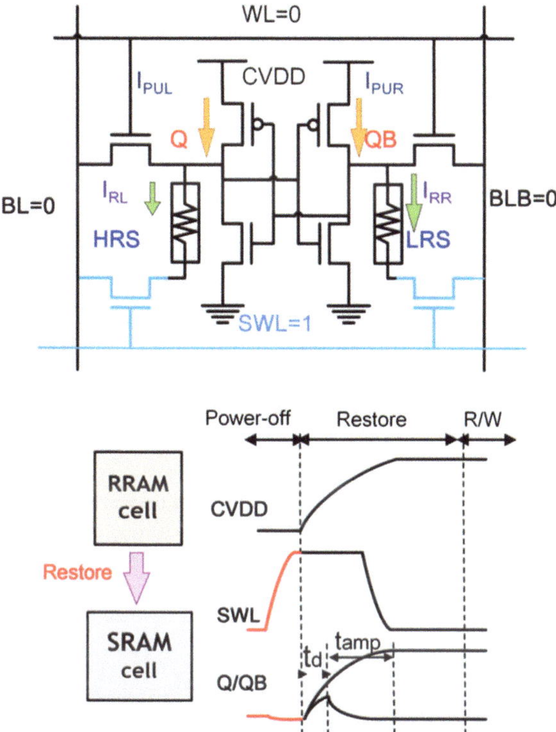

Fig. 4.28 Restore operation of the Rnv8T cell

4.4.2.4 Cell Stability and Read/Write VDDmin

Cell stability limits the VDDmin of the proposed Rnv8T cell during read operations. Fortunately, the greatly improved WM provides sufficient room to trade the WM for increased cell stability. Based on this improvement in WM, a read-favored sizing scheme can be used to reduce the read/write VDDmin. Under the same area constraints, the sizing of Rnv8T transistors can be adjusted to provide better read stability performance by (1) enlarging the size ratio of PDL (PDR) over PGL (PGR) to reduce read disturbance and (2) raising the trip point of the latch (inv1 and inv2) to prevent data flipping.

Figure 4.29 compares the read static noise margin (RSNM) [44], WM, and cell VDDmin for Rnv8T cells with and without the read-favored sizing scheme. This analysis evaluates process variations using a 10,000-point Monte Carlo simulation. The VDDmin of the nominal 6T cell and Rnv8T cell without the read-favored sizing scheme are limited to 0.55 V, because of poor RSNM performance. Applying a BL-CL sharing scheme to the Rnv8T cell improves the WM by 4.08x at 0.5 V. However, this does not lower the VDDmin because the RSNM does not benefit from this structure (i.e., the bottleneck remains).

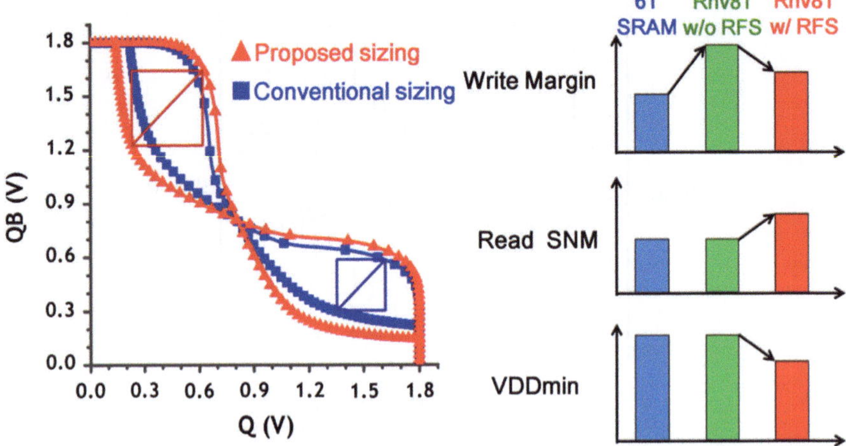

Fig. 4.29 Trade-offs between WM, RSNM, and VDDmin in the Rnv8T structure

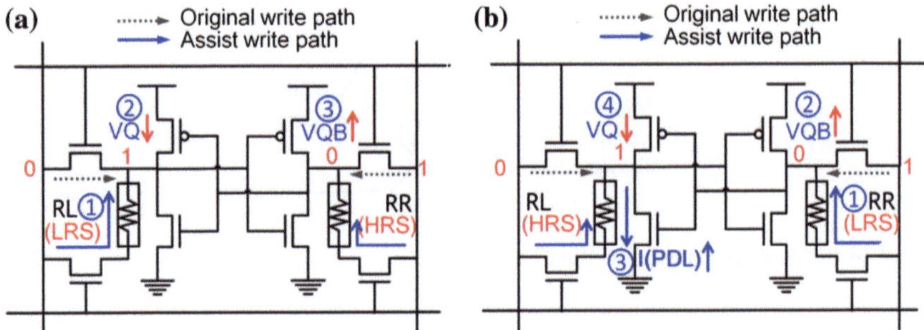

Fig. 4.30 **a** Best-case condition and **b** worst-case condition of WM improvement

Applying the read-favored sizing scheme decreases the improvement in WM slightly to 3.32x, whereas the RSNM produces an 11.1x improvement at VDD = 0.6 V. The Rnv8T cell with read-favored sizing scheme achieved a VDDmin 150 mV lower than without the read-favored sizing scheme. Consequently, the Rnv8T cell with BL-CL sharing and the read-favored sizing technique achieved superior cell stability and a lower SRAM read/write VDDmin than nominal 6T SRAM or other nvSRAM cells.

Figure 4.30a shows the best-case conditions for an improvement in the WM, in which the write '0' operation is conducted with a low-resistance RL and a high-resistance RR. The BL-CL sharing scheme provides (1) an additional write path on the write '0' side, which helps to (2) pull Q to ground and then (3) charge QB to VDD by PUR and PGR. In the worst case as shown in Fig. 4.30b, the WM improvement occurs for a write '0' operation with a high-resistance RL and a low-resistance RR. All the WM simulations in this paper are conducted under these conditions. Although the RL could be considered an 'off' state that provides negligible write current to help the WM, the RR (low resistance) side can still influence write performance. As the figure shows, (1) RR provides an additional write current to charge QB. Therefore, (2) QB increases to a rate beyond that achieved without the Rnv8T scheme. (3) When QB exceeds the threshold voltage of PDL, PDL can provide the strength in (4) to pull Q to ground. As a result, even though the low resistance is not located on the write '0' side, it can still improve the write performance. Under one specific condition (VDD = 1.8 V, TT corner, 25 °C), the best case can provide a 55 % improvement in the WM, while the worst case can still provide a 25 % improvement in the WM. This improvement could be much more significant in cases of device mismatch and low-voltage operation, as shown in Fig. 4.26. Figure 4.22 shows a comparison of the worst-case WM improvement and the resistance of memristors. The improvement in WM decreases as the HRS resistance increases for writing a "0" when Q = 1, particularly for more advanced process nodes. Nevertheless, the minimum improvement in the WM exceeds 67 % for 65-nm process nodes. Therefore, the write operation of the Rnv8T cell is still better than 6T SRAM cells even with an extremely large HRS–LRS ratio.

4.4.2.5 Restore Yield

The restore function of the Rnv8T cell can be viewed as a sensing operation in which the latch structure works like a sense amplifier. This restore operation uses a differential sensing scheme with two resistive devices written to different states, which greatly increases the sensing margin. Figure 4.31 shows that the restore yield can reach 99.6 % when the resistance ratio exceeds 3 and a higher yield when the resistance ratio exceeds 5, under device mismatch demonstrated by a 40,000-point Monte Carlo simulation. The memristor/ReRAM devices employed in this study achieved a resistive ratio exceeding 10, ensuring 100 % restore yield in the Rnv8T cell.

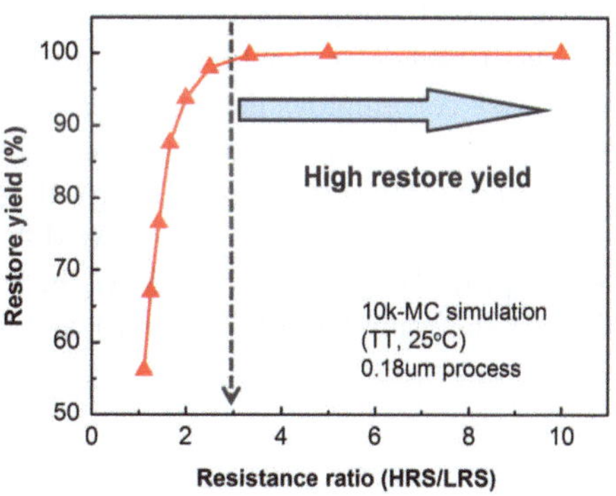

Fig. 4.31 Restore yield of Rnv8T cell

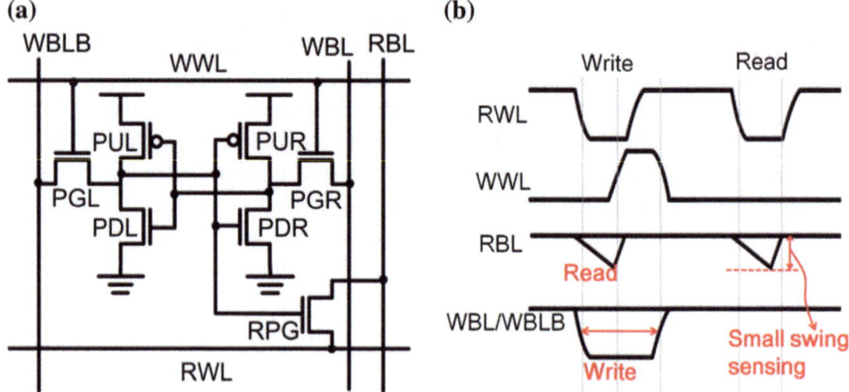

Fig. 4.32 **a** Proposed L7T SRAM cell. **b** Operation waveform of L7T cell

However, because of limited read stability, the Rnv8T cell is only capable of achieving 0.45 V VDDmin for processes with a small variation in threshold voltage (V_{TH}) (i.e., 0.18 um). For nanometer processes, which have larger V_{TH} variations, the VDDmin of Rnv8T cells degrades (increases). This highlights the need for a new nvSRAM cell structure for low-VDD applications and particularly for nanometer chips.

4.4.2.6 Low-Voltage nvSRAM Example–9T2R nvSRAM

To overcome read and write failure at low VDD, a nonvolatile 9T2R SRAM cell (Rnv9T) based on resistive memory was proposed [42]. The Rnv9T cell combines the features of the previously proposed Rnv8T nvSRAM and low-voltage L-shaped 7T (L7T) SRAM [83] cells. Figure 4.32 shows the schematic and waveform of the low-voltage 7T SRAM cell, which was developed from the Zigzag 8T (Z8T) cell [10].

The Rnv9T cell in Fig. 4.33 uses 1T-decoupled read port and pseudo read word line (RWL) schemes to achieve read-disturb-free performance. Like the Rnv8T cell, the Rnv9T cell employs the 2T memristor switch to oversee memristor control and improve the WM. Because of this read-disturb-free feature, the Rnv9T cell is not constrained to the read-favored sizing (RFS) scheme in the Rnv8T cell to improve the read stability. This provides a larger WM than that achieved by the Rnv8T cell. Thus, with improved read stability and WM, the Rnv9T cell achieves a lower VDDmin than 6T, L7T (i.e., sub-360 mV at 65 nm), and Rnv8T cells.

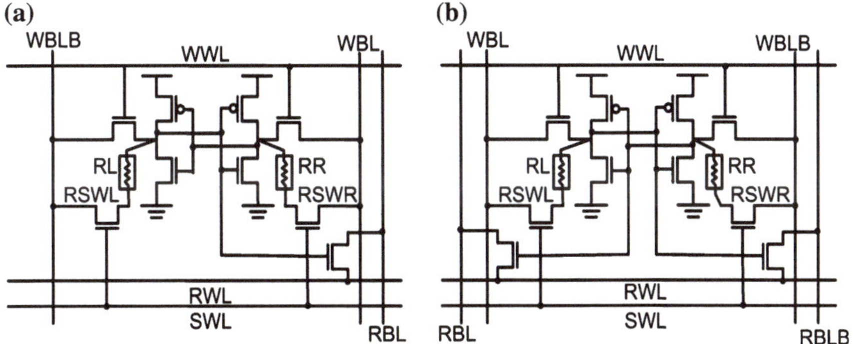

Fig. 4.33 Structure of **a** proposed Rnv9T cell and **b** proposed Rnv10T cell

4.4.3 Comparison of nvSRAM Cells

Figure 4.34 presents a comparison of the WMs of Rnv8T cell and other nvSRAM cells using a 10,000-point Monte Carlo simulation for various VDDs. Previous nvSRAM cells were reconstructed according to [37–40, 82, 84] with the same transistor sizing (pull-down, pass-gate, and pull-up transistors) as conventional a 6T SRAM cell. The Rnv8T cell achieved an improvement in WM of 3.3x compared to other nvSRAM cells. Thanks to the WM improvement, the Rnv8T cell also achieves faster write time than other nvSRAM cells, as shown in Fig. 4.35.

Table 4.3 summarize the performance of various nvSRAM cell. Unlike previous 6T2R cells, the proposed Rnv8T cell does not suffer degraded cell stability because of DC leakage current [82]. The VDDmin of the Rnv8T cell is 500 mV lower than that of other nvSRAM and 6T SRAM cell designs for a variety of process considerations (standard deviation of threshold voltage, $\sigma - V_{\mathrm{th}}$), as shown in Fig. 4.36.

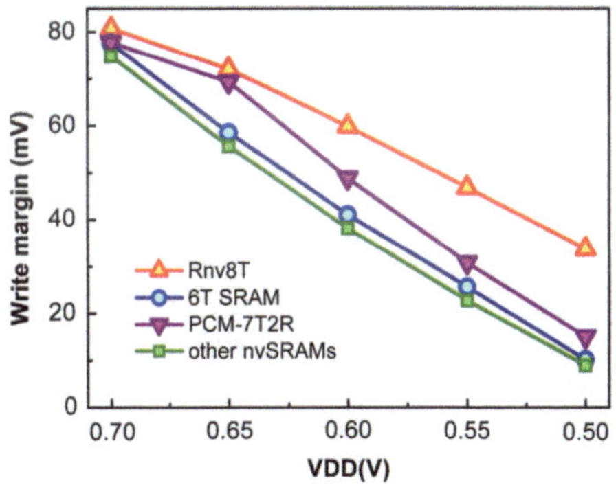

Fig. 4.34 Write margin of nvSRAM cells

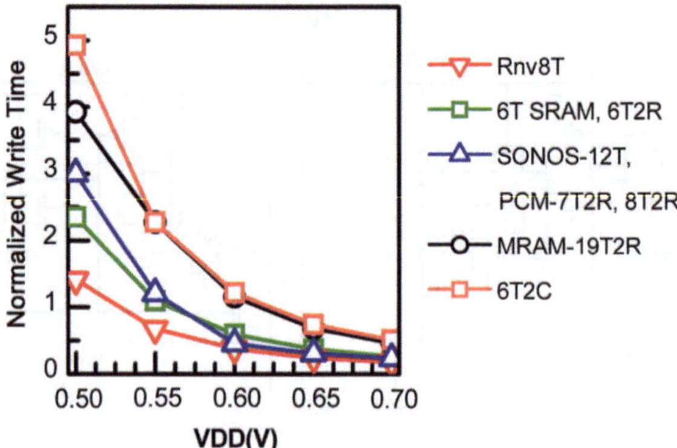

Fig. 4.35 Comparison of SRAM-mode write time between nvSRAM cells

4.5 Energy Efficient Systems Using Nonvolatile-SRAM and ReRAM

4.5.1 Overview of Recent ReRAM Macros

The current widespread development of resistive memory devices has recently led to dramatic improvements in random read–write access speed, operation voltage, and array density. Figure 4.37 compares the read speed and macro VDD of recently reported ReRAM macros.

4.5.1.1 High-speed ReRAM Macros

Table 4.3 Performance of various nvSRAM cells

Structure	9T2R (Rnv9T)	ST2R (Rnv8T)	12T	19T2R	6T2C	7T2R	8T2R	6T2R
NV device	ReRAM	ReRAM	SONOS	MRAM	Fe. C	PCM	ReRAM	ReRAM
Cell area	S	S	M	L	S	S	S	S
Write speed[1]	+	+	–	–	–	X	–	–
VDDmin[1]	++	+	X	X	X	X	X	–
T_{STORE}	10 ns[2]	10 ns[2]	4 ms	6 ns	200 ns	200 ns	10 ns	100 ns
V_{STORE}	1.8/-1.6	1.8/-1.6	-10/11	N/A	3V	3V	3V	N/A
I_{STORE}	<25 uA	<25 uA	N/A	1 mA	N/A	600 uA	100 uA	N/A
Endurance	$> 2 \times 10^8$	$> 2 \times 10^8$	10^6	Inf.	10^8	10^6	N/A	N/A

[1]compared with 6T SRAM, *S* small, *M* medium, *L* large, *plus* improved, *minus* degraded, *X* unchanged
[2]10 ns = SET + RESET time

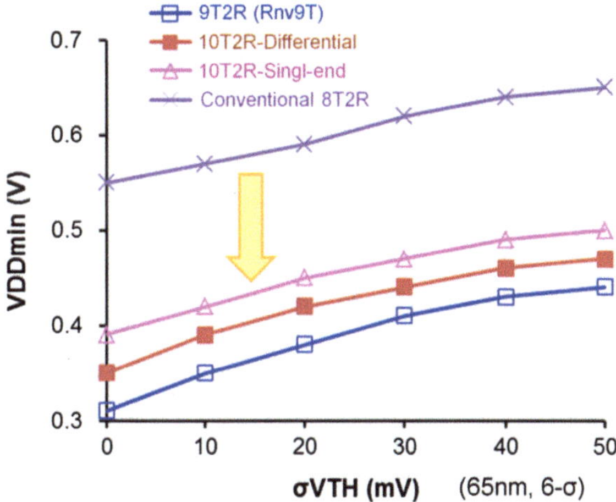

Fig. 4.36 Comparison of SRAM-mode VDDmin for nvSRAM cells

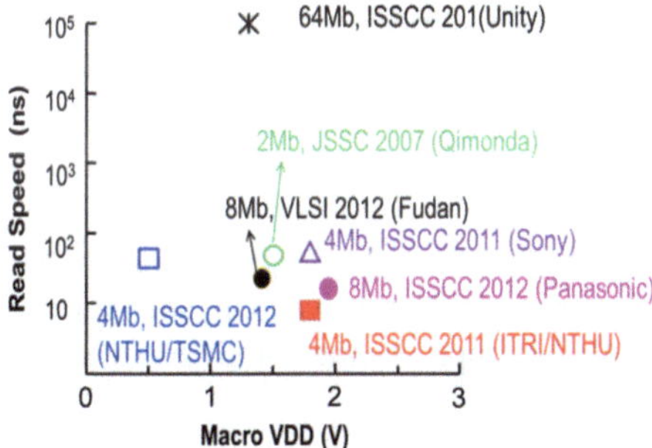

Fig. 4.37 Performance of published ReRAM macros

Voltage sensing or current sensing can be applied to the RRAM macro during the read operation. The demonstrated 1-Kb-RRAM macro (Fig. 4.38) [85] uses a reference cell to generate a reference current, which is $1/k$ of the read cell current (I_{CELL}) of the LRS RRAM cell. This reference current generates a reference voltage on the reference bit line (RBL) for a voltage-mode differential-input sense amplifier. However, a critical design challenge is to limit the BL voltage to under 0.3 V throughout the sensing period to prevent read disturbance of the RRAM cell.

A novel current-mode sensing scheme [86] for ReRAM was proposed to achieve fast random access time and high sensing yield against wide ReRAM-resistance distribution and process variations. This high-speed ReRAM macro uses the parallel-series reference-cell (PSRC) scheme to narrow the reference current distribution to achieve a high read yield against resistance variation. This design also employs a process-temperature-aware dynamic BL bias (PTADB) scheme to achieve small BL bias voltage fluctuations, preventing read disturbance while maintaining a fast BL pre-charge speed. A fabricated 4 Mb-SLC/8 Mb-MLC dual-mode ReRAM macro confirmed the effectiveness of proposed schemes for both SLC and MLC operations. This is the first study to present an embedded mega-bit-scale NVM macro with sub-8ns read–write random access time. This sensing scheme also enables robust read operation against power integrity degradation in a 3D-IC with a large number of stacked layers.

4.5.1.2 Logic-Process-Based ReRAM Macros

Two logic-process-based ReRAM macros were proposed: a contact-based ReRAM (C-ReRAM) [87] and a Cu_XSi_YO-based ReRAM [88].

The contact-based ReRAM macro use low-voltage circuit techniques to achieve 0.5 V read and write operation while offering a read speed similar to other NVM designs operated at nominal VDD. This low-voltage ReRAM uses body-drain-driven CSA (BDD-CSA) with a dynamic BL bias voltage (V_{BL}) scheme to achieve sub-0.5 V current-mode operation with a high sensing yield and fast read speed. This study also develops low-voltage level shifters and a low-voltage charge pump to achieve 0.5 V write (reset/set) operations. A fabricated 65 nm CMOS logic-compatible 4Mb C-ReRAM macro using the BDD-CSA achieves 0.5 V VDDmin. This BDD-CSA achieved 0.32 V VDDmin. This study is the first to present a 0.5 V ReRAM macro and a sub-0.4 V CSA.

The Cu_XSi_YO-based ReRAM macro employs voltage-mode sensing and two self-adaptive approaches for yield improvement and power reduction. The self-adaptive write mode (SAWM) improves the R-ratio by overcoming large write speed

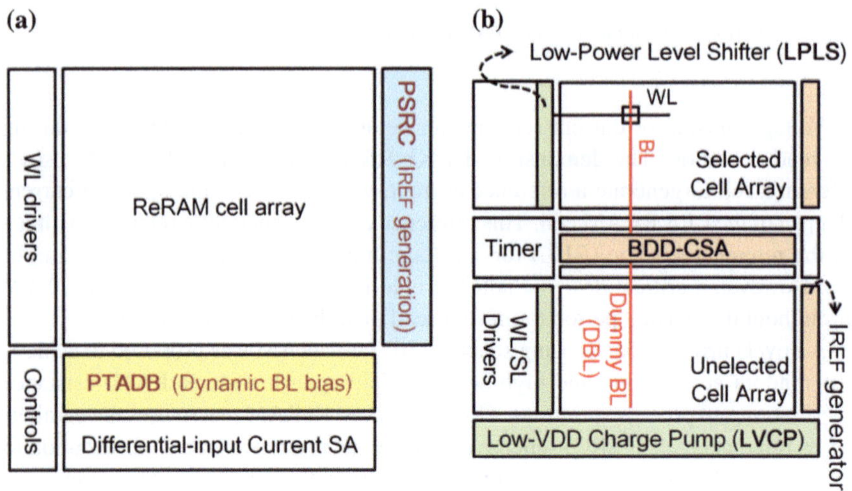

Fig. 4.38 Structures of high-speed and low-voltage ReRAM macros

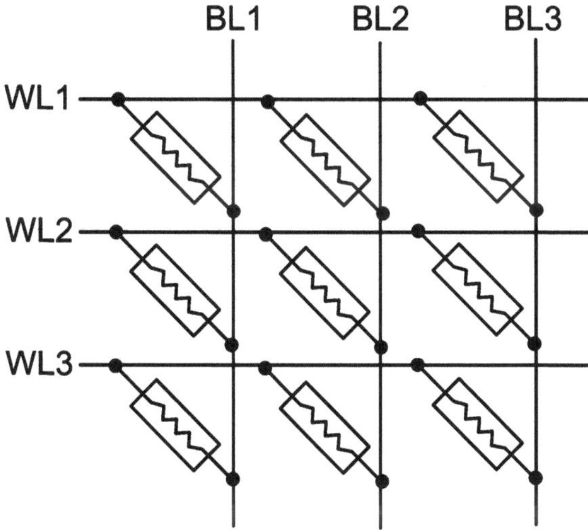

Fig. 4.39 Structures of cross-point ReRAM

variations. To improve read yield, the self-adaptive read mode (SARM) adds dummy cells in a regular array to track the ReRAM cell and BL load variation for voltage-mode operation. A fabricated 130 nm 8 Mb Cu_XSi_YO-based ReRAM device achieves a high yield and a 21ns read access time.

4.5.1.3 Cross-Point ReRAMt Macro

Researchers have proposed several cross-point-based 3D ReRAM macros [89] (Fig. 4.39)

A 0.13 μm 64 Mb conductive metal oxide (CMOx)-based ReRAM [86] macro was implemented by stacking an insulating oxide material on top of a conductive oxide material [90]. This macro uses a 0.17 μm^2 cell with multiple layer stacking capability for high-density applications. This design achieves sub-100 nA sensing but suffers from a slow read speed (exceed 50 us).

Another study presents a 0.18 um 8 Mb TaOx-based bipolar-type multilayered cross-point ReRAM macro. This macro adopts the following three techniques to reduce the sneak current in cross-point cell array structure: (1) use a bidirectional diode as a memory cell select element, (2) a hierarchical BL structure, and (3) a multibit write architecture. This 8 Mb macro achieves 443 MB/s write throughput (with 64b parallel write operations per 17.2 ns cycle). This design is applicable to high-density, stand-alone, and embedded memory with more stacked memory arrays and or scaling memory cells.

4.5.2 Standby-Mode Energy Reduction Using nvSRAM

Because of high data store/restore energy consumption of the two-macro scheme, the memory power-off (or deep sleep mode) scheme offers no advantage in power reduction for applications with a short standby/idle period (with frequent wake-up operations). This prevents several mobile chips, which require a fast wake-up response time, from employing memory power-off schemes to suppress standby current, because of the long operation time required to recall data from eFlash to the SRAM macro. For mobile chips requiring a fast wake-up time, lowering the SRAM VDD to its data-retention voltage (DRV) is the only option for suppressing SRAM standby power. Unfortunately, as the process is scaled down, growing demand for increased memory capacity on chips and bias SRAM in DRV still leads to considerable power consumption because of leakage.

The nvSRAM structure also plays a minor role in the store time and energy consumption. The store operation is determined by the characteristics of the resistive switching device. There is an obvious trend that memristor devices are moving toward a fast switching time and low switching current. Because of the low energy consumption and fast access time for store/restore operations, the proposed Rnv8T makes the memory power down approach applicable not only to long standby periods, but also to fast wake-up applications.

Figure 4.40 compares the standby-mode energy consumption of data-retention-voltage (DRV)-biased SRAM with the memory power down approach for two-macro schemes and nvSRAM based on Flash/RRAM. The standby energy of the two-macro scheme includes the energy consumed in data backup operations (SRAM read out + RRAM write in) and the restore operation (ReRAM read out + SRAM write in). Similarly, the standby energy of Rnv8T includes the energy consumed by the store and restore operations. Compared with a 0.18 μm SRAM macro holding at a DRV voltage of 0.4 V [91], the memory power down approach (one power-on operation and one power-off operation) with the Rnv8T/Rnv9T macro consumes 2.24x less

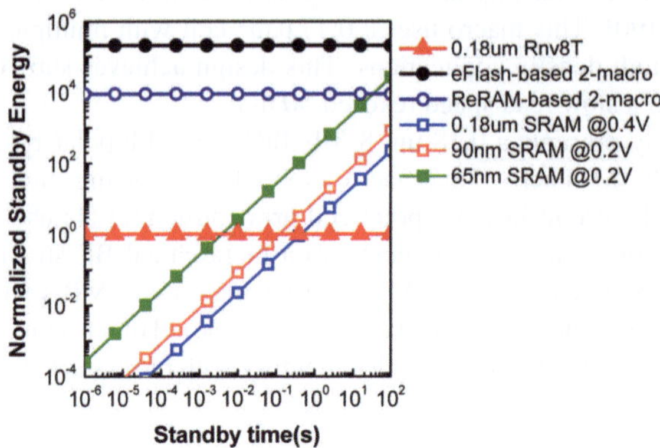

Fig. 4.40 Comparison of standby-mode energy consumption

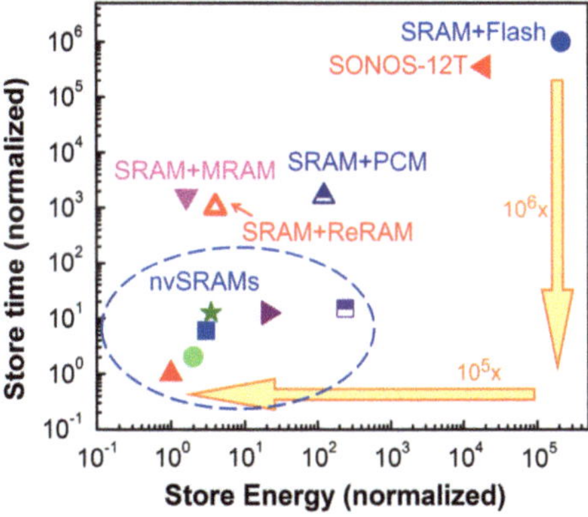

Fig. 4.41 Comparison of the store time and store energy

standby-mode energy when the standby period exceeds 1 s. The nvSRAM scheme offers additional benefits in more advanced technologies because of an increase in SRAM leakage current. Thus, the Rnv8T macro saves more energy than DRV-biased SRAM for applications with a standby period exceeding 4/160 ms for 65/90-nm process nodes. This makes the Rnv8T/Rnv9T macro a good candidate for suppressing standby energy in mobile chips across a wide range of standby durations.

4.5.2.1 Store Time

Figure 4.41 presents a comparison of the store time and store energy of Rnv8T//Rnv9T cells, two-macro solutions, and other nvSRAM cells. The word-by-word data transfer of two-macro solutions leads to an extremely long store time compared with the parallel data transfer of nvSRAMs. The fast switching time and low write current of the HfO_2-based memristor/ReRAM devices allow the Rnv8T//Rnv9T cell to achieve a faster store speed and lower store energy than other nvSRAM cells. The low-voltage and low-current switching characteristics of this memristor/ReRAM device enable multiple Rnv8T//Rnv9T cells to perform store operations simultaneously, further reducing the store time.

The widespread development of resistive memory devices has recently led to dramatic improvements in the SET/RESET current, and the reported minimum RESET/SET current is currently below 1 uA [30, 92]. The use of a memristor device with a smaller store current enables a further reduction in the inrush current for store operations, facilitating large-size block-store operations. Figure 4.42 presents a comparison of restore time. The restore time of the two-macro solution includes the time to read data from NVM and store it to SRAM word by word. In contrast, the Rnv8T SRAM restores all data simultaneously in a mass restore operation as the CVDD recovers.

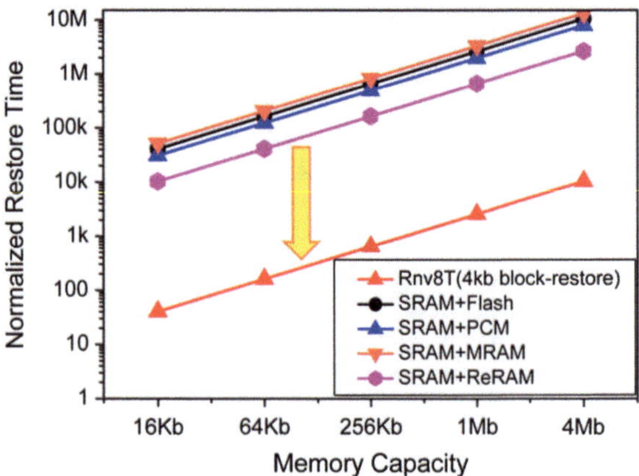

Fig. 4.42 Comparison of restore time

4.6 Summary and Conclusions

Resistive memory (memristor) devices offer the advantages of a short write time, low write voltage, and small write power. However, current devices suffer from limited endurance. Applying memristors to nvLogics and nvSRAMs suppresses the power consumption required for store operations and achieves fast power-on/power-off performance for the chip. This chapter explores various circuit structures associated with nvLogic and nvSRAM and considers account memristor endurance, particularly for low-voltage applications.

Acknowledgments The authors would like to thank ITRI, NTHU, and TSMC for their support.

References

1. Sasagawa, R., Fukushi, I., Hamaminato, M., & Kawashima, S. (1999). *High speed cascode sensing scheme for 1.0 V contact-programming Mask ROM* (pp. 95–96). Symposium on VLSI Circuits Digest of Technical Papers.
2. Chang, M.-F., Chiou, L.-Y., & Wen, K.-A. (2006). A full code-pattern coverage high-speed embedded ROM using dynamic virtual guardian technique. *IEEE Journal of Solid-State Circuits, 41*(2), 496–506.
3. Chang, M.-F., & Yang, S.-M. (2009). Analysis and reduction of supply noise fluctuations induced by embedded ROM. *IEEE Transactions on Very Large Scale Integration (VLSI) Systems, 17*(6), 758–769.
4. Seok, M., Hanson, S., Seo, J.-S., Sylvester, D., & Blaauw, D. (2008). Robust ultralow voltage ROM design (pp. 423–426). *IEEE CICC Digest of Technical Papers*.
5. Chang, M.-F., et al. (2010, Feb) A 0.29V embedded NAND-ROM in 90nm CMOS for ultra-low-voltage applications (pp. 266–267). *IEEE International Solid-State Circuits Conference (ISSCC) Digest of Technical Papers*.

6. Chang, M.-F., et al. (2010, November). Embedded non-volatile memory circuit design technologies for mobile low-voltage SoC and 3D-IC (p. 13). *10th IEEE International Solid-State and Integrated Circuit Technology (ICSICT)*.
7. Chang, M.-F., et al. (2011, January). Circuit design challenges in embedded memory and resistive RAM (RRAM) for mobile SoC and 3D-IC (pp. 197–203). In *Proceedings of 2011 IEEE Asia South Pacific Design Automation Conference (ASP-DAC)*.
8. Chang, M.-F., et al. (2011, September). *Challenges and trends of resistive memory (Memristor) based circuits for 3D-IC applications* (pp. 1053–1054). 2011 Symposium on Solid State Devices and Materials (SSDM).
9. Chang, M.-F., Chang, S.-W., Chou, P.-W., & Wu, W.-C. (2011). A 130 mV SRAM with expanded write and read margins for subthreshold applications. *IEEE Journal of Solid-State Circuits*, *46*(2), 520–529.
10. Wu, J.-J., et al. (2011). A large σVTH/VDD tolerant ZigZag 8T SRAM with area-efficient decoupled differential sensing and fast write-back scheme. *IEEE Journal of Solid-State Circuits*, *46*(4), 815–827.
11. Chang, M.-F., et al. (2011, October). Challenges and trends in low-power 3D die-stacked IC designs using RAM, memristor logic, and resistive memory (ReRAM) (pp. 327–330). *The IEEE 9th International Conference on ASIC (ASICON)*.
12. Chiu, P.-F., Chang, M.-F., et al. (2010, June). *A low store energy, low VDDmin, nonvolatile 8T2R SRAM with 3D stacked RRAM devices for low power mobile applications* (pp. 229–230). Symposium on VLSI Circuits Digest of Technical Papers.
13. Ohno, H., et al. (2010). *Magnetic tunnel junction for nonvolatile CMOS logic* (pp. 9.4.1–9.4.4). IEEE International Electron Devices Meeting (IEDM).
14. Yoon, S.-M., et al. (2009). Phase-change-driven programmable switch for nonvolatile logic applications. *IEEE Electron Device Letters*, *30*(4), 371–373.
15. Burd, T. D., et al. (2000). A dynamic voltage scaled microprocessor system. *IEEE Journal of Solid-State Circuits*, *35*(11), 1571–1580.
16. Nowka, K. J., et al. (2002). A 32-bit PowerPC system-on-a-chip with support for dynamic voltage scaling and dynamic frequency scaling. *IEEE Journal of Solid-State Circuits*, *37*(11), 1441–1447.
17. Ultra-low-voltage circuit design forum. *IEEE International Solid-State Circuits Conference (ISSCC)* (2009, Feb).
18. Zhai, B., Blaauw, D., Sylvester, D., & Hanson, S. (2007). A sub-200mV 6T SRAM in 0.13μm CMOS (pp. 332–333). *IEEE International Solid-State Circuits Conference (ISSCC) Digest of Technical Papers*.
19. Calhoun, B. H., & Chandrakasan, A. P. (2006). A 256kb subthreshold SRAM in 65nm CMOS (pp. 2592–2601). *IEEE International Solid-State Circuits Conference (ISSCC) Digest of Technical Papers*.
20. Chua, L. O. (1971). Memristor-the missing circuit element. *IEEE Transactions on Circuit Theory*, *18*(5), 507–519.
21. Chen, A., et al. (2005). Non-volatile resistive switching for advanced memory applications (pp. 746–749). *IEDM*.
22. Tseng, Y. H., et al. (2010). Electron trapping effect on the switching behavior of contact RRAM devices through random telegraph noise analysis (pp. 28.5.1–28.5.4). *IEDM*.
23. Tsunoda, K., et al. (2007). Low power and high speed switching of Ti-doped NiO ReRAM under the unipolar voltage source of less than 3V (pp. 767–770). *IEDM*.
24. Strukov, Dmitri B., et al. (2008). The missing memristor found. *Nature*, *453*, 80–83.
25. Lee, H. Y., et al. (2008). Low power and high speed bipolar switching with a thin reactive Ti buffer layer in robust HfO_2 based RRAM (pp. 1–4). *IEDM*.
26. Wei, Z., et al. (2008). Highly reliable TaOx ReRAM and direct evidence of redox reaction mechanism (pp. 293–296). *IEDM*.
27. Chien, W. C., et al. (2010). A forming-free Wox resistive memory using a novel self-aligned field enhancement feature with excellent reliability and scalability (pp. 440–443). *IEDM*.

28. Kawabata, S., et al. (2010). CoOx-RRAM memory cell technology using recess structure for 128Kbits memory array (pp. 1–2). In Proceedings of the International Memory Workshop (IMW).
29. Kim, M. J., et al. (2010). Low power operating bipolar TMO ReRAM for sub 10 nm Era (pp. 444–447). *IEDM*.
30. Kim, W., et al. (2011). *Forming-free nitrogen-doped AlOx RRAM with sub-*μA *programming current* (pp. 22–23). Symposium on VLSI Technology.
31. Kim, Y.-B., et al. (2011). *Bi-layered RRAM with unlimited endurance and extremely uniform switching* (pp. 52–53). Symposium on VLSI Circuits Digest of Technical Papers.
32. Lee, H., et al. (2010). Evidence and solution of over-RESET problem for HfOx based resistive memory with sub-ns switching speed and high endurance (pp. 19.7.1–19.7.4). *IEDM*.
33. Lee, H. Y., et al. (2008). *Low power and high speed bipolar switching with a thin reactive Ti buffer layer in robust* HfO_2 *based RRAM* (pp. 297–300). In International Electron Devices Meeting (IEDM) Technical Digest Papers.
34. Chen, Y. S., et al. (2009). *Highly scalable hafnium oxide memory with improvements of resistive distribution and read disturb immunity* (pp. 105–108). In International Electron Devices Meeting (IEDM) Technical Digest Papers.
35. Lee, H. Y., et al. (2010). *Evidence and solution of Over-RESET problem for HfOx based resistive memory with sub-ns switching speed and high endurance* (pp. 460–463). In International Electron Devices Meeting (IEDM) Technical Digest Papers.
36. Lee, H. Y., et al. (2010). Low-power and nanosecond switching in robust Hafnium Oxide resistive memory with a thin Ti cap. *IEEE Electron Device Letters*, *31*(1), 44–46.
37. Fliesler, M., et al. (2008). A 15ns 4Mb NVSRAM in 0.13u SONOS Technology (pp. 83–86). Non-Volatile Semiconductor, Memory Workshop (NVSMW).
38. Sakimura, N., et al. (2009). Nonvolatile magnetic flip-flop for standby-power-free SoCs. *IEEE Journal of Solid-State Circuits*, *44*(8), 2244–2250.
39. Miwa, T., et al. (2001). NV-SRAM: a nonvolatile SRAM with backup ferroelectric capacitors. *IEEE Journal of Solid-State Circuits*, *36*(3), 522–527.
40. Takata, M., et al. (2006). Nonvolatile SRAM based on Phase Change (pp. 95–96). Non-Volatile Semiconductor, Memory Workshop (NVSMW).
41. Wang, W., et al. (2006). *Nonvolatile SRAM cell* (pp. 1–4). International Electron Device Meeting (IEDM).
42. Chang, M.-F., et al. (2012). Endurance-aware circuit designs of nonvolatile logic and nonvolatile SRAM using resistive memory (Memristor) device (pp. 329–334). In *2012 IEEE Asia South Pacific Design Automation Conference (ASP-DAC)*.
43. Chiu, P.-F., Chang, M.-F., Wu, C.-W., Chuang, C.-H., Sheu, S.-S., Chen, Y.-S., et al. (2012). Low store energy, low VDDmin, 8T2R nonvolatile latch and SRAM with vertical-stacked resistive memory (Memristor) devices for low power mobile applications. *IEEE Journal of Solid-State Circuits*, *47*(6), 1483–1496.
44. Seevinck, E., List, F. J., & Lohstroh, J. (1987). Static-noise margin analysis of MOS SRAM cells. *IEEE Journal Solid-State Circuits*, *22*(5), 748–754.
45. Agarwal, A., Paul, B. C., Mukhopadhyay, S., & Roy, K. (2005). Process variation in embedded memories: failure analysis and variation aware architecture. *IEEE Journal of Solid-State Circuits*, *40*(9), 1804–1814.
46. Calhoun, B. H., & Chandrakasan, A. P. (2006). Static noise margin variation for sub-threshold SRAM in 65 nm CMOS. *IEEE Journal of Solid-State Circuits*, *41*(7), 1673–1679.
47. Yamaoka, M., et al. (2005). Low-power embedded SRAM modules with expanded margins for writing (pp. 480–611). In *IEEE International Solid-State Circuits Conference (ISSCC) Digest Technical Papers*.
48. Zhang, K., et al. (2006). A 3-GHz 70-Mb SRAM in 65-nm CMOS technology with integrated column-based dynamic power supply. *IEEE Journal of Solid-State Circuits*, *41*(1), 146–151.
49. Pilo, H., et al. (2006). *An SRAM design in 65 and 45 nm technology nodes featuring read and write-assist circuits to expand operating voltage* (pp. 15–16). Symposium on VLSI Circuits Digest Technical Papers.

50. Nii, K., et al. (2008). A 45 nm bulk CMOS embedded SRAM with improved immunity against process and temperature variations. *IEEE Journal of Solid-State Circuits, 43*(1), 180–191.
51. Ohbayashi, S., et al. (2007). A 65 nm SoC embedded 6T-SRAM designed for manufacturability with read and write operation stabilizing circuits. *IEEE Journal of Solid-State Circuits, 42*(4), 820–829.
52. Sohn, K., et al. (2008). A 100 nm double-stacked 500Mhz 72Mb separate-I/O synchronous SRAM with automatic cell-bias scheme and adaptive block redundancy (pp. 386–622). In *IEEE International Solid-State Circuits Conference (ISSCC) Digest of Technical Papers.*
53. Hirabayashi, O., et al. (2009). A process-variation-tolerant dual-power-supply SRAM with 0.179μm2 cell in 40 nm CMOS using level-programmable wordline driver (pp. 458–459). In *IEEE International Solid-State Circuits Conference (ISSCC) Digest of Technical Papers.*
54. Bhavnagarwala, A. et al. (2007). *A sub-600 mV, fluctuation tolerant 65nm CMOS SRAM array with dynamic cell biasing* (pp. 78–79). Symposium on VLSI Circuits Digest of Technical Papers.
55. Suzuki, T., et al. (2008). A stable 2-port SRAM cell design against simultaneously read/write-disturbed accesses. *IEEE Journal of Solid-State Circuits, 43*(9), 2109–2119.
56. Verma, N., & Chandrakasan, A. P. (2007). A 65 nm 8T subthreshold SRAM employing sense-amplifier redundancy (pp. 328–606). ISSCC Digest of Technical Papers.
57. Chang, I. J., Kim, J. J., Park, S. P., & Roy, K. (2008). A 32kb 10T subthreshold SRAM array with bit-interleaving and differential read scheme in 90 nm CMOS (pp. 388–622). In *IEEE International Solid-State Circuits Conference (ISSCC) Digest of Technical Papers.*
58. Chen, Y. H., et al. (2008). A 0.6 $V 45 nm$ adaptive dual-rail SRAM compiler circuit design for lower $VDD_{m}in$ VLSIs (pp. 210–211). Symposium on VLSI Circuits Digest of Technical Papers.
59. Khellah, M., et al., (2006). A 4.2GHz 0.3 mm 2 256Kb Dual-VCC SRAM building block in 65 nm CMOS (pp. 2572–2581). In *IEEE International Solid-State Circuits Conference (ISSCC) Digest of Technical Papers.*
60. Davis, J., et al. (2006). A 5.6GHz 64KB dual-read data cache for the POWER6TM processor (pp. 2564–2571). In *IEEE International Solid-State Circuits Conference (ISSCC) Digest of Technical Papers.*
61. Pille, J., et al. (2007). Implementation of the CELL broadband engine in a 65 nm SOI technology featuring dual-supply SRAM arrays supporting 6 GHz at 1.3 V (pp. 322–606). In *IEEE International Solid-State Circuits Conference (ISSCC) Digest of Technical Papers.*
62. Yamaoka, M., Osada, K., & Ishibashi, K. (2004). 0.4 V logic-library-friendly SRAM array using rectangular-diffusion cell and delta-boosted-array voltage scheme. *IEEE Journal of Solid-State Circuits, 39*(6), 934–940.
63. Lai, F.-S., & Lee, C.-F. (2007). On-chip voltage down converter to improve SRAM read/write margin and static power for sub-nano CMOS technology. *IEEE Journal of Solid-State Circuits, 42*(9), 2061–2070.
64. Shibata, N., et al. (2006). A 0.5 V 25 MHz 1 mW 256-kb MTCMOS/SOI SRAM for solar-power-operated portable personal digital equipment–sure write operation by using step-down negatively overdriven bitline scheme. *IEEE Journal of Solid-State Circuits, 41*(3), 728–742.
65. Khellah, M., et al. (2006). *Wordline and bitline pulsing schemes for improving SRAM cell stability in low-Vcc 65 nm CMOS designs* (pp. 9–10). Symposium on VLSI Circuits Digest of Technical Papers.
66. Kushida, K., et al. (2009). A 0.7 V single-supply SRAM with 0.495umP2P cell in 65nm Technology utilizing self-write-back sense amplifier and cascaded bit line scheme. *IEEE Journal of Solid-State Circuits, 44*(4), 1192–1198.
67. Takeda, K., et al. (2006). A read-static-noise-margin-free SRAM cell for low-VDD and high-speed applications. *IEEE Journal of Solid-State Circuits, 41*(1), 113–121.
68. Chang, L., et al. (2005). *Stable SRAM cell design for the 32 nm node and beyond* (pp. 128–129). Symposium on VLSI Technology Digest of Technical Papers.
69. Frustaci, F., Corsonello, P., Perri, S., & Cocorullo, G. (2006). Techniques for leakage energy reduction in deep submicrometer cache memories. *IEEE Transactions on Very Large Scale Integration (VLSI) Systems, 14*(11), 1238–1249.

70. Ishikura, S., et al. (2007). Symposium on VLSI Circuits Digest of Technical Papers (pp. 254–255).
71. Joshi, R., et al. (2007). 6.6+GHz *low Vmin, read and half select disturb-free 1.2Mb SRAM* (pp. 250–251). Symposium on VLSI Circuits Digest of Technical Papers.
72. Chang, L., et al. (2008). An 8T-SRAM for variability tolerance and low-voltage operation in high-performance caches. *IEEE Journal of Solid-State Circuits, 43*(4), 956–963.
73. Kim, T.-H., Liu, J., & Kim, C. H. (2009). A voltage scalable 0.26 V, 64 kb 8T SRAM with Vmin lowering techniques and deep sleep mode. *IEEE Journal of Solid-State Circuits, 44*(6), 1785–1795.
74. Liu, Z., & Kursun, V. (2008). Characterization of a novel nine-transistor SRAM Cell. *IEEE Transactions on Very Large Scale Integration (VLSI) Systems, 16*(4), 488–492.
75. Verkila, S. A., Bondada, S. K., & Amrutur, B. S. (2008). A 100 MHz to 1 GHz, 0.35-1.5 V Supply 256 × 64 SRAM block using symmetrized 9T SRAM cell with controlled read (pp. 560–565). In *Proceedings of Conference on, VLSI Design.*
76. Calhoun, B. H., & Chandrakasan, A. P. (2006). A 256kb Sub-threshold SRAM in 65 nm CMOS (pp. 2592–2601). ISSCC Digest of Technical Papers.
77. Noguchi, H., et al. (2007). *A 10T non-precharge two-port SRAM for 74% power reduction in video processing* (pp. 107–112). In IEEE Computer Society Annual Symposium VLSI Circuits (ISVLSI) Digest of Technical Papers.
78. Kim, T.-H., Liu, J., Keane, J., & Kim, C. H. (2007). A high-density subthreshold SRAM with data-independent bitline leakage and virtual ground replica scheme (pp. 330–606). In *IEEE International Solid-State Circuits Conference (ISSCC) Digest of Technical Papers.*
79. Kulkarni, J. P., Kim, K., & Roy, K. (2007). A 160 mV robust Schmitt trigger based subthreshold SRAM. *IEEE Journal of Solid-State Circuits, 42*(10), 2303–2313.
80. Chen, J., Clark, L. T., & Chen, T.-H. (2006). An ultra-low-power memory with a subthreshold power supply voltage. *IEEE Journal of Solid-State Circuits, 41*(10), 2344–2353.
81. Kawasumi, A., et al. (2008). A single-power-supply 0.7 V 1 GHz 45 nm SRAM with an asymmetrical unit-ß-ratio memory Cell (pp. 382–622). ISSCC Digest of Technical Papers.
82. Wang, W., et al. (2006). Nonvolatile SRAM Cell (pp. 27–30). In *International Electron Devices Meeting (IEDM) Technical Digest Papers.*
83. Chen, L.-F. (2011). A 7T SRAM circuit design for low voltage applications. M.S. thesis, Institute of Electronics Engineering, National Tsing Hua Univ., Hsinchu, Taiwan.
84. Yamamoto, S., et al. (2009). Nonvolatile SRAM (NV-SRAM) using functional MOSFET merged with resistive switching devices (pp. 531–534). In *IEEE Custom Integrated Circuits Conference (CICC) Technical Digest Papers.*
85. Sheu, S.-S., et al. (2009). *A 5ns fast write multi-level non-volatile 1 K bits RRAM memory with advance write scheme* (pp. 82–83). IEEE Symposium VLSI Circuits Digest of Technical Papers.
86. Sheu, S.-S., & Chang, M.F., et al. (2011). A 4Mb embedded SLC resistive-RAM macro with 7.2ns read-write random-access time and 160ns MLC-access capability (p. 200). IEEE *International Solid-State Circuits Conference (ISSCC) Digest of Technical Papers.*
87. Chang, M.-F., et al. (2012). A 0.5 V 4Mb logic-process compatible embedded resistive RAM (ReRAM) in 65 nm CMOS using low-voltage current-mode sensing scheme with 45ns random read time (pp. 434–436). *IEEE International Solid-State Circuits Conference (ISSCC) Digest of Technical Papers.*
88. Xue, X.Y., Jian, W.X., Yang, J.G., Xiao, F.J., Chen, G., Xu, X.L., Xie, Y.F., Lin, Y.Y., Huang, R., Zhou, Q.T., & Wu, J.G. (2012). Fudan University, A 0.13μm 8Mb Logic Based CuxSiyO Resistive Memory with Self-Adaptive Yield Enhancement and Operation Power Reduction", 2012 Symposium on VLSI Circuits, June 13, CIRCUITS SESSION 6 - TAPA I -6.1.
89. Kawahara, A., et al. (2012). Panasonic, Moriguchi, Japan, An 8Mb Multi-Layered Cross-Point ReRAM Macro with 443MB/s Write Throughput (pp. 432–434). *IEEE International Solid-State Circuits Conference (ISSCC) Digest of Technical Papers.*
90. Chevallier, C.J., et al. (2010). A 0.13μm 64Mb multi-layered conductive metal-oxide memory (pp. 260–261). *IEEE International Solid-State Circuits Conference (ISSCC) Digest of Technical Papers.*

91. Chang, M.-F., et al. (2007). Experiments on reducing standby current for compilable SRAM using hidden clustered source line control (pp. 1038–1041). In *Proceedings of IEEE International Conference of ASIC.*
92. Cheng, C. H., et al. (2010). *Novel ultra-low power RRAM with good endurance and retention* (pp. 85–86). In IEEE Symposium VLSI Technology Digest of Technical Papers.

81. Chang, M.-F., et al. (2009). [illegible] standby current [illegible] SRAM using hidden clustered source line control (pp. [illegible]). In [illegible] International Conference of [illegible].
82. Chang, C.-H., et al. (2011). [illegible] (pp. [illegible]). In IEEE Symposium VLSI [illegible] Digest of Technical Papers.

Chapter 5
Asymmetry in STT-RAM Cell Operations

Yaojun Zhang, Wujie Wen and Yiran Chen

Abstract Spin-transfer torque random access memory (STT-RAM) has emerged as a promising technology to replace SRAM and DRAM in embedded memory applications. In STT-RAM, the data are stored in a magnetic device (magnetic tunneling junction or MTJ) as different resistance states. The unique data storage mechanism of STT-RAM introduces the different design optimization concerns from conventional memories. As one important characteristic, programming "1" and "0" into an STT-RAM cell is very asymmetric in terms of performance, power, and reliability. In this chapter, we will review this asymmetry and analyze its sources. The impacts of this asymmetry on the STT-RAM cell optimization will be also discussed, followed by the introduction on a model to simulate the STT-RAM cell asymmetry.

5.1 Introduction

The conventional memory technologies such as SRAM, DRAM, and Flash have achieved a remarkable success in modern electronic designs. Following technology scaling, the shrunk feature size and the increased process variations impose serious concerns on the power and reliability of these conventional memory technologies. Many new memory technologies, including spin-transfer torque random access memory (STT-RAM), have emerged above the horizon. By leveraging a good combination of the non-volatility of Flash, the comparable cell density to DRAM, and the nanosecond programming time like SRAM, many applications of STT-RAM in

Y. Zhang
University of Pittsburgh, Pittsburgh, USA
e-mail: yaz24@pitt.edu

W. Wen
e-mail: wuw2@pitt.edu

Y. Chen (✉)
e-mail: yic52@pitt.edu

Y. Xie (ed.), *Emerging Memory Technologies*,
DOI: 10.1007/978-1-4419-9551-3_5,

embedded memory and on-chip cache designs have been successfully demonstrated [16, 20, 23].

Conventional memory technologies rely on electrical charges to store the data. In STT-RAM, the data are represented as the resistance states of a magnetic tunneling junction (MTJ) device, which can be switched by applying a programming current with different polarizations. Compared to the charge-based storage mechanism, the dependency of this mechanism on the device volume is much less, introducing a better scalability. Nonetheless, STT-RAM still suffers from the process variation issue as the technology scales. Also, an intrinsic random process called thermal fluctuations may cause the intermittent errors in the read and write operations of STT-RAM.

As technology scales, the STT-RAM density and power consumption improve. However, process variations on STT-RAM cell designs, including MOS transistor device variations, MTJ geometry variations, and resistance variations, become prominent. Many researches have been conducted to simulate the impacts of process variations and thermal fluctuations on STT-RAM reliability. A common simulation flow is as follows: First, a macro-magnetic model runs extensively to characterize the MTJ switching behaviors under different MTJ device variations; after that, the derived statistical MTJ electrical properties, i.e., the resistance distributions, together with the CMOS device variations, are sent to SPICE simulations to get the MTJ programming current distributions. In some recent works, thermal fluctuations are handled as either the random magnetic field in macro-magnetic models or the resistance noise in SPICE simulations to obtain the MTJ switching time variation or the read disturbance.

In this chapter, the impacts of the device parametric variations in MTJ and transistors and intrinsic MTJ operating uncertainties on the performances and reliability of STT-RAM cells are systematically analyzed. A fast, scalable, and portable statistical STT-RAM reliability analysis methodology, namely "PS3-RAM", is introduced to simulate the impacts of multidimensional variations on the STT-RAM read and write operations. We reveal that the write mechanism of STT-RAM cells is highly asymmetric at different switching directions, i.e., '0'→'1' and '1'→'0'. Specifically, the switching of '0'→'1' takes longer time than '1'→'0' at the same switching current while suffering from the larger variations. This asymmetry is further aggravated by the different biasing conditions of the driving NMOS transistor at different switching directions and the device variations in both MTJ and NMOS transistors. An example of minimizing the design space pessimism that required to tolerate the asymmetry and variations in STT-RAM write performance is also shown.

5.2 STT-RAM Basics

Spin-transfer torque random access memory (STT-MRAM) uses MTJ device to store the information. An MTJ has two ferromagnetic layers (FL) and one oxide barrier layer (BL). The resistance of MTJ depends on the relative magnetization directions (MDs) of the two FLs. When their MDs are parallel or anti-parallel, the MTJ is in

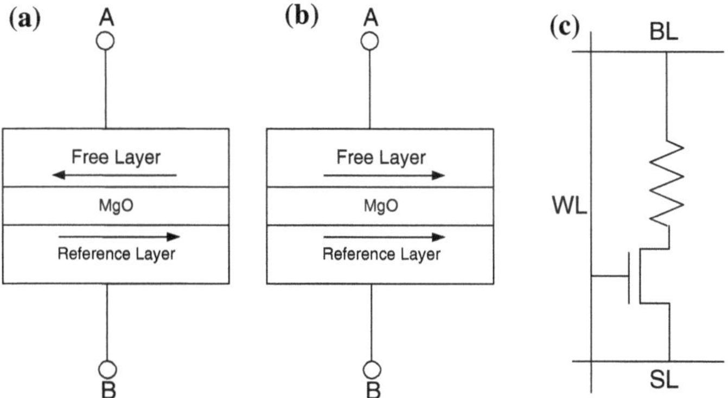

Fig. 5.1 MTJ structure. **a** Anti-parallel (high-resistance state). **b** Parallel (low-resistance state). **c** 1T1J STT-RAM cell structure

its low or high-resistance state, as illustrated in Fig. 5.1. R_h and R_l are usually used to denote the high and the low MTJ resistance, respectively. Tunneling magnetoresistance (TMR) is defined as $(R_h - R_l)/R_l$, which presents the distinction between the two resistance states.

In an MTJ, the MD of one FL (reference layer) is pinned, while the one of the other free layer (FL) can be flipped by applying a polarized write current through the MTJ. For example, the switching from low-resistance state ("0") to high-resistance state ("1") can be realized by applying a current from B to A, as shown in Fig. 5.1. A larger write current can shorten the MTJ switching time by paying the additional memory cell area overhead: In the popular "1T1J" (one-transistor-one-MTJ) cell structure (see Fig. 5.1c), the MTJ write current is supplied by the NMOS transistor. Increasing the write current requires a larger NMOS transistor. Also, the increased write current raises the breakdown possibility of the MTJ device.

5.3 Persistent and Non-persistent Variations

The persistent errors in an STT-RAM cell include the write errors incurred by the insufficient MTJ write current, the variation in MTJ critical switching current, and the insufficient read sense margin. They are mainly caused by the process variations in the NMOS transistors and the MTJ devices.

5.3.1 Persistent Errors

The persistent errors in an STT-RAM cell include the write errors incurred by the insufficient MTJ write current, the variation in MTJ critical switching current, and the insufficient read sense margin. They are mainly caused by the process variations in the NMOS transistors and the MTJ devices.

5.3.1.1 Transistor and MTJ Device Variation

The CMOS process variations that contribute to the variability of the driving strength of the NMOS transistor in an "1T1J" STT-RAM cell structure include random dopant fluctuations (RDFs), line-edge roughness (LER), shallow trench isolation (STI) stress, and the geometry variations in transistor channel length/width. Besides the geometry variations, most of the CMOS process variations are reflected as the threshold voltage deviations. The random variation of the threshold voltage is prominent in the scaled CMOS technology and can severely affect circuit stability and performance.

CMOS process variations affect not only the driving strength of the MOS transistor but also its equivalent resistance. The relative deviations of MOS transistor parameters are reduced when the transistor size increases.

The major sources of MTJ device variations include (1) MTJ shape variations; (2) MgO thickness variations; and (3) normally distributed localized fluctuation of magnetic anisotropy $K = M_s \cdot H_k$ [11]. The first two factors cause the variations in the MTJ resistance and the MTJ switching current by changing the bias conditions of the NMOS transistor. The third factor is the intrinsic variation in magnetic material that affects the MTJ switching threshold current density (Eq. 5.1) and the magnetization stability barrier height (Eq. 5.2) [11].

$$J_{C0} = \left(\frac{2e}{\hbar}\right)\left(\frac{\alpha}{\eta}\right)(t_F M_s)(H_k \pm H_{\text{ext}} + 2\pi M_s) \tag{5.1}$$

$$\Delta = \frac{K_u V}{k_B T} = \frac{M_s H_k V \cos^2(\theta)}{k_B T} \tag{5.2}$$

Here, the switching threshold current density J_{C0} is the minimal current density that causes the MTJ resistance flipping in the absence of any external magnetic field at 0K; e is the electron charge; α is the damping constant; M_s is the saturation magnetization; t_F is the thickness of the free layer; $\hbar$ is the reduced Planck's constant; H_k is the effective anisotropy field including magnetocrystalline anisotropy and shape anisotropy; H_{ext} is the external field; η is the spin-transfer efficiency; T is the working temperature; K_B is Boltzmann constant; and V is MTJ element volume.

Without considering any power rail fluctuations, during the write operation of an STT-RAM cell, the write current through the MTJ is mainly determined by the size of CMOS transistor and the MTJ resistance. The channel width, length, and threshold voltage of the NMOS transistor are the main parameters that affect the CMOS transistor performance. The standard variation in threshold voltage V_{th} decreases when the transistor size increases as $\sigma(V_{\text{th}}) \propto 1/\sqrt{WL}$. The MTJ resistance $R_{\text{MTJ}} \propto e^t/A$, where t is the tunneling oxide thickness and A is the MTJ surface area. The variations in both t and A follow Gaussian distribution [7]. Since $V_{\text{ds}} = V_{\text{dd}} - \Delta V_{\text{MTJ}}$, where $\Delta V_{\text{MTJ}} = I_{\text{MTJ}} \cdot R_{\text{MTJ}}$ is the voltage drop across the MTJ, V_{ds} should be a function of I_{MTJ}.

Normally, the variations in tunneling oxide thickness (t) and cross-sectional area (A) follow Gaussian distributions with a standard deviations of 2 or 5 % of their means, respectively [7]. Also, the MTJ variations are independent from the CMOS process variations since they are fabricated at different layers with different processes.

5.3.1.2 Fluctuation of Magnetic Anisotropy

Different from the CMOS process variations and the MTJ geometry variations that directly affecting the MTJ write current, the localized fluctuation of the MTJ magnetic anisotropy results in the variations in the switching threshold current density J_{C0}. In the concerned MTJ switching time range (a few ns to hundreds of ns), our magnetic model shows that the fluctuation of the MTJ magnetic anisotropy causes a standard deviation of the MTJ switching threshold current density that is about 2 % of the nominal value.

5.3.2 *Non-persistent Errors*

Device variations are introduced by the uncertainties during the manufacturing process. After the device is fabricated, the device parameters are fixed and their impacts on the circuit performance are deterministic. Besides the device variations of MOS transistor and MTJ, the MTJ switching performance is also affected by the intrinsic thermal fluctuations. The thermal-induced MTJ switching variation is a purely random process and cannot be deterministically repeated. It is the major source of the non-persistent errors in STT-RAM operations.

5.3.2.1 Thermal Fluctuation

In general, the impact of thermal fluctuations can be modeled by the thermal-induced random field h_{fluc} in stochastic Landau–Lifshitz–Gilbert (LLG) equation (Eq. 5.3) [1, 3, 5] as

$$\frac{d\vec{m}}{dt} = -\vec{m} \times (\vec{h}_{\text{eff}} + \vec{h}_{\text{fluc}}) + \alpha\vec{m} \times (\vec{m} \times (\vec{h}_{\text{eff}} + \vec{h}_{\text{fluc}})) + \frac{\vec{T}_{\text{norm}}}{M_s}. \quad (5.3)$$

Here, $\vec{m}$ is the normalized magnetization vector. Time t is normalized by γM_s. γ is the gyromagnetic ratio, and M_s is the magnetization saturation. $\vec{h}_{\text{eff}} = \frac{\vec{H_{\text{eff}}}}{M_s}$ is the normalized effective magnetic field. $\vec{h}_{\text{fluc}}$ is the normalized thermal agitation fluctuating field at finite temperature, which represents the thermal fluctuation. α is the LLG damping parameter. $\vec{T}_{\text{norm}} = \frac{\vec{T}}{M_s V}$ is the spin-torque term with units of magnetic field. The net spin torque $\vec{T}$ can be obtained through microscopic quantum electronic spin transport model. Under the intrinsic thermal fluctuations, the MTJ switching time becomes unrepeatable and follows a distribution. As we shall show later, this distribution is also affected by the MTJ and NMOS transistor device variations and causes the asymmetric STT-RAM cell switching at two switching directions.

Based on the switching time, the switching process of an MTJ can be categorized into three working regions:

In a relatively long switching time (> 10 ns), we have

$$I_{C1}(t_w) = I_{C0}(1 - (1/\Delta)\ln(t_w/\tau_0)). \tag{5.4}$$

Here, t_w is switching time; τ_0 is relaxation time; and T is the working temperature. In a ultrashort switching time (< 3 ns):

$$I_{C3}(t_w) = I_{C0} + C\ln(\pi/2\theta). \tag{5.5}$$

Here, C is a fitting parameter and θ is the initial angle between the magnetization vector and the easy axis.

When the MTJ switching time is in the middle (3 ns $< t_w <$ 10 ns), a dynamic reversal that combines the precessional and thermally activated switching occurs. Based on the simulation results of our macro-magnetic model, we derive a fitting function of the critical MTJ switching current I_{C2} as

$$I_{C2}(t_w) = 30(I_{C3}(3n) - I_{C1}(10n))/t_w + (10I_{C3}(3n) - 3I_{C1}(10n))/7. \tag{5.6}$$

Here, n is a fitting parameter.

Figure 5.2 shows the MTJ switching current versus the mean and the SDMR of the MTJ switching time. The device parameters are extracted from a 45 $\times$ 90 nm elliptical MTJ device, which have been calibrated with the measurement data of a real fabricated device from a leading magnetic recording company. The results of both switching directions ('1'→'0' and '0'→'1') are depicted.

Figure 5.2a shows the means of MTJ switching current and switching time in both 1→0 (negative) and '0'→'1' (positive) switchings for the same MTJ. As Eq. 5.3, thermal fluctuation influences the magnetic process in MTJ switching and causes the variations in MTJ switching time. When MTJ working in a relatively long time region (>10 ns), the thermal fluctuation is dominated by the thermal component of internal energy; when MTJ working in a short time region (<10 ns), the thermal fluctuation is dominated by the thermally active initial angle of procession [21]. This uncertainty causes the unsuccessful writes if the MTJ does not switch before the write pulse width is removed or the read overwrite errors if the MTJ resistance switches before the read voltage/current is removed.

Figure 5.2b shows the distribution of MTJ switching time at both '1'→'0' and '0'→'1' switchings. When the mean of MTJ switching time decreases, its standard deviation decreases first and then increases. The minimal SDMR of the MTJ switching time occurs around $t_w = 10$ ns. Noted that when the MTJ is working at a small

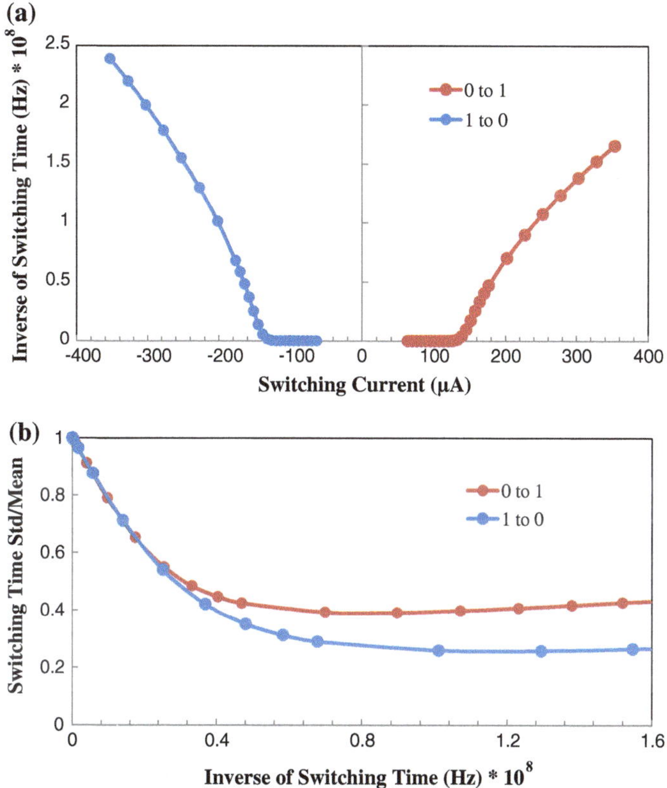

Fig. 5.2 **a** Switching current versus switching time mean. **b** Switching time mean versus switching time standard deviation/mean ratio (SDMR)

current region, the MTJ switching time follows Poisson's distribution; when the MTJ is working at a large current region, the MTJ switching time follows Gaussian distribution; and the MTJ switching time is a mixed distribution when it is in between the two above working regions.

The distinction between the means of the MTJ switching time of two switching directions under the same switching current can be explained as the asymmetric impacts of tunneling spin polarization P as

$$\frac{J_{C0}^{0\to 1}}{J_{C0}^{1\to 0}} = \frac{1+P^2}{1-P^2}. \tag{5.7}$$

Here, $J_{C0}^{0\to 1}$ and $J_{C0}^{1\to 0}$ denote the MTJ switching threshold current densities at the switchings of '0'→'1' and '1'→'0,' respectively.

The different standard deviations of the MTJ switching time at two switching directions, however, are caused by the asymmetric influences of the thermal agitation fluctuating field $\vec{h}_{\text{fluc}}$. Noticed that when the MTJ works at long switching time (>40 ns) under a low switching current, the standard deviation of the MTJ switching time for both switching directions is almost the same. However, following the decrease in the MTJ switching time, the standard deviation difference of the MTJ switching time becomes prominent. It is due to the reduced thermal impacts and the increased asymmetry of the spin-torque term $\vec{T}_{norm}$ in MTJ switching under a high switching current. In general, MTJ switching time has a larger mean and a wider distribution in '0'→'1' switching than '1'→'0' switching under the same switching current.

5.3.2.2 Temperature Dependency

The switching performance of an MTJ can be improved by raising the working temperature. Higher temperature degrades the magnetization stability barrier height (Eq. 5.2) and reduces the critical MTJ switching current and/or the switching time. Figure 5.3 shows the critical MTJ switching currents versus switching time under different temperatures. The impacts of temperature variations are more significant in long working time region: The impact of thermal fluctuations on the MTJ switching performance is more prominent when the MTJ switching current is low, compared to the impact of spin torque.

The temperature sensitivity of the nominal switching time of the MTJ driven by the NMOS transistors with different sizes is shown in Fig. 5.4. The MTJ switching time increases when the temperature rises. It actually indicates that the improvement in MTJ magnetic switching cannot compensate the driving ability loss of the NMOS transistor when the working temperature rises up.

5.4 PS3-RAM Method

"PS3-RAM" is a fast, scalable, and portable statistical STT-RAM reliability analysis methodology. Figure 5.5 depicts the overview of "PS3-RAM" methods, including the sensitivity analysis for MTJ switching current (I), the I sample recovery,

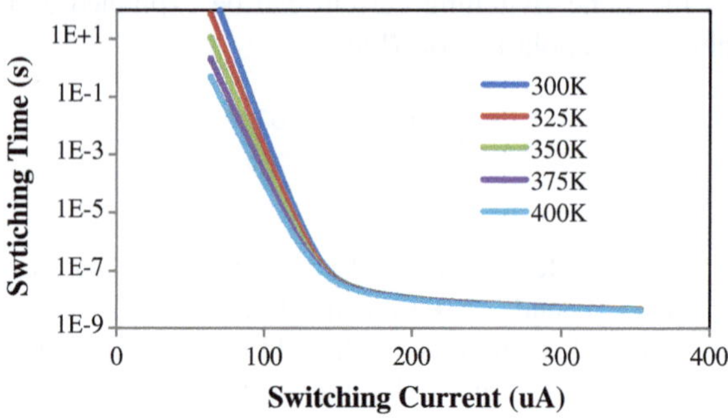

Fig. 5.3 MTJ Critical switching current versus switching time under varying temperature

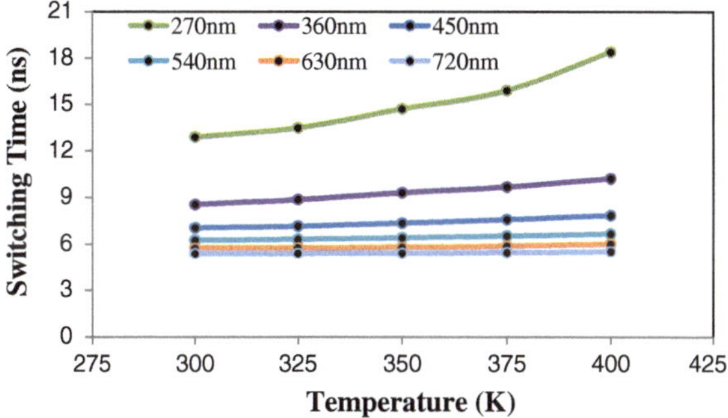

Fig. 5.4 Threshold switching time against temperature

and the statistical thermal analysis of STT-RAM. Array-level analysis and design optimizations can be also conducted using PS3-RAM.

5.4.1 Sensitivity Analysis on MTJ Switching

In this section, the sensitivity model is used for to characterize the MTJ switching current distribution. The contributions of different variation sources to the distribution of the MTJ switching current will be discussed. The definitions of the variables used in this section are summarized in Table 5.1.

5.4.1.1 Threshold Voltage Variation

The variations in channel length (L), width (W), and threshold voltage (V_{th} are three major factors inducing the variations in transistor driving ability. V_{th} variation mainly comes from RDF and LER, which is also the source of some geometry variations (i.e., L and W) [14, 21]. It is known that the V_{th} variation is also correlated with L and W and its variance decreases as the transistor size increases. The deviation of the V_{th} from the nominal value following the change in L (ΔL) can be modeled by [22]

Table 5.1 Simulation parameters and environment setting

Parameters	Mean	Standard deviation
Channel length	$\overline{L} = 45\,\text{nm}$	$\sigma_L = 0.05\overline{L}$
Channel width	$\overline{W} = 90 \sim 1800\,\text{nm}$	$\sigma_W = 0.05\overline{L}$
Threshold voltage	$\overline{V}_{\text{th}} = 0.466\,\text{V}$	compact model
Mgo thickness	$\overline{\tau} = 2.2\,\text{nm}$	$\sigma_\tau = 0.02\overline{\tau}$
MTJ surface area	$\overline{A} = 45 \times 90\,\text{nm}^2$	$\sigma_A = 0.05\overline{A}$
Resistance low	$R_L = 1000\,\Omega$	By calculation
Resistance high	$R_H = 2000\,\Omega$	By calculation

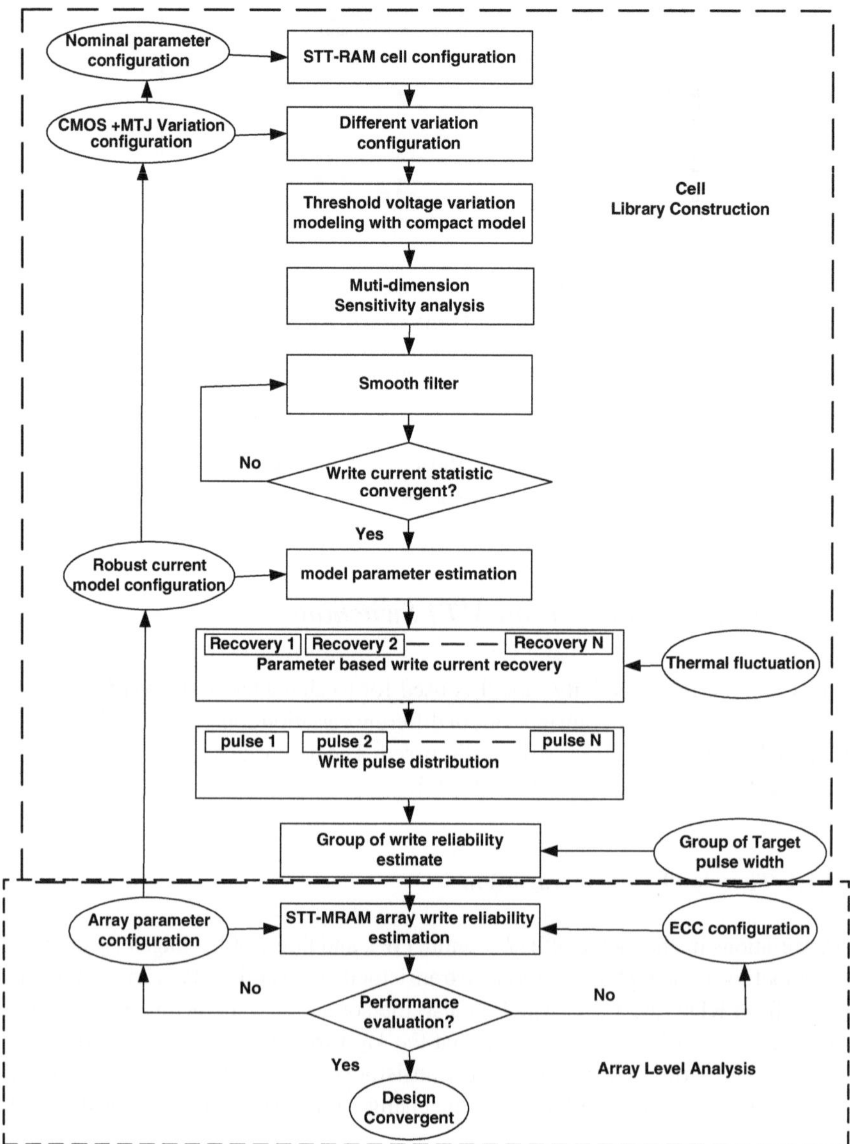

Fig. 5.5 Overview of PS3-RAM

$$\Delta V_{\text{th}} = \Delta V_{\text{th0}} + V_{\text{ds}}\exp(-\frac{L}{l'}) \cdot \frac{\Delta L}{l'}. \tag{5.8}$$

The standard deviation of V_{th} can be calculated as

$$\sigma_{V_{\mathrm{th}}}^2 = \frac{C_1}{WL} + \frac{C_2}{\exp\left(\mathrm{L}/l'\right)} \cdot \frac{W_c}{W} \cdot \sigma_L^2. \tag{5.9}$$

Here, W_c is the correlation length of non-rectangular gate (NRG) effect, which is caused by the randomness in sub-wavelength lithography. C_1, C_2, and l' are technology-dependent coefficients. The first term at the right side of Eq. (5.9) describes the RDF's contribution to $\sigma_{V_{\mathrm{th}}}$. The second term in Eq. (5.9) represents the contribution from NRG, which is heavily dependent on L and W. Following technology scaling, the contribution of this term becomes prominent due to the reduction in L and W.

5.4.1.2 Sensitivity Analysis on Variations

Although the contributions of MTJ and CMOS parameters to the MTJ switching current distribution cannot be explicitly expressed, it is still possible for us to conduct a sensitivity analysis to obtain the critical characteristics of the distribution. Without the loss of generality, the MTJ switching current I can be modeled by a function of W, L, V_{th}, A, and τ. A and τ are the MTJ surface area and MgO layer thickness, respectively. The first-order Taylor expansion of I around the mean values of every parameter is

$$\begin{aligned} &I\left(W, L, v_{\mathrm{th}}, A, \tau\right) \approx I\left(\overline{W}, \bar{L}, \bar{V_{\mathrm{th}}}, \bar{A}, \bar{\tau}\right) + \tfrac{\partial I}{\partial W}\left(W - \overline{W}\right) \\ &+ \tfrac{\partial I}{\partial L}\left(L - \bar{L}\right) + \tfrac{\partial I}{\partial V_{\mathrm{th}}}\left(V_{\mathrm{th}} - \bar{V_{\mathrm{th}}}\right) + \tfrac{\partial I}{\partial A}\left(A - \bar{A}\right) + \tfrac{\partial I}{\partial \tau}\left(\tau - \bar{\tau}\right). \end{aligned} \tag{5.10}$$

As aforementioned, W, L, A, and τ generally follow Gaussian distribution [8]. V_{th} is correlated with W and L, as shown in Eq. (5.8) and (5.9). Because the MTJ resistance $R \propto \frac{e^{\tau}}{A}$ [8], we have

$$\frac{\partial I}{\partial A}\Delta A + \frac{\partial I}{\partial \tau}\Delta\tau = \frac{\partial I}{\partial R}\left(\frac{\partial R}{\partial A}\Delta A + \frac{\partial R}{\partial \tau}\Delta\tau\right) = \frac{\partial I}{\partial R}\Delta R. \tag{5.11}$$

It indicates that the combined contribution of A and τ is the same as the impact of MTJ resistance R. The difference between the actual I and its mathematical expectation μ_I can be calculated by

$$\begin{aligned} &I\left(W, L, V_{\mathrm{th}}, R\right) - E\left(I\left(\overline{W}, \bar{L}, \bar{V_{\mathrm{th}}}, \overline{R}\right)\right) \approx \\ &\frac{\partial I}{\partial W}\Delta W + \frac{\partial I}{\partial L}\Delta L + \frac{\partial I}{\partial V_{\mathrm{th}}}\Delta V_{\mathrm{th}} + \frac{\partial I}{\partial R}\Delta R. \end{aligned} \tag{5.12}$$

Here, $\mu_I \approx E\left(I\left(\overline{W}, \bar{L}, \bar{V_{\mathrm{th}}}, \overline{R}\right)\right) = I\left(\overline{W}, \bar{L}, \bar{V_{\mathrm{th}}}, \overline{R}\right)$ and the mean of MTJ resistance $\overline{R} \approx R\left(\bar{A}, \bar{\tau}\right)$. Combining Eqs. (5.8), (5.9), and (5.12), the standard deviation of I (σ_I) can be calculated as

$$
\begin{aligned}
\delta_I^2 = &\left(\frac{\partial I}{\partial W}\right)^2 \sigma_W^2 + \left(\frac{\partial I}{\partial L}\right)^2 \sigma_L^2 + \left(\frac{\partial I}{\partial R}\right)^2 \sigma_R^2 \\
&+ \left(\frac{\partial I}{\partial V_{\text{th}}}\right)^2 \left(\frac{C_1}{WL} + \frac{C_2}{\exp\left(\mathrm{L}/l'\right)} \cdot \frac{W_c}{W} \cdot \sigma_L^2\right) \\
&+ 2\frac{\partial I}{\partial L}\frac{\partial I}{\partial v_{\text{th}}}\rho_1 \frac{1}{\sqrt{WL}}\sigma_L + 2\frac{\partial I}{\partial W}\frac{\partial I}{\partial V_{\text{th}}}\rho_2 \frac{1}{\sqrt{WL}}\sigma_W \\
&+ 2\frac{\partial I}{\partial L}\frac{\partial I}{\partial V_{\text{th}}} V_{\text{ds}} \exp\left(-\frac{L}{l'}\right)\frac{\sigma_L^2}{l'}.
\end{aligned}
\tag{5.13}
$$

Here, $\rho_1 = \frac{\mathrm{cov}(V_{\text{th0}}, L)}{\sqrt{\sigma_{V_{\text{th0}}}^2 \sigma_L^2}}$ and $\rho_2 = \frac{\mathrm{cov}(V_{\text{th0}}, W)}{\sqrt{\sigma_{V_{\text{th0}}}^2 \sigma_W^2}}$ are the correlation coefficients between V_{th0} and L or W, respectively [21]. $\sigma_{V_{\text{th0}}}^2 = \frac{C_1}{WL}$. The last three terms at the right side of Eq. (5.13) are significantly smaller than other terms and can be safely ignored in the simulations of normal STT-RAM operations.

The accuracy of the coefficient in front of the variances of every parameter at the right side of Eq. (5.13) can be improved by applying window-based smooth filtering. Take W as an example:

$$
\left(\frac{\partial I}{\partial W}\right)_i = \frac{I\left(\overline{W} + i\Delta W, L, V_{\text{th}}, R\right) - I\left(\overline{W} - i\Delta W, L, V_{\text{th}}, R\right)}{2i\Delta W}, \tag{5.14}
$$

where $i = 1, 2, ...K$. Different $\frac{\partial I}{\partial W}$ can be obtained at the different step i. K samples can be filtered out by a window-based smooth filter to balance the accuracy and the computation complexity as

$$
\overline{\frac{\partial I}{\partial W}} = \sum_{i=1}^{K} \omega_i \left(\frac{\partial I}{\partial W}\right)_i. \tag{5.15}
$$

Here, ω_i is the weight of sample i, which is determined by the window type, i.e., Hamming window or rectangular window [6].

5.4.1.3 Variation Contribution Analysis

The variations' contributions to I are mainly represented by the first four terms at the right side of Eq. (5.13) as

$$
\begin{aligned}
S_1 &= \left(\frac{\partial I}{\partial W}\right)^2 \sigma_W^2,\ S_2 = \left(\frac{\partial I}{\partial L}\right)^2 \sigma_L^2,\ S_3 = \left(\frac{\partial I}{\partial R}\right)^2 \sigma_R^2 \\
S_4 &= \left(\frac{\partial I}{\partial V_{\text{th}}}\right)^2 \left(\frac{C_1}{WL} + \frac{C_2}{\exp\left(\mathrm{L}/l'\right)} \cdot \frac{W_c}{W} \cdot \sigma_L^2\right).
\end{aligned}
\tag{5.16}
$$

As pointed out by many prior arts [22], an asymmetry exists in STT-RAM write operations: the switching time of '0'→'1' is longer than that of '1'→'0' and suffers from a larger variance. Also, the switching time variance of '0'→'1' is more sensitive to the transistor size changes than '1'→'0'. As we shall show later, these phenomena can be well explained by using our sensitivity analysis.

As shown in Fig. 5.1, when writing '0', the word line (WL) and bit line (BL) are connected to V_{dd}, while the source line (SL) is connected to ground. $V_{gs} = V_{dd}$ and $V_{ds} = V_{dd} - IR$. The NMOS transistor is working in saturation region when the W is small or in triode region when W is large. Based on short-channel BSIM model, the MTJ switching current supplied by the NMOS transistor working in saturation region can be calculated by

$$I = \frac{\beta \cdot \left[(V_{dd} - V_{th})(V_{dd} - \mathrm{IR}) - \frac{a}{2}(V_{dd} - \mathrm{IR})^2 \right]}{1 + \frac{1}{v_{sat} L}(V_{dd} - \mathrm{IR})}. \tag{5.17}$$

Here, $\beta = \frac{\mu_0 C_{ox}}{1+U_0(V_{dd}-V_{th})} \frac{W}{L}$. U_0 is the vertical field mobility reduction coefficient, μ_0 is electron mobility, C_{ox} is gate oxide capacitance per unit area, a is body effect coefficient, and v_{sat} is carrier velocity saturation. Based on short-channel PTM model [10] and short-channel BSIM model [2, 13], we derive $\left(\frac{\partial I}{\partial W}\right)^2$, $\left(\frac{\partial I}{\partial L}\right)^2$, $\left(\frac{\partial I}{\partial R}\right)^2$, and $\left(\frac{\partial I}{\partial V_{th}}\right)^2$ as

$$\left(\frac{\partial I}{\partial W}\right)_0^2 \approx \frac{1}{(A_1 W + B_1)^4}, \quad \left(\frac{\partial I}{\partial L}\right)_0^2 \approx \frac{1}{\left(\frac{A_2}{W} + B_2 W + C\right)^2}$$
$$\left(\frac{\partial I}{\partial R}\right)_0^2 \approx \frac{1}{\left(\frac{A_3}{W} + B_3\right)^4}, \quad \left(\frac{\partial I}{\partial V_{th}}\right)_0^2 \approx \frac{1}{\left(\frac{A_4}{\sqrt{W}} + B_4\sqrt{W}\right)^4}. \tag{5.18}$$

Here, R is the high-resistance state of the MTJ, or R_H.

$$A_1 = \sqrt{\frac{\mu_0 C_{ox} V_{dd}(V_{dd} - V_{th})}{L}} R$$
$$B_1 = \sqrt{\frac{L}{\mu_0 C_{ox} V_{dd}(V_{dd} - V_{th})}}$$
$$A_2 = \frac{L^2}{\mu_0 C_{ox} V_{dd}(V_{dd} - V_{th})}$$
$$B_2 = R^2 \mu_0 C_{ox} \frac{V_{dd} - V_{th}}{V_{dd}}$$
$$A_3 = \frac{L}{\mu_0 C_{ox} \sqrt{V_{dd}}(V_{dd} - V_{th})}, \quad B_3 = \frac{R}{\sqrt{V_{dd}}}, \quad C = \frac{2LR}{V_{dd}}$$

$$A_4 = \sqrt{\frac{L}{\mu_0 C_{ox} V_{\mathrm{dd}}}},\ B_4 = \sqrt{\frac{\mu_0 C_{ox}}{L V_{\mathrm{dd}}}} R\left(V_{\mathrm{dd}} - V_{\mathrm{th}}\right). \tag{5.19}$$

For a NMOS transistor working in triode region at '0'→'1' switching, the MTJ switching current becomes

$$I = \frac{\beta}{2a}\left[\left(V_{\mathrm{dd}} - IR - V_{\mathrm{th}}\right) - \frac{I}{W C_{ox} v_{\mathrm{sat}}^2}\right]^2. \tag{5.20}$$

where R is the low-resistance state of the MTJ, or R_L. We have

$$\begin{aligned}
\left(\frac{\partial I}{\partial W}\right)_1^2 &\approx \frac{1}{(A_5 W + B_5)^4},\quad \left(\frac{\partial I}{\partial L}\right)_1^2 \approx \frac{1}{\left(\frac{A_6}{W} + B_6\right)^2} \\
\left(\frac{\partial I}{\partial R}\right)_1^2 &\approx \frac{1}{\left(\frac{A_7}{W} + B_7\right)^4},\quad \left(\frac{\partial I}{\partial V_{\mathrm{th}}}\right)_1^2 \approx \frac{1}{\left(\frac{A_8}{W} + B_8\right)^2}.
\end{aligned} \tag{5.21}$$

Here,

$$\begin{aligned}
A_5 &= \sqrt{\frac{2C_{ox} v_{\mathrm{sat}} \mu_0}{La + \mu_0\left(V_{dd} - V_{\mathrm{th}}\right)}} R \\
B_5 &= \frac{\mu_0}{2C_{ox} v_{\mathrm{sat}}\left[La + \mu_0\left(V_{dd} - V_{\mathrm{th}}\right)\right]} \\
A_6 &= \frac{\mu_0}{2aC_{ox} v_{\mathrm{sat}}^2},\ B_6 = \frac{R\mu_0}{a v_{\mathrm{sat}}} \\
A_7 &= \frac{1}{2C_{ox} v_{sat}} \sqrt{\frac{\mu_0}{Lav_{sat} + \mu_0\left(V_{\mathrm{dd}} - V_{\mathrm{th}}\right)}} \\
B_7 &= \sqrt{\frac{\mu_0}{Lav_{\mathrm{sat}} + \mu_0\left(V_{\mathrm{dd}} - V_{\mathrm{th}}\right)}} R \\
A_8 &= \frac{1}{2C_{ox} v_{\mathrm{sat}}},\ B_8 = R.
\end{aligned} \tag{5.22}$$

In general, a large S_i, $i = 1...4$, corresponds to a large contribution to I variation. When W is approaching infinity, only S_3 is non-zero at '1'→'0' switching, while both S_2 and S_3 are non-zero at '0'→'1' switching. It indicates that the residual values of S_1–S_4 at '0'→'1' switching are larger than that at '1'→'0' switching when $W \to \infty$. In other words, '0'→'1' switching suffers from a larger MTJ switching current variation than '1'→'0' switching when NMOS transistor size is large.

5.4.1.4 Simulation Results of Sensitivity Analysis

Sensitivity analysis [4] can be used to obtain the statistical parameters of MTJ switching current, i.e., the mean and the standard deviation, without running the costly SPICE and Monte Carlo simulations. It can be also used to analyze the contributions of different variation sources to I variation in details. The normalized contributions (P_i) of variation resources W, L, V_{th}, and R are defined as

$$P_i = \frac{S_i}{\sum_{i=1}^{4} S_i}, i = 1, 2, 3, 4. \tag{5.23}$$

Here,

$$S_1 = \left(\frac{\partial I}{\partial W}\right)^2 \sigma_W^2, S_2 = \left(\frac{\partial I}{\partial L}\right)^2 \sigma_L^2, S_3 = \left(\frac{\partial I}{\partial R}\right)^2 \sigma_R^2,$$
$$S_4 = \left(\frac{\partial I}{\partial V_{\text{th}}}\right)^2 \left(\frac{C_1}{WL} + \frac{C_2}{\exp\left(\text{L}/l'\right)} \cdot \frac{W_c}{W} \cdot \sigma_L^2\right). \tag{5.24}$$

Figures 5.6 and 5.7 show the normalized contributions of every variation source at '0'→'1' and '1'→'0' switchings, respectively, at different transistor sizes. We can see that L and V_{th} are two major contributors to I variation at both switching directions when W is small. At '1'→'0' switching, the contribution of L ramps up until reaching its maximum value when W increases and then quickly decreases when W further increases. At '0'→'1' switching, however, the contribution of L monotonically decreases, but keep being the dominant factor over the simulated W range. At both switching directions, the contributions of R rise up when W increases.

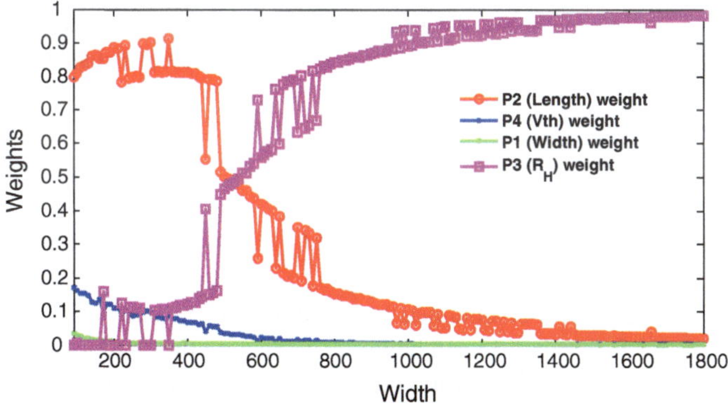

Fig. 5.6 The normalized contributions under different W at '1'→'0' switching

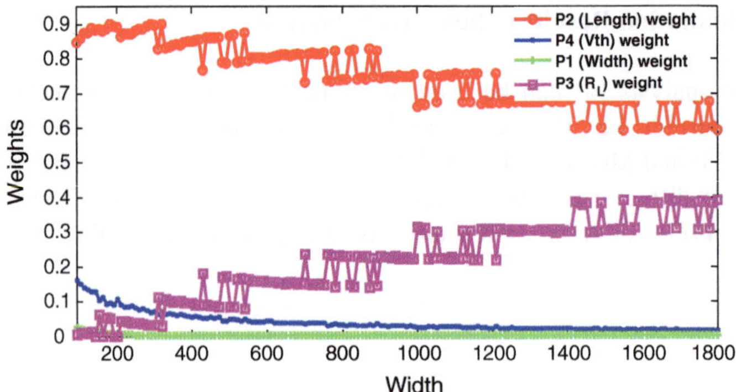

Fig. 5.7 The normalized contributions under different W at '0'→'1' switching

At '1'→'0' switching, the normalized contribution of R becomes almost 100 % when W is really large.

5.4.2 Write Current Distribution Recovery

After the I distribution is characterized by the sensitivity analysis, the next question becomes how to recover the distribution of I from the characterized information in the statistical analysis of STT-RAM reliability. It is found that dual-exponential function can provide an excellent accuracy of the typical distributions of I in modeling and recovering these distributions. The dual-exponential function for the I distributions is shown below:

$$f(I) = \begin{cases} a_1 e^{b_1(I-u)} & I \leq u \\ a_2 e^{b_2(u-I)} & I > u. \end{cases} \tag{5.25}$$

Here, a_1, b_1, a_2, b_2, and u are the fitting parameters, which can be calculated by matching the first- and second-order momentums of the actual I distribution and the dual-exponential function as

$$\begin{aligned} &\int f(I)dI = 1 \\ &\int I f(I)dI = E(I) \\ &\int I^2 f(I)dI = E(I)^2 + \sigma_I^2. \end{aligned} \tag{5.26}$$

Here, $E(I)$ and σ_I^2 can be obtained from the sensitivity analysis.

The recovered I distribution can be used to generate the MTJ switching current samples, as shown in Fig. 5.8. At the beginning of the sample generation flow, the confidence interval for STT-RAM design is determined, e.g., $[\mu_I - 6\sigma_I, \mu_I + 6\sigma_I]$ for a six sigma confidence interval. For example, if N samples are needed to generate within the confidence interval, at the point of $I = I_i$, a switching current sequence

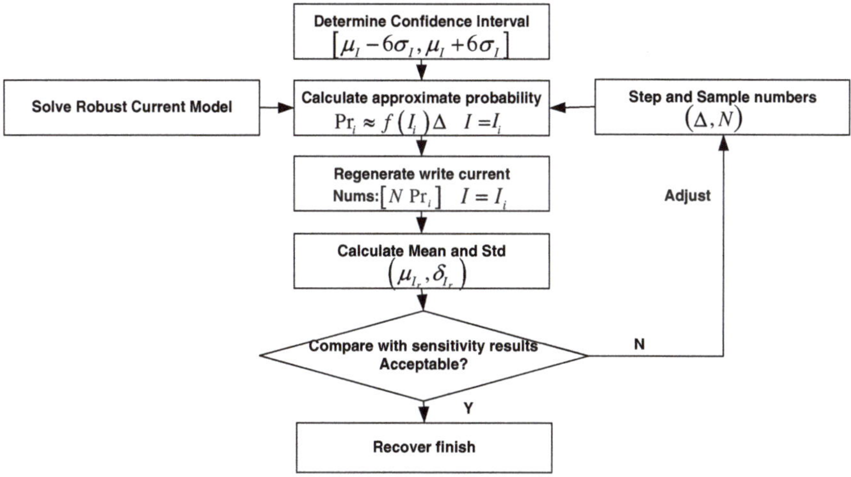

Fig. 5.8 Basic flow for MTJ switching current recovery

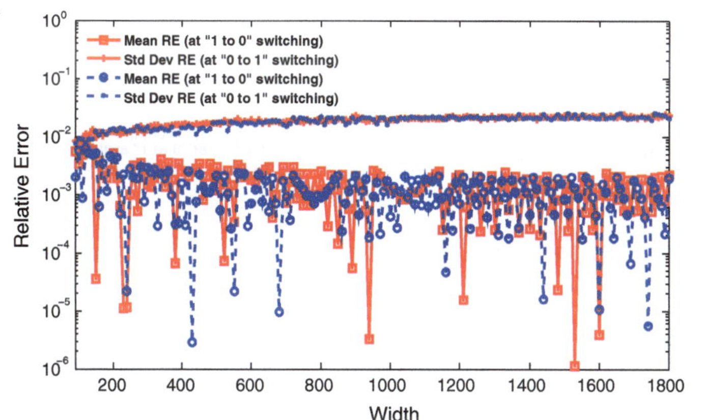

Fig. 5.9 Relative error of the recovered I w.r.t. the result from sensitivity analysis

of $[N\Pr_i]$ samples must be generated. Here, $\Pr_i \approx f(I_i)\,\Delta$. Δ equals $\frac{12\sigma_I}{N}$, or the step of sampling generation. $f(I_i)$ is the dual-exponential function.

Figure 5.9 shows the relative error of the mean and the standard deviation of the recovered I distribution w.r.t. the results directly from the sensitivity analysis (see Eqs. (5.12) and (5.13)). The maximum relative error $<10^{-2}$, which proves the accuracy of our dual-exponential model.

Figures 5.10 and 5.11 compare the probability distribution functions (PDF's) of I from SPICE and Monte Carlo simulations and from the recovery process based on our sensitivity analysis at two switching directions. This method achieves good

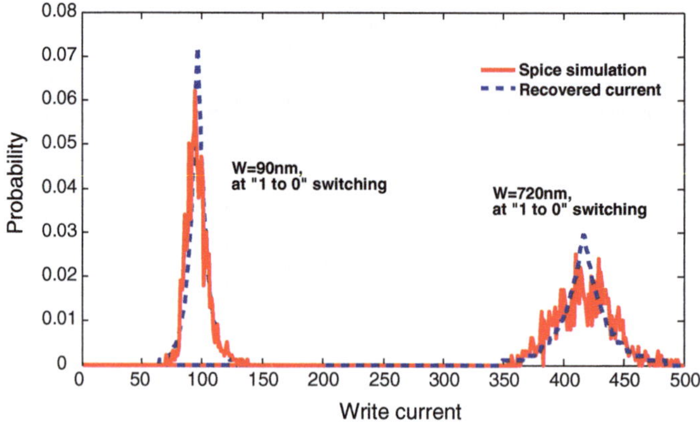

Fig. 5.10 Recovered I versus Monte Carlo result at '1'→'0'

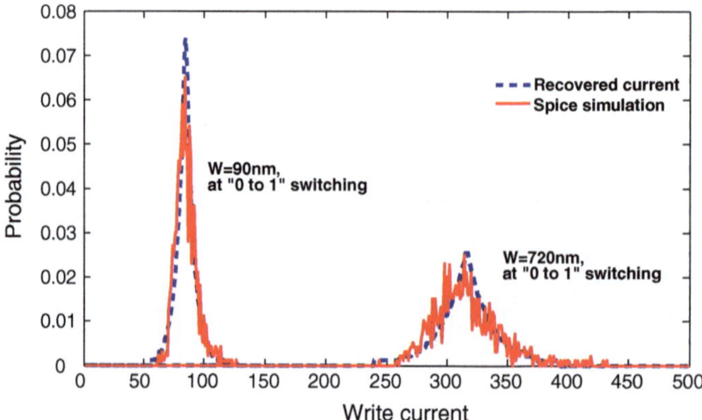

Fig. 5.11 Recovered I versus Monte Carlo result at '0'→'1'

accuracy at both simulated representative transistor channel widths ($W = 90\,\text{nm}$ or $= 720\,\text{nm}$).

5.4.3 Statistical Thermal Analysis

The variation in the MTJ switching time (τ_{th}) incurred by the thermal fluctuations follows Gaussian distribution when τ_{th} is below $10 \sim 20\,\text{ns}$, as Sect. 5.3.2 shows [22]. In this range, the distribution of τ_{th} can be easily constructed after the I is determined. The distribution of MTJ switching performance can be obtained by combining the τ_{th} distributions of all I samples.

5.5 Write Reliability Analysis

In this section, the statistical analysis is conducted on the write reliability of STT-RAM cells by leveraging our PS3-RAM method. Both device variations and thermal fluctuations are considered.

5.5.1 Reliability Analysis of STT-RAM Cells

The write failure rate P_{WF} of an STT-RAM cell can be defined as the probability that the actual MTJ switching time τ_{th} is longer than the write pulse width T_w or $P_{WF} = P\left(\tau_{\text{th}} > T_w\right)$. τ_{th} is impacted by the MTJ switching current, MTJ and MOS device variations, MTJ switching direction, and thermal fluctuations. The simulation of P_{WF} can be conducted by PS3-RAM without incurring the costly Monte Carlo runs with hybrid SPICE and macro-magnetic modeling steps.

Figures 5.12 and 5.13 show the P_{WF}'s simulated with PS3-RAM for both switching directions at 300 K. The simulation environment is summarized in Table 5.1. For comparison purpose, the Monte Carlo simulation results are also presented. Different T_w's are selected at either switching directions due to the asymmetric MTJ switching performances [22]: ($T_w = 10, 15, 20$ ns at '0'→'1' and $T_w = 6, 8, 10, 12$ ns at '1'→'0'). The PS3-RAM results are in excellent agreement with the ones from Monte Carlo simulations.

Since '0'→'1' is the limiting switching direction for STT-RAM reliability, in Fig. 5.14, the P_{WF} of different STT-RAM cell designs under different temperatures is also compared at this switching direction based on the result in Sect. 5.3.2. The results show that PS3-RAM can provide very close but pessimistic results compared to those of the conventional simulations. PS3-RAM is also capable to precisely

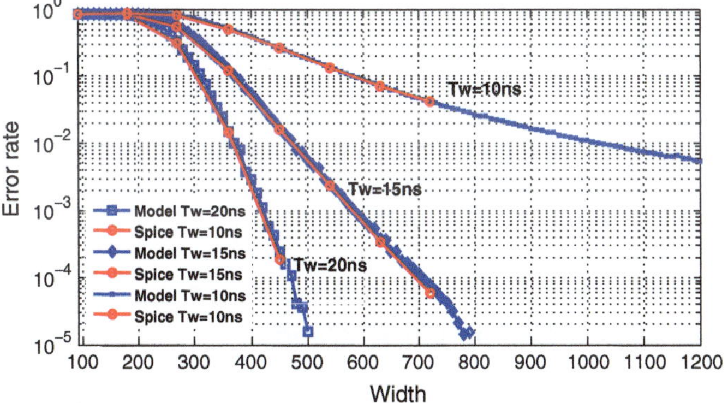

Fig. 5.12 Write failure rate at '0'→'1' when T = 300 K

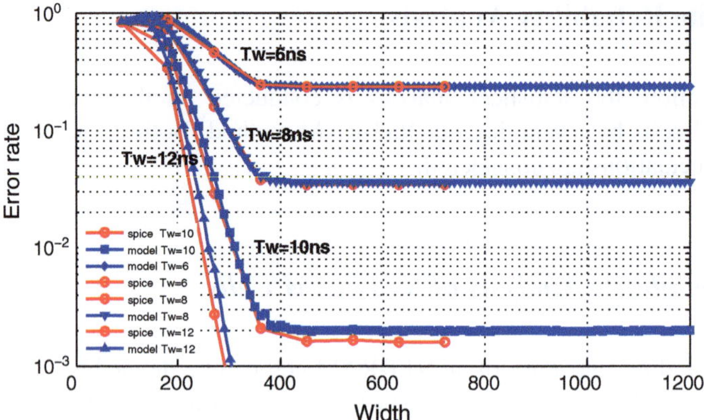

Fig. 5.13 Write failure rate at '1'→'0' when T = 300 K

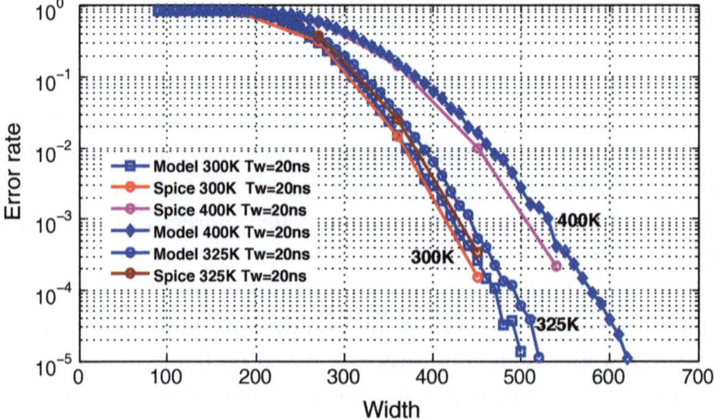

Fig. 5.14 P_{WF} under different temperatures at '0'→'1'

capture the small error rate change due to a little temperature shift (from T = 300 K to T = 325 K).

Figure 5.15 is one of the examples that PS3-RAM is used to explore the STT-RAM design space: The trade-off curves between P_{WF} and T_w are simulated at different W's. The corresponding trade-off between W and T_w can be easily identified on Fig. 5.15.

5.5.2 Computation Complexity Evaluation

We can also compare the computation complexity of our proposed PS3-RAM with the conventional simulation method. Assume the number of variation sources is M; for a statistical analysis of an STT-RAM design, the numbers of SPICE simulations

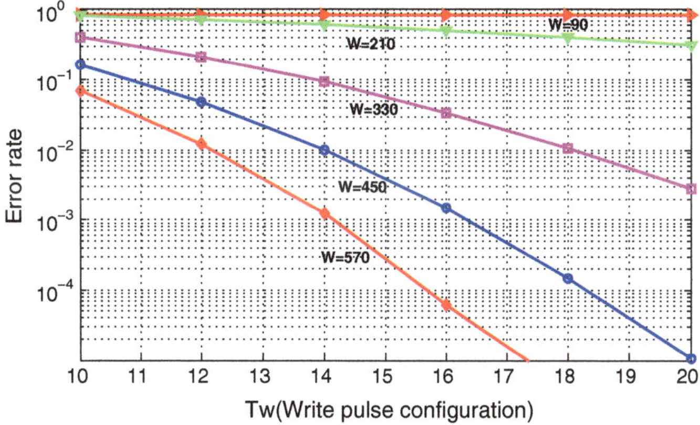

Fig. 5.15 STT-RAM design space exploration at '0'→'1'

required by conventional flow and PS3-RAM are $N_{\text{std}} = {N_s}^M$ and $N_{PS3-\text{RAM}} = 2KM + 1$, respectively. Here, K denotes the sample numbers for window-based smooth filter in sensitivity analysis. N_s is average sample numbers of every variation in the Monte Carlo simulations in conventional method and $K \ll N_s$. The speedup $X_{\text{speedup}} \approx \frac{N_s^M}{2KM}$ can be up to multiple orders of magnitude: For example, if we set $N_s = 100$, $M = 4$, (Note: V_{th} is not an independent variable) and $K = 50$, the speedup is around 2.5×10^5.

5.6 Design Space Exploration

Device variations and thermal fluctuations on the write performance and reliability have been analyzed in the previous section. Based on the statistical analysis in Sect. 5.5, in this section, the outcomes of different design methodologies will be presented, followed by the exploration on the approaches to minimize the design pessimism for the STT-RAM write operations.

A corner design methodology is usually used to overcome the impacts of device variations. The design corner can be set up as the combinations of device parameters. In STT-RAM cell design, the design corner can be set up as follows:

Based on the impacts of the major sources of device variations, the worst corner happens when L, V_{th}, and τ show positive deviations from their nominal values and W shows a negative deviation from its nominal value. However, the worst corner of A is difficult to determined: A large MTJ surface raises the magnitude of MTJ switching threshold current, while it causes the reduction in MTJ resistance, which can improve the NMOS transistor driving ability and vice versa. Two sub-worst corners need to be created for both the positive and the negative deviations of A. Table 5.2 lists the

Table 5.2 3σ worst case parameters

W	L	V_{th}	τ	A
−15 %	+15 %	+15 %	+6 %	±15 %

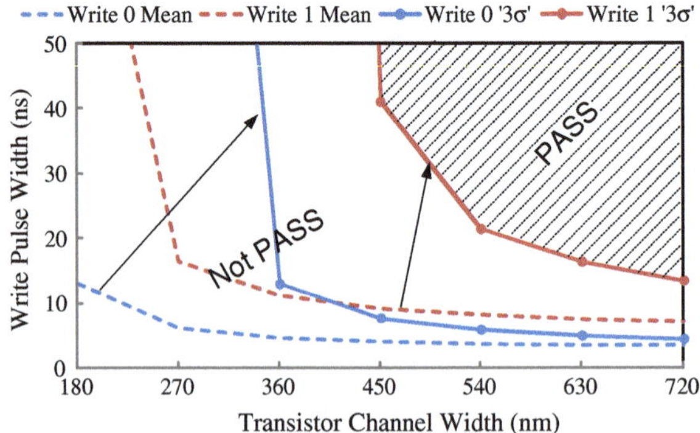

Fig. 5.16 Design space based on device parameters corner (Corner-P-I)

parameter deviations we used in the 3σ worst corner of both NMOS transistor and MTJ devices.

The simulated relationship between the NMOS transistor size and the required write pulse width is shown in Fig. 5.16. Here, only the device variations are considered and the thermal fluctuations are neglected. The required write pulse width value is calculated from the nominal relationship curve between the MTJ write current and the switching time (see Fig. 5.2a), while the MTJ write current is calculated based on the 3σ device parameter corner. The solid blue and red lines denote the results of '1'→'0' switching and '0'→'1' switching, respectively. The worst result is obtained when the MTJ surface area A is 15 % less than the nominal value. Simulation shows that '0'→'1' switching is the limiting switching direction, which requires larger transistor size and/or longer write pulse width. The pass region is constrained by the solid red line. This design method here is called as "Corner-P-I." For comparison purpose, the nominal design result that does not consider any device variations or thermal fluctuations is also plotted as the dash blue and red lines. A larger pass region is observed though it is an optimistic result.

5.6.1 Process Variation Aware Only Corner Design

There is another way to create design corner, i.e., directly using the 3σ value of the MTJ write current distribution to compute the required write pulse width. This method equals characterizing the MTJ write current corner by conventional statistical CMOS circuit design method and then deriving the MTJ switching time with the nominal MTJ switching curve. We refer to this design method as "Corner-P-II." The

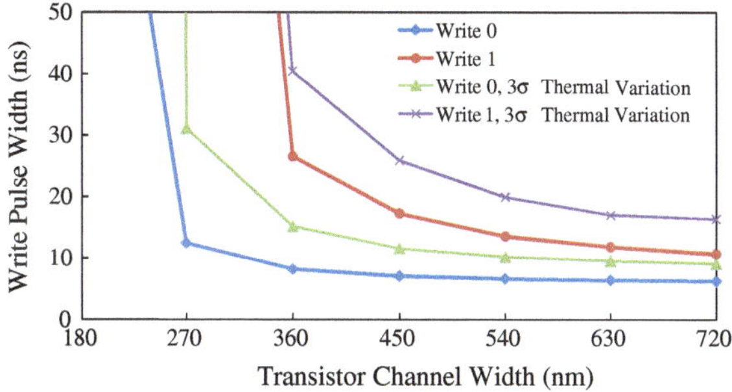

Fig. 5.17 Design space based on driving current corner (Corner-P-II & Corner-PT-II)

corresponding results are shown in Fig. 5.17. The pass region is relaxed from the "Corner-P-I" result by accurately estimating the 3σ corner value of the MTJ write current. However, this result may become optimistic as the thermal fluctuation is ignored.

5.6.2 Process Variation and Thermal Fluctuation Aware Corner Design

As aforementioned, thermal fluctuations cause the variation in MTJ switching time even the MTJ write current is fixed. If thermal fluctuations are considered, a corner representing MTJ switching time variation must be also created in the corner design of STT-RAM cell. For example, the distribution of MTJ switching time under certain MTJ write current can be obtained by macro-magnetic model. Then, the required MTJ write pulse width can be selected as the one corresponding to the $+3\sigma$ deviation of the MTJ switching time from its nominal value. The simulations results of the required write pulse width at different transistor size are also shown in Fig. 5.17. Compared to the result of "Corner-P-II," additional pessimism is added into the pass region because of the consideration on thermal fluctuations. Here, the same current corner of "Corner-P-II" is used, and this corner design is called as "Corner-PT-II."

5.6.3 Process Variation and Thermal Fluctuation Aware Statistical Design

It is well known that the combination of the worst corners of all device parameters may derive very pessimistic design since the worst cases seldom happen simultaneously. To reduce the design pessimism introduced by conventional corner designs, we established our macro-magnetic–SPICE design platform to simulate the statistical

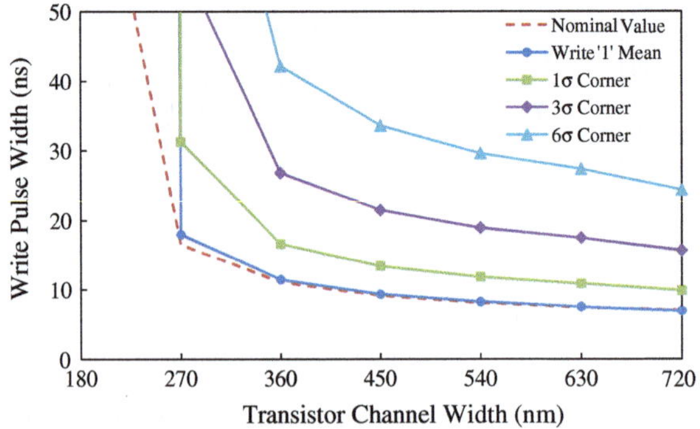

Fig. 5.18 Required write pulse width at various σ's in statistical design

property of STT-RAM cell operations. Monte Carlo simulations are run on both macro-magnetic MTJ model and SPICE transistor model to obtain the overall MTJ switching time distributions when both device variations and thermal fluctuations are considered.

Figure 5.18 shows the pass regions of the STT-RAM cell at different σ's of MTJ switching time. The pass region of the STT-RAM cell at $+3\sigma$ corner of MTJ switching time is between the results of "Corner-P-II" and "Corner-PT-II" designs, indicating the optimism of "Corner-P-II" and the pessimism of "Corner-PT-II." In any cases, '0'→'1' switching continues to be the limiting direction. It actually means that we should avoid the '0'→'1' switching in the operation of STT-RAM, as pointed out by many other studies [9].

5.7 Word-line Override Designs and Statistical Optimization Flow

5.7.1 Word-line Override Designs

As shown in Table 5.3, the effectiveness of increasing NMOS transistor size is degraded by the reduced V_{gs}. Word-line override, which compensates the loss of V_{gs} by adding additional voltage on WL for higher V_{gs}, was proposed to improve the driving ability of the NMOS transistor [17].

We assume that the WL voltage is raised to 1.1V rather than the normal 1V in '0'→'1' switching in WL override scheme. Table 5.3 compares the NMOS transistor driving abilities of the original design and the WL override design. Substantial improvement in MTJ write current is achieved in WL override design, while the incurred current variation is minimal. As a result, the total error rates reduce, as shown in Fig. 5.19.

Table 5.3 Driving current distribution with and without override voltage

Transistor size (nm)	Original design		Override design	
	Mean (μA)	Std. dev. (μA)	Mean (μA)	Std. dev. (μA)
180	148.28	14.35	169.07	14.39
270	194.75	18.11	222.21	18.19
360	230.18	20.68	262.80	20.85
450	258.18	22.76	294.89	23.01
540	280.79	24.51	320.83	24.87
630	299.91	26.15	342.69	26.62
720	315.41	27.31	360.49	27.88

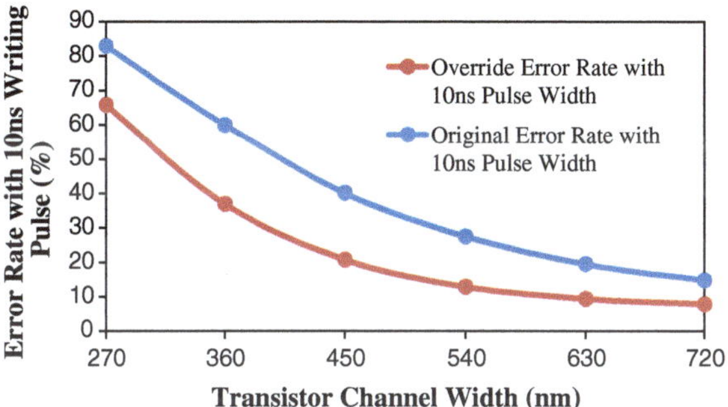

Fig. 5.19 Error rate for 10 ns writing pulse width

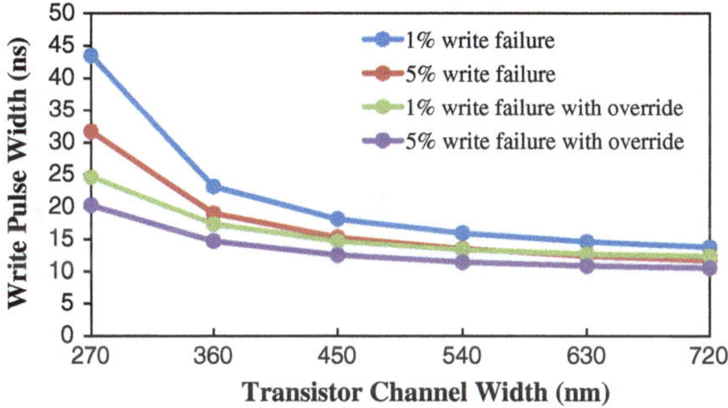

Fig. 5.20 1 and 5 % error rate for required writing pulse width

Figure 5.20 depicts the required write pulse widths by both original and WL override designs for certain total write error rates, in both '0'→'1' and '1'→'0' switches. Substantial reductions in the required writing pulse width are observed in WL override design. However, the effectiveness of WL override design is degraded by the reduced V_{gs} when the NMOS transistor size increases.

5.7.2 STT-RAM Cell Design Optimization Flow

Figure 5.21 illustrates an STT-RAM cell design optimization flow to estimate and minimize the operation errors. After the MTJ device parameters are given, the NMOS transistor sizes are calculated accordingly, based on the designed (nominal) values of both MTJ and CMOS parameters. Meanwhile, a reasonable operation pulse width will be calculated based on the nominal design. In the second step, the device parameter samples, including both the geometry and the material parameters, are generated based on the process variations in both NMOS transistor and MTJ. These samples are sent to the SPICE simulations to collect the write current samples through the MTJ. The third step takes into account the thermal fluctuation effects and the fluctuation of magnetic anisotropy under the given operation pulse width to calculate the distribution of the MTJ switching time and the write errors. Based on the specific write performance and the write error rate requirements, the optimal design points for both the NMOS transistor and the MTJ are found. If the result leads to a design

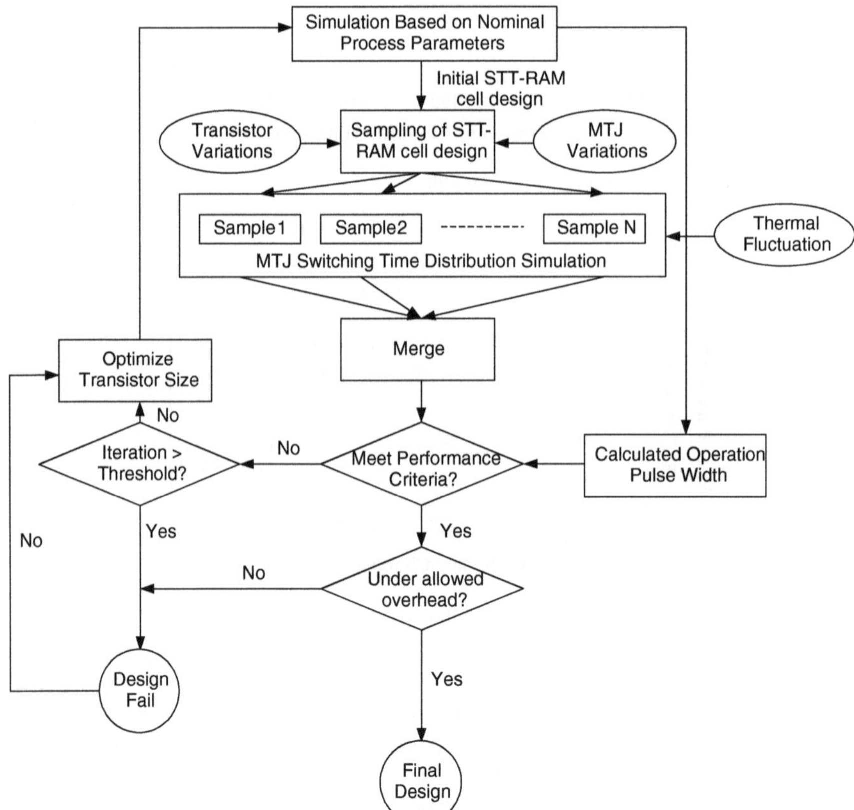

Fig. 5.21 Precess variation aware STT-RAM design flow

failure, then the override design will be applied to meet the performance. Similar design flow can be applied to the read error optimization or take it into the overall STT-RAM error rate optimization design flow.

5.8 Conclusion

In this chapter, we conduct a comprehensive discussion on the major variation sources in the STT-RAM designs and quantitatively analyze their impacts on the STT-RAM cell read and write operations. Both process variations and thermal fluctuations, which cause the significantly unbalanced write reliability at the switchings of '1'→'0' and '0'→'1', are considered in the analysis. After that, a fast and scalable statistical STT-RAM reliability analysis method named PS3-RAM is introduced. PS3-RAM is able to simulate the impact of the concerned variation sources on the statistical STT-RAM write performance, without running the costly Monte Carlo simulations on SPICE and macro-magnetic models. The effectiveness of different design methodologies are also evaluated, including nominal design, corner designs (with only device variations and with both device variations and thermal fluctuations) and full statistical design.

Acknowledgments This work was supported by National Science Foundation grants CNS-1116171 and CCF-1217947, and 49th Design Automation Conference A. Richard Newton Scholarship.

References

1. Berger, L. (Oct 1996). Emission of Spin waves by a magnetic multilayer traversed by a current. *Physical Review B*, *54*, 9353–9358.
2. BSIM.http://www-device.eecs.berkeley.edu/bsim3/. UC Berkeley.
3. Diao, Z., Li, Z., Wang, S., Ding, Y., Panchula, A., Chen, E., et al. (2007). Spin-transfer torque switching in magnetic tunnel junctions and Spin-transfer torque random access memory. *Journal of Physics: Condensed Matter*, *19*, 165209.
4. Doubilet, P., Begg, C. B., Weinstein, M. C., Braun, P., McNeil, B. J. (1985). A Practical approach: Probabilistic sensitivity analysis using Monte Carlo Simulation.
5. Gilbert, T. L. (1955). A lagrangian formulation of the gyromagnetic equation of the magnetization field. *Physics Review*, *100*(1243).
6. Harris, F. J. (Jan. 1978). On the use of windows for Harmonic analysis with the discrete fourier transform. *Proceedings of the IEEE*, *66*(1), 51–83.
7. Li, J., Augustine, C., Salahuddin, S., Roy, K. (2008). Modeling of failure probability and statistical design of spin-torque transfer magnetic random access memory (STT MRAM) array for yield enhancement. In 45th Design Automation Conference, pp. 278–283, june 2008.
8. Li, J., Liu, H., Salahuddin, S., Roy, K. (2008). Variation-tolerant Spin-Torque transfer (STT) MRAM array for yield enhancement. In CICC, pp. 193–196, Sep. 2008.
9. Nigam, A., Smullen, C. W., Mohan, V., Chen, E., Gurumurthi, S., Stan, M. R. (2011). Delivering on the promise of universal memory for Spin-Transfer Torque RAM (STT-RAM), *International Symposium on Low Power Electronics and Design (ISLPED)*, pp. 121–126, aug 2011.

10. Predictive Technology Model (PTM). http://www.eas.asu.edu/ptm/. ASU.
11. Raychowdhury, A., Somasekhar, D., Karnik, T., De V. (2009). Design space and scalability exploration of 1T–1STT MTJ memory arrays in the presence of variability and disturbances. In IEEE International Electron Devices Meeting (IEDM), pp. 1–4, dec. 2009.
12. Raychowdhury, A., Somasekhar, D., Karnik, T., De, V. (2009). Design space and scalability exploration of 1t–1stt MTJ memory arrays in the presence of variability and disturbances. In IEDM, pp. 1–4, Dec. 2009.
13. Sheu, B. J., Scharfetter, D. L., Ko, P.-K., & Jeng, M.-C. (Aug 1987). BSIM: Berkeley short-channel IGFET model for MOS transistors. *JSSC*, *22*(4), 558–566.
14. Singha, R., Balijepalli, A., Subramaniam, A., Liu, F., Nassif, S. (2007). Modeling and analysis of non-rectangular gate for post-lithography circuit simulation. In 44th DAC, pp. 823–828, June 2007.
15. Smullen, C. W., Nigam, A., Gurumurthi, S., Stan, M. R. (2011). The STeTSiMS STT-RAM simulation and modeling system. In ICCAD, pp. 318–325, Nov 2011.
16. Sun, G., Dong, X., Xie, Y., Li, J., Chen, Y. (2009). A novel architecture of the 3D stacked MRAM L2 cache for CMPs. In 15th HPCA, pp. 239–249. IEEE, 2009.
17. Wang, X., Zheng, Y., Xi, H., Dimitrov, D. (2008). Thermal fluctuation effects on Spin Torque induced switching: Mean and variations. *JAP*, *103*(3):034507–034507-4, Feb. 2008.
18. Xu, W., Chen, Y., Wang, X., Zhang, T. (2009). Improving STT MRAM storage density through smaller-than-worst-case transistor sizing. In 46th DAC, pp. 87–90, July 2009.
19. Xu, C., Niu, D., Zhu, X., Kang, H. S, Nowak, M., Yuan, X. (2011). Device architecture co-optimization of STT-RAM based memory for low power embedded systems. In ICCAD, p. 463–470, Nov 2011.
20. Xu, W., Sun, H., Chen, Y., Zhang, T. (2011). Design of last-level on-chip cache using Spin-Torque Transfer RAM (STT-RAM). In *IEEE Transactions on VLSI System*, pp. 483–493. IEEE, 2011.
21. Ye, Y., Liu, F., Nassif, S., Cao, Y. (2008). Statistical modeling and simulation of threshold variation under dopant fluctuations and line-edge roughness. In 45th DAC, pp. 900–905, June 2008.
22. Zhang, Y., Wang, X., Chen, Y. (2011). STT-RAM Cell Design Optimization for Persistent and Non-Persistent Error rate Reduction: A statistical Design View. In ICCAD, pp. 471–477, Nov. 2011.
23. Zhou, P., Zhao, B., Yang, J., Zhang, Y. (2009). Energy Reduction for STT-RAM Using Early Write Termination. In ICCAD, pp. 264–268. ACM, 2009.

Chapter 6
An Energy-Efficient 3D Stacked STT-RAM Cache Architecture for CMPs

Guangyu Sun, Xiangyu Dong, Yiran Chen and Yuan Xie

Abstract In this chapter, we introduce how to adopt spin-transfer torque random access memory (STT-RAM) as on-chip L2 caches to achieve better performance and lower energy consumption, compared to traditional L2 cache designs. STT-RAM is a promising memory technology for on-chip cache design because of its fast read access, high density, and non-volatility. Using 3D heterogeneous integrations, it becomes feasible and cost-efficient to stack STT-RAM atop conventional chip multiprocessors (CMPs). However, one disadvantage of STT-RAM is its long write latency and its high write energy. In this chapter, we first stack STT-RAM-based L2 caches directly atop CMPs and compare it against SRAM counterparts in terms of performance and energy. We observe that the direct STT-RAM stacking might harm the chip performance due to the aforementioned long write latency and high write energy. To solve this problem, we then propose two architectural techniques: *read-preemptive write buffer* and *SRAM–STT-RAM hybrid L2 cache*. The simulation result shows that our optimized STT-RAM L2 cache improves performance by 4.91 % and reduces power by 73.5 % compared to the conventional SRAM L2 cache with the similar area.

G. Sun
Peking University, Beijing, China
e-mail: gsun@pku.edu.cn

X. Dong
Qualcomm Research Lab, San Diego, USA
e-mail: rioshering@gmail.com

Y. Chen
University of Pittsburgh, Pittsburgh, USA
e-mail: yic52@pitt.edu

Y. Xie (✉)
Pennsylvania State University, University Park, PA, USA
e-mail: yuanxie@cse.psu.edu

Y. Xie (ed.), *Emerging Memory Technologies*,
DOI: 10.1007/978-1-4419-9551-3_6,

6.1 Introduction

The diminishing return of endeavors to increase clock frequencies and exploit instruction-level parallelism in a single processor has led to the advent of chip multiprocessors (CMPs) [5]. The integration of multiple cores on a single chip is expected to accentuate the already daunting "memory wall" problem [3], and it becomes a major challenge of supplying massive multicore chips with sufficient memories.

The introduction of the three-dimensional (3D) integration technology [6, 35] provides the opportunity of stacking memories atop compute cores and therefore alleviates the memory bandwidth challenge of CMPs. Recently, active research [1, 10, 24] has targeted SRAM caches or DRAM memories stacking.

Spin-transfer torque random access memory (STT-RAM) is a promising memory technology with attractive features such as fast read access, high density, and non-volatility [11, 36]. How to integrate STT-RAM into compute cores on planular chips is the key obstacle since the STT-RAM fabrication involves hybrid magnetic-CMOS processes. Fortunately, 3D integrations enable the cost-efficient integration of heterogeneous technologies, which is ideal for STT-RAM stacking atop compute cores. Some recent work [7, 9] has evaluated the benefits of STT-RAM as a universal memory replacement for L2 caches and main memories in single-core chips.

In this chapter, we further evaluate the benefits of stacking STT-RAM L2 caches atop CMPs. We first develop a cache model for stacking STT-RAM and then compare the STT-RAM-based L2 cache against its SRAM counterpart with the similar area in terms of performance and energy. The comparison shows that (1) for applications that have moderate write intensities to L2 caches, the STT-RAM-based cache can reduce the total cache power significantly because of its zero standby leakage and achieve considerable performance improvement because of its relatively larger cache capacity and (2) for applications that have high write intensities to L2 caches, the STT-RAM-based cache can cause performance and power degradations due to the long latency and the high energy of STT-RAM write operations.

These two observations imply that STT-RAM-based caches might not work efficiently if we directly introduce them into the traditional CMP architecture because of their disadvantages on write latency and write energy. In light of this concern, we propose two architectural techniques, *read-preemptive write buffer* and *SRAM–STT-RAM hybrid L2 cache*, to mitigate the STT-RAM write-associated issues. The simulation result shows that performance improvement and power reduction can be achieved effectively with our proposed techniques even under the write-intensive workloads.

6.2 Related Work

Some previous research focused on the performance improvement by stacking DRAM main memories on top of processors. 3D cache model has been developed to facilitate architectural-level analysis [15, 34]. Performance analysis of 3D stacking memory was studied by Loi et al. [23]. Li et al. [22] have also reported performance improvement by using stacked SRAM L2 caches for CMPs. Black et al. studied the benefits of stacking a large DRAM or SRAM cache on a Intel Core 2 Duo processor and achieved considerable performance improvement [1]. Loh [24–26] presented an aggressive 3D DRAM integration as on-chip caches and main memories. Kgil et al. [17] have implemented an aggressive CMP method by replacing all the L2 caches with in-order simple processor cores and used 3D stacking DRAM to satisfy the memory capacity and bandwidth requirements. Ghosh et al. [10] proposed a new method to reduce the power consumption in systems where the DRAM is stacked on top of the processor cores. A prototype of the 80-core teraflop processor with an SRAM layer stacked on top of the processor cores, which was designed and fabricated by Intel, also demonstrated the benefits of stacking SRAM memories on CMPs [2].

In order to overcome the memory wall, extensive research has been done to find alternatives of the traditional SRAM and DRAM technologies. Various emerging memory technologies, such as STT-RAM, phase change memory (PCM), floating body DRAM (FBDRAM), have been proposed to replace SRAM/DRAM in different levels of the memory hierarchy [21, 30, 32, 33, 37]. STT-RAM is normally employed as a competitive replacement of SRAM as on-chip memories because of its advantages of fast read speed, low leakage power, and high density [30, 33]. PCM is widely studied as a potential candidate of the main memory because it has higher density and lower standby power compared to DRAM [21, 32, 37]. As an emerging memory compatible with CMOS technology, FBDRAM is also attracting more attention, recently [19, 28]. With these emerging memory technologies, prior research has shown improvements in performance, power consumption, and reliability.

These emerging memories, however, have some common limitations such as long write latency and high write energy. Although the memory capacity is increased after using these high-density memories, the long write latency may offset the benefits and degrade the performance for workloads with intensive write operations. Similarly, although the standby power is reduced by using these emerging memories, the dynamic power can be significantly increased due to high write energy. Some research has been done to hide the long write latency by stalling the write and preempting the blocked read [31]. The replacement policy can also be tailored to reduce the write intensity [38]. Recently, Smullen et al. proposed to scarify the non-volatility to improve the write speed [33].

6.3 Background

This section briefly introduces the background of STT-RAM and 3D integration technologies.

6.3.1 STT-RAM Background

The basic difference between the STT-RAM and the conventional RAM technologies (such as SRAM/DRAM) is that the information carrier of STT-RAM is magnetic tunnel junctions (MTJs) instead of electric charges [36]. As shown in Fig. 6.1, each MTJ contains *a pinned layer* and *a free layer*. The *pinned layer* has fixed magnetic direction, while the *free layer* can change its magnetic direction by spin torque transfers [11]. If the free layer has the same direction as the pinned layer, the MTJ resistance is low and indicates state “0”; otherwise, the MTJ resistance is high and indicates state “1”.

The latest STT-RAM technology (spin torque transfer ram, STT-RAM) changes the magnetic direction of the free layer by directly passing spin-polarized currents through MTJs. Compared to the previous generation of MRAM using external magnetic fields to reverse the MTJ status, STT-RAM has the advantage of scalability, which means that the *threshold current* to make the status reversal will decrease as the size of the MTJ becomes smaller.

The most popular structure of STT-RAM cells is composed of one NMOS transistor as the access device and one MTJ as the storage element (“1T1J” structure) [11]. As illustrated in Fig. 6.1, the storage element, MTJ, is connected in series with the NMOS transistor. The NMOS transistor is controlled by the word line (WL) signal. The detailed read and write operations for each STT-RAM cell are described as follows:

- **Read Operation**: When a *read operation* happens, the NMOS is turned on and a small voltage difference (-0.1 V as demonstrated in [11]) is applied between the bit line (BL) and the source line (SL). This voltage difference causes a current through the MTJ whose value is determined by the status of MTJs. A sense amplifier compares this current to a reference current and then decides whether a “0” or a “1” is stored in the selected STT-RAM cell.
- **Write Operation**: When a *write operation* happens, a large positive voltage difference is established between SLs and BLs for writing “0”s or a large negative one for writing “1”s. The current amplitude required to ensure a successful status

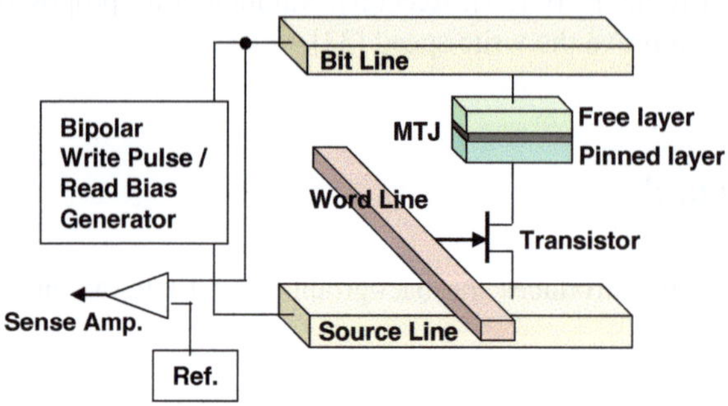

Fig. 6.1 An illustration of one STT-RAM cell

reversal is called threshold current. The current is related to the material of the tunnel barrier layer, the writing pulse duration, and the MTJ geometry [8].

In this work, we use the writing pulse duration of 10 ns [36], below which the writing threshold current will increase exponentially. In addition, we scale the STT-RAM size of previous work [11] down to 65 nm technology node. Assuming the size of MTJs is 65×90 nm, the derived threshold current for magnetic reversal is about 195 μA.

6.3.2 3D Integration Overview

The 3D integration technology has recently emerged as a promising means to mitigate interconnect-related problems. By using the vertical through silicon via (TSV), multiple active device layers can be stacked together (through wafer stacking or die stacking) in the third dimension [35].

3D integrations offer a number of advantages over traditional two-dimensional (2D) designs [6]: (1) shorter global interconnects because the vertical distance (or the length of TSVs) between two layers is usually in the range of 10–100 μm [35], depending on manufacturing processes; (2) higher performance because of reducing the average interconnect length; (3) lower interconnect power consumption due to the wire length reduction; (4) denser form factor and smaller footprint; and (5) support for the cost-efficient integration of heterogenous technologies.

In this chapter, we rely on the 3D integration technology to stack a massive amount of L2 caches (2 MB for SRAM caches and 8 MB for STT-RAM caches) on top of CMPs. Furthermore, the heterogenous technology integration enabled by 3D makes it feasible to fabricate STT-RAM caches and CMP logics as two separate dies and then stack them together in a vertical way. Therefore, the magnetic-related fabrication process of STT-RAM will not affect the normal CMOS logic fabrication and keep the integration cost-efficient.

6.4 STT-RAM and Non-uniform Cache Access Models

In this section, we describe an STT-RAM circuit model and a non-uniform cache access (NUCA) model which is implemented with network on chip (NoC).

6.4.1 STT-RAM Modeling

To model STT-RAM, we first estimate the area of STT-RAM cells. As shown in Fig. 6.1, each STT-RAM cell is composed of one NMOS transistor and one MTJ. The size of MTJs is only limited by manufacturing techniques, but the NMOS transistor

Table 6.1 STT-RAM cell specifications

Technology	65 nm
Write pulse duration	10 ns
Threshold current	195 μA
Cell size	$40F^2$
Aspect ratio	2.5

Table 6.2 Comparison of area, access time, and energy consumption (65 nm technology)

Cache size	128 KB SRAM	512 KB STT-RAM
Area	3.62 mm^2	3.30 mm^2
Read latency	2.252 ns	2.318 ns
Write latency	2.264 ns	11.024 ns
Read energy	0.895 nJ	0.858 nJ
Write energy	0.797 nJ	4.997 nJ

has to be sized properly so that it can drive sufficiently large current to change the MTJ status. The current driving ability of NMOS transistor is proportional to its W/L ratio. Using HSPICE simulation, we find that the minimum W/L ratio for the NMOS transistor under 65 nm technology node is around 10 to drive the threshold writing current of 195 μA. We further assume the width of the source or drain regions of an NMOS transistor is $1.5F$, where F is the feature size. Therefore, we estimate the STT-RAM cell size is about $10F \times 4F = 40F^2$. The parameters of our targeted STT-RAM cell are tabulated in Table 6.1.

Despite the difference in storage mechanisms, STT-RAM and SRAM have the similar peripheral interfaces from the circuit designers' point of view. By simulating with a modified version of CACTI [13], our result shows that the area of a 512 KB STT-RAM cache is similar to a 128 KB SRAM cache whose cell is about $146F^2$ (this value is extracted from CACTI). Table 6.2 lists the comparison between a 512 KB *STT-RAM cache bank* and a 128 KB *SRAM cache bank*, which are used later in this chapter, in terms of area, access time, and access energy.

6.4.2 Modeling 3D NUCA Cache

As the caches' capacity and area increase, the wire delay has made the NUCA architecture [18] more attractive than the conventional uniform cache access (UCA) one. In NUCA, the cache is divided into multiple banks with different access latencies according to their locations relative to cores and these banks can be connected through a mesh-based NoC.

Extending the work of CACTI [13], we develop our NoC-based 3D NUCA model. The key concept is to use NoC routers for communications within planular layers, while using a specific through silicon bus (TSB) for communications among different layers. Figure 6.3a illustrates an example of the 3D NUCA structure. There are four cores located in the *core layer* and 32 cache banks in each *cache layer*, and all layers are connected by TSB, which is implemented with TSVs. This interconnect style has the advantage of short connections provided by 3D integrations. It has been reported that the vertical latency of traversing a 20-layer stack is only 12 ps [27], thus the latency of TSB negligible compared to the latency of 2D NoC routers. Consequently, it is feasible to have single-hop vertical communications by utilizing TSBs. In addition, hybridization of 2D NoC routers with TSBs requires one (instead of two) additional link on each NoC router, because TSB can move data both upward and downward [22].

As shown in Fig. 6.3a, cache layers are on top of core layers and they can either be SRAM or STT-RAM caches. Figure 6.3b shows a detailed 2D structure of cache layers. Every four cache banks are grouped together and routed to other layers via TSBs.

Similar to prior approaches [4, 22], the proposed model supports *data migration*, which moves data closer to their accessing core. For set-associative cache, the cache ways belonging to the set should be distributed into different banks so that data migration can be implemented. In our 3D NUCA model, each cache layer is equally divided into several zones. The number of zones is equal to the number of cores, and each zone has a TSB located at its center. The cache ways of each set are uniformly distributed into these zones. This architecture promises that, within each cache set, there are several ways of cache lines close to the active core. Figure 6.2 gives an illustration of distributing eight ways into four zones. Figure 6.3a shows an example of data migration after which the core in the upper left corner can access the data faster. In this chapter, this kind of data migrations is called *inter-migration* to differentiate another kind of migration policy introduced later.

The advantages of this 3D NUCA cache are as follows: (1) Placing L2 caches in separate layers makes it possible to integrate STT-RAM with traditional CMOS process technology and (2) separating cores from caches simplifies the design of TSBs and routers because TSBs are now connected to cache controllers directly, and there is no direct connection between routers and cache controllers.

We provide one TSB for each core in the model. Considering that the TSV pitch size is reported to be only 4–10 μm [27]; thus, even a 1024-bit bus (much wider than our proposed TSB) would only incur an area overhead of 0.32 mm^2. In our study, the die area of an 8-core CMP is estimated to be 60 mm^2 (discussed later). Therefore, it is feasible to assign one TSB for each core and the TSV area overhead is negligible.

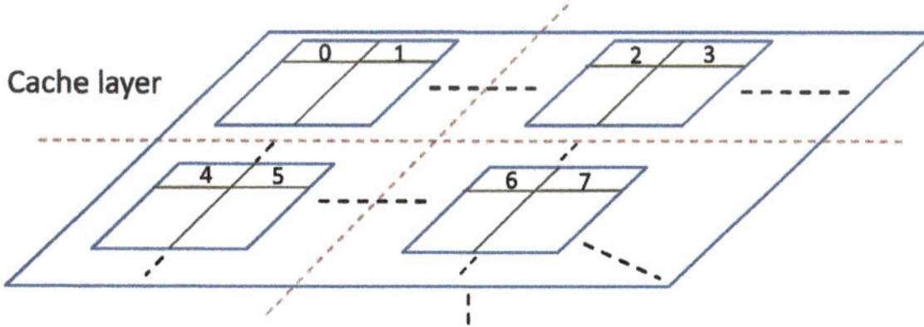

Fig. 6.2 Eight cache ways are distributed in four banks. Assume four cores and accordingly four zones each layer

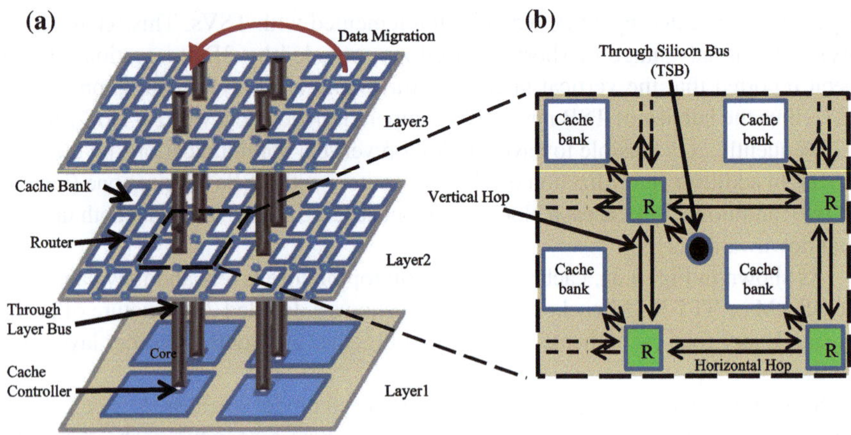

Fig. 6.3 **a** An illustration of the proposed 3D NUCA structure, which includes 1 core layer, 2 cache layers. There are 4 processing cores per core layer, 32 cache banks per cache layer, and 4 through-layer-bus across layers. **b** Connections among routers, caches banks, and through-layer-buses

6.4.3 Configurations and Assumptions

Our baseline configuration is an 8-core in-order processor using the UltraSPARC III ISA. In order to predict the chip area, we investigate some die photos, such as cell processor [16], Sun UltraSPARC T1 [20], etc., and estimate the area of an 8-core CMP without caches to be 60 mm^2. By using our modified version of CACTI [13], we further learn that one cache layer fits to either a 2 MB SRAM or an 8 MB STT-RAM L2 cache, assuming that each cache layer has the similar area to that of core layer (60 mm^2). The configurations are detailed in Table 6.3. Note that the power of processors is estimated based on the data sheet of real designs [16, 20].

We use the Simics toolset [29] for performance simulations. Our 3D NUCA architecture is implemented as an extended module in Simics. We use a few multithreaded benchmarks from *OpenMP2001* [14] and *PARSEC* [12] suites. Since the performance and power of STT-RAM caches are closely related to transaction intensity, we select some simulation workloads as listed in Table 6.4 so that we have a wide range of transaction intensities to L2 caches. The average numbers of total transactions (TPKI)[1] and write transactions (WPKI) of L2 caches are listed in Table 6.4. For each simulation, we fast forward to warm up the caches and then run 3 billion cycles. We use the total IPC of all the cores as the performance metric.

[1] TPKI is the number of total transactions per 1K instructions, and WPKI is the number of write transactions per 1K instructions.

Table 6.3 Baseline configuration parameters

Processors	
# of cores	8
Frequency	3 GHz
Power	6 W/core
Issue width	1 (in order)
Memory parameters	
L1 cache	Private, 16 + 16 KB, 2-way, 64 B line, 2-cycle, write-through, 1 read/write port
SRAM L2	Shared, 2 MB (16 × 128 KB), 32-way, 64 B line, read/write per bank: 7-cycle, write-back, 1 read/write port
STT-RAM L2	Shared, 8 MB (16 × 512 KB), 32-way, 64 B line, read penalty per bank: 7-cycle, write penalty per bank: 33-cycle, write-back, 1 read/write port
Write buffer	4 entry, retire-at-2
Main memory	4 GB, 500-cycle latency
Network parameters	
# of layers	2
# of TSB	8
Hop latency	TSB 1 cycle, V_hop 1 cycle H_hop 1 cycle
Router latency	2-cycle

Table 6.4 L2 transaction intensities

Name	TPKI	WPKI
Galgel	1.01	0.31
Apsi	4.15	1.85
Equake	7.94	3.84
Fma3d	8.43	4.00
Swim	19.29	9.76
Streamcluster	55.12	23.326

6.4.4 SNUCA and DNUCA

Static NUCA (SNUCA) and dynamic NUCA (DNUCA) are two different implementations of the NUCA architecture proposed by Kim et al. [18]. SNUCA statically partitions the address space across cache banks, which are connected via NoC; DNUCA dynamically migrates frequently accessed blocks to the closest banks. These two NUCA implementations result in different access patterns and variable write intensities. In our later simulations, we use both SNUCA-SRAM and DNUCA-SRAM L2 caches as our baselines when evaluating the performance and power benefits of STT-RAM caches.

6.5 Direct Replacing SRAM with STT-RAM as L2 Caches

In this section, we directly replace SRAM L2 caches with STT-RAM ones that have the comparable area and show that without any optimization, a naive STT-RAM replacement will harm both performance and power when the workload write intensity is high.

6.5.1 Same Area Replacement

As shown in Table 6.2, a 128 KB SRAM bank has the similar area as a 512 KB STT-RAM bank does. Thereby, in order to keep the area of cache layers unchanged, it becomes reasonable to replace SRAM L2 caches with STT-RAM ones whose capacity is three times larger. We call this replacement strategy as "same area replacement."

Using this strategy, we integrate as many caches in the cache layers as possible. Considering our baseline SRAM L2 cache has 16 banks and each cache bank has the capacity of 128 KB, we keep the number of banks unchanged but replace each 128 KB SRAM L2 cache bank with a 512 KB STT-RAM cache bank. The read/write access time and read/write energy consumption are tabulated in Table 6.2 for both SRAM and STT-RAM.

6.5.2 Performance Analysis

Because the number of banks remains the same and our modified CACTI shows that 128 KB SRAM bank and 512 KB STT-RAM bank have similar read latencies (2.252 vs. 2.318 ns in Table 6.2), the read latencies of the 2 MB SRAM cache and the 8 MB STT-RAM cache are similar as well. Since the STT-RAM cache capacity is three times larger, the access miss rate to the L2 cache decreases as shown in Fig. 6.4. On average, the miss rates are reduced by 19.0 % and 12.5 % for SNUCA STT-RAM cache and DNUCA STT-RAM cache, respectively.

The IPC comparison is illustrated in Fig. 6.5. Caused by the large STT-RAM cache capacity, the L2 cache miss rate decrease improves the performance of the first two workloads ("galgel" and "apsi"); however, the performance of the rest four workloads is not improved as expected. On average, the performance *degradation* of *SNUCA STT-RAM* and *DNUCA STT-RAM* is 3.09 % and 7.52 %, respectively, compared to their SRAM counterparts.

This performance degradation of direct STT-RAM replacement can be explained by Table 6.4, where we can observe the write operation intensity (presented by WPKI) of "equake," "fma3d," "swim," and "streamcluster" is much higher than that of "galgel" and "apsi." Due to the long latency of STT-RAM write operations, the high write intensity is reflected by the performance loss. When the write intensity is sufficiently high, the resulting performance loss overwhelms the performance

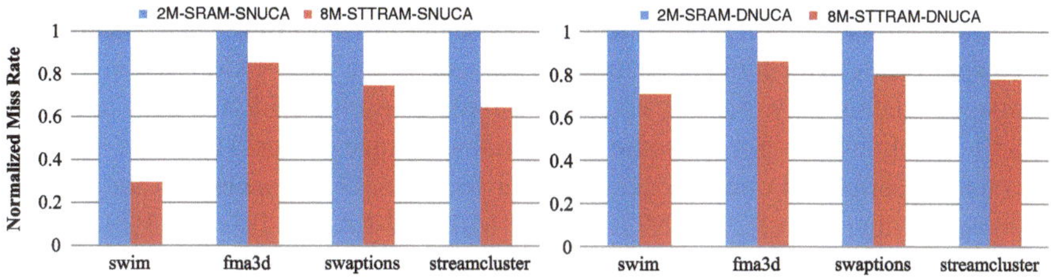

Fig. 6.4 The comparison of L2 caches access miss rates for SRAM L2 cache and STT-RAM L2 cache that have similar area. Larger capacity of STT-RAM cache results in smaller cache miss rates

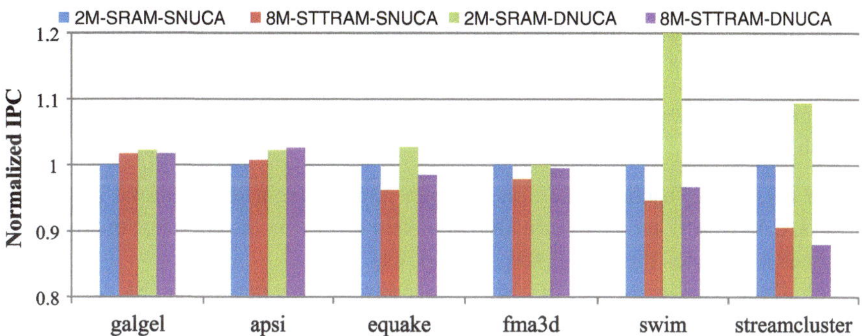

Fig. 6.5 IPC comparison of SRAM and STT-RAM L2 caches (normalized by 2M SNUCA-SRAM cache)

gain achieved by reduced L2 cache miss rate. This observation is further supported by comparison between SNUCA and DNUCA. From Fig. 6.5, one can observe that performance degradation is more significant when we use DNUCA STT-RAM caches because data migrations in DNUCA initiate more write operations than SNUCA does and thus cause high write intensities.

To summarize, we conclude our first observation of using STT-RAM caches as

Observation 1 *Replacing SRAM L2 caches directly with STT-RAM, which has the similar area but with a large capacity, can reduce the access miss rate of the L2 cache. However, the long latency associated with the write operations to the STT-RAM cache has a negative impact on the performance. When the write intensity is high, the benefits caused by miss rate reductions could be offset by the long latency of STT-RAM write operations and eventually result in performance degradation.*

6.5.3 Power Analysis

The major contributors of the total power consumption in caches are leakage power and dynamic power:

- *Leakage Power*: When process technology scales down to sub-90 nm, the leakage power in CMOS technology becomes dominant. Since STT-RAM is a non-volatile memory technology, there is no power supply to each STT-RAM cell, and then, STT-RAM cells do not consume any standby leakage power. Therefore, we only consider peripheral circuit leakage power for STT-RAM caches and the leakage power comparison of 2 MB SRAM and 8 MB STT-RAM is listed in Table 6.5.
- *Dynamic Power*: The dynamic power estimation for the NUCA cache is described as follows. For each transaction, the total dynamic power is composed of the memory cell access power, the router access power, and the power consumed by wire connections. In this chapter, these values are either simulated by HSPICE or obtained from our modified version of CACTI. The access number of routers and the length of wire connections vary from the location of the requesting core and the requested cache lines.

Figure 6.6 shows the power comparison of SRAM and STT-RAM L2 caches. One can observe that

- For SRAM L2 caches, since the leakage power dominates, the total power for SNUCA-SRAM and DNUCA-SRAM is very close. On the contrary, the dynamic power dominates the STT-RAM cache power.
- For all the workloads, STT-RAM caches consume less power than SRAM caches do. The average power savings across all the workloads are about 78 and 68 % for SNUCA and DNUCA, respectively. The power saving for DNUCA STT-RAM is smaller because of the high write intensity caused by data migrations. It is obvious that the "low leakage power" feature makes STT-RAM more attractive to be used

Table 6.5 Leakage power of SRAM and STT-RAM caches at 80 °C

Cache configurations	Leakage power (W)
2 MB 16 × 128 KB SRAM cache	2.089
8 MB 16 × 512 KB STT-RAM cache	0.255

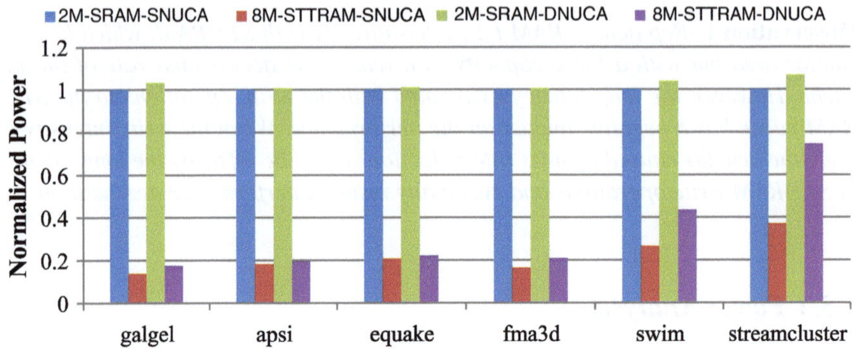

Fig. 6.6 Power comparison of SRAM and STT-RAM L2 caches (normalized by 2 MB SNUCA-SRAM cache)

as large on-chip memory, especially when SRAM leakage power becomes worse with technology scaling.

- The average power savings for the first four workloads are more than 80 %. However, for the workload "streamcluster," the total power saving is only 63 and 30 % for SNUCA and DNUCA, respectively, due to its much higher L2 cache write intensity (see Table 6.4).

To summarize, our second conclusion of direct STT-RAM cache replacement is

Observation 2 *Direct replacing of the SRAM L2 cache with a STT-RAM cache, which has similar area but with larger capacity, can greatly reduce the leakage power. However, when the write intensity is high, the dynamic power increases significantly because of the high energy associated with the STT-RAM write operation and the amount of total power saving could be reduced.*

These two conclusions show that if we directly replace SRAM caches with STT-RAM caches using "same area strategy," the long latency and high energy consumption of STT-RAM write operations can offset the performance and power benefit brought by STT-RAM cache when the cache write intensity is high.

6.6 Novel 3D-Stacked Cache Architecture

In this section, we propose two techniques to mitigate the write operation problem of using STT-RAM caches: *Read-preemptive write buffer* is employed to reduce the stall time caused by the STT-RAM long write latency; *SRAM–STT-RAM hybrid L2 cache* is proposed to reduce the number of STT-RAM write operations and thereby improve both performance and power. Finally, we combine these two techniques together as an optimized STT-RAM cache architecture.

6.6.1 Read-Preemptive Write Buffer

The first observation in Sect. 6.5 shows that the long STT-RAM write latency has a serious impact on the performance. In the scenario where a write operation is followed by several read operations, the ongoing write operation may block the upcoming read operations and cause performance degradations. Although the write buffer design in modern processors works well for SRAM caches, our experiment result in Sect. 6.5.2 shows that this write buffer does not fit for STT-RAM caches due to the large variation between STT-RAM read latency and write latency. In order to make STT-RAM caches work efficiently, we explore the proper write buffer size and propose a "read-preemptive" management policy for it.

6.6.1.1 The Exploration of the Buffer Size

The choice of the buffer size is important. The larger the buffer size is, the more write operations can be hidden. Thereby, the number of stall cycles decreases. However, on the other hand, the larger the buffer size is, the longer time it takes to check whether there is a "hit" in the buffer and then to access it. Furthermore, the design complexity and the area overhead also increase with the buffer size growth. Figure 6.7 shows the relative IPC improvement by using different buffer sizes for workloads "streamcluster" and "swim." Observing the simulation result, we choose the size of 20 entries as the optimal STT-RAM write buffer size. Compared to the SRAM write buffer, which has only 4 entries (as listed in Table 6.3), the STT-RAM write buffer size is much larger and we use 20-entry write buffer for STT-RAM caches in the later simulations.

6.6.1.2 Read-Preemptive Policy

Since the L2 cache can receive requests from the upper level memory (L1 cache) and the write buffer, a priority policy is necessary to solve the conflict that a read request and a write request compete for the execution right. For STT-RAM caches, write operation latencies are much larger than read latencies; thus, our objective is to prevent write operations from blocking read operations. As a result, we have our first rule:

Rule 1 *The read operation always has the higher priority in a competition for the execution right.*

Additionally, consider there is a read request blocked by a write operation that is already in process, the STT-RAM write latency is so large that its retirement may block one or more read request for a long period and further cause performance degradations. In order to mitigate this problem, we propose another read-preemptive rule as follows:

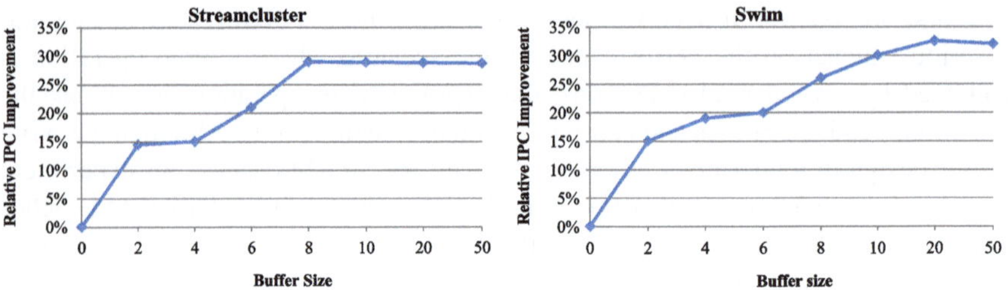

Fig. 6.7 The impact of buffer size. The IPC improvement is normalized by that of 8M STT-RAM cache without write buffer

Rule 2 *When a read request is blocked by a write retirement and the write buffer is not full, the read request can trap and stall the write retirement if the preemption condition (discussed later) is satisfied. Then, the read operation obtains the right of the execution to the cache. The stalled write retirement will retry later.*

Our proposed read-preemptive policy tries to execute STT-RAM read requests as early as possible, but the drawback is that some write retirements need to be re-executed and the possibility of full buffer increases. The pivot is to find a proper *preemption condition*. One extreme method is to stall the write retirement as long as there is a read request, which means that read requests can always be executed immediately. Theoretically, if the write buffer size is large enough, no read request will be blocked. However, since the buffer size is limited, the increased possibility of full buffer could also harm the performance. In some other cases, stalling write retirements for read requests is not always good. For example, if a write retirement almost finishes, no read request should stall the retirement process. Consequently, we propose to use the *retirement accomplishment degree*, denoted as α, as the preemption condition. The *retirement accomplishment degree* is the accomplishment percentage of the ongoing write retirement, below which no preemption will occur.

Figure 6.8 compares the IPC of using different α in our read-preemptive policy. Note that $\alpha = 0\,\%$ represents the non-conditional preemption policy and $\alpha = 100\,\%$ represents the traditional write buffer. We can find that for the workloads with low write intensities, such as "galgel" and "apsi," the performance improves as α increases and the non-conditional preemption policy works the best. However, for the benchmark with high write intensities, like "streamcluster," the performance improves at the beginning but then degrades as α increases. Generally, in this chapter, we set $\alpha = 50\,\%$ to make our read-preemptive policy effective for all the workloads.

A counter is required in order to make the accomplishment degree aware of the cache controller. The counter resets to zero and begins to count the number of cycles when a retirement begins. The cache controller checks the counter and

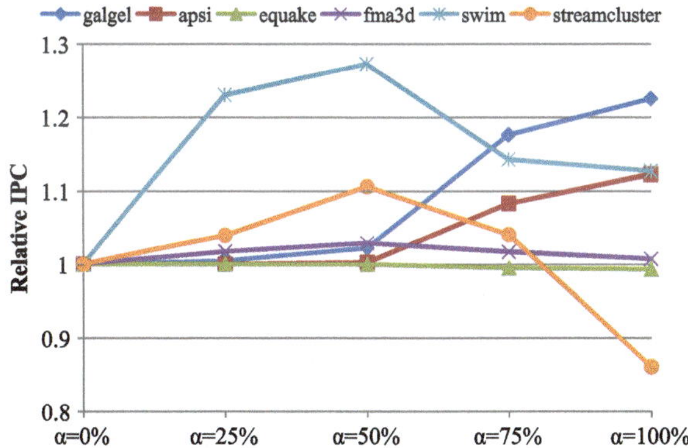

Fig. 6.8 The impact of α on the performance. The IPC values are normalized by that of using the traditional policy

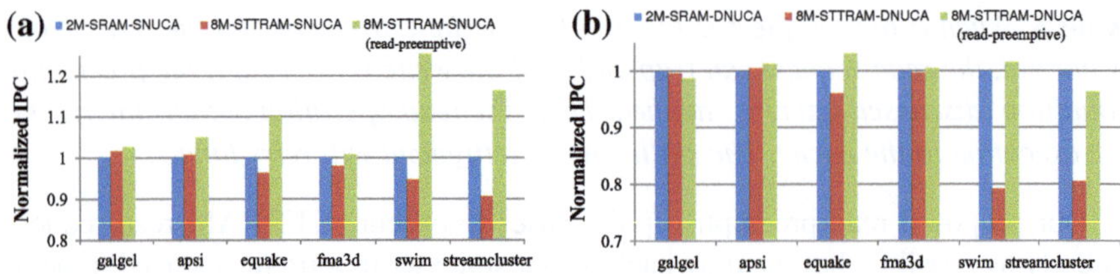

Fig. 6.9 The comparison of IPC among 2M SRAM, 8M STT-RAM with traditional write buffer, and 8M STT-RAM with read-preemptive write buffer (normalized by that of SRAM)

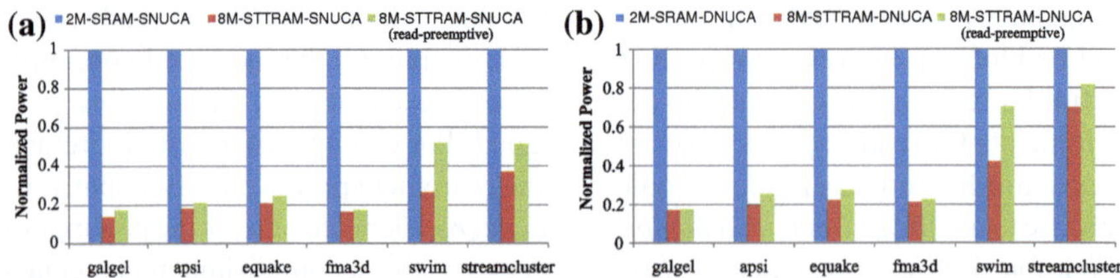

Fig. 6.10 The comparison of total power of 2M SRAM, 8M STT-RAM with traditional write buffer, and 8M STT-RAM with read-preemptive write buffer (normalized by that of SRAM)

decides whether to stall the retirement for the read request. The area of 20 buffer entries can be evaluated as a cache whose size is 20×64 byte (less than 2 KB). We use a 7-bit counter to record the retirement accomplishment degree. Since the area of each 3D-stacked layer is around 60 mm^2, the area overhead of our proposed *read-preemptive write buffer* is less than 1 %. Similarly, the leakage power increase caused by this buffer is also negligible.

Figures 6.9 and 6.10 illustrates the performance and power improvement gained by our proposed *read-preemptive write buffer*. Compared to the IPC of SRAM baseline configurations, the average performance improvements are 9.93 % and 0.41 % for SNUCA and DNUCA, respectively. The average power reductions are 67.26 % and 59.3 % for SNUCA and DNUCA, respectively. Compared to the result of direct STT-RAM replacement shown in Figs. 6.5 and 6.6, the performance degradation is eliminated, but the amount of power savings decrease. It is because using our read-preemptive write buffer causes some re-executions of write operations which consume more power.

6.6.2 SRAM–STT-RAM Hybrid L2 Cache

The aforementioned read-preemptive write buffer hides the STT-RAM long write latency, but the total number of write operations remains the same. In order to reduce the number of write operations to STT-RAM cells, we propose another technique called *SRAM–STT-RAM Hybrid Cache* and show how this technique can further reduce the dynamic power as well as improve the performance.

6.6.2.1 SRAM–STT-RAM Hybrid Cache Implementation

The proposed hybrid cache implementation is that, instead of building a pure STT-RAM cache, we compose the ways in each cache set with a majority of STT-RAM cache lines and a minority of SRAM ones. The main purpose is to keep as many write-intensive data in the SRAM part as possible and hence reduce the number of write operations to the STT-RAM part. In this work, we design an SRAM–STT-RAM hybrid L2 cache with 31 ways of STT-RAM and 1 way of SRAM (*31M1S*).

After having these hybrid cache lines, the second step is to distribute STT-RAM cache lines and SRAM ones into separate cache banks. Considering the SRAM part is the minority in the proposed *31M1S* cache, one partitioning alternative is to distribute these SRAM cache lines into different banks so that there are several SRAM cache lines close to each processing core. However, this method requires each cache bank to be a heterogenous memory array with SRAM and STT-RAM cells and increases the complexity of the cache design. In addition, this distributed partitioning of SRAM cells implies that the SRAM and STT-RAM cells have to be fabricated together. Considering the specialization of the STT-RAM fabrication process, this method also eliminates the cost advantages of stacking STT-RAMs on top of processing cores.

Therefore, we use another alternative that we reduce the number of cache lines in some STT-RAM cache banks compared to the pure STT-RAM cache structure (as shown in Fig. 6.11a that the STT-RAM banks at four corners are smaller than other

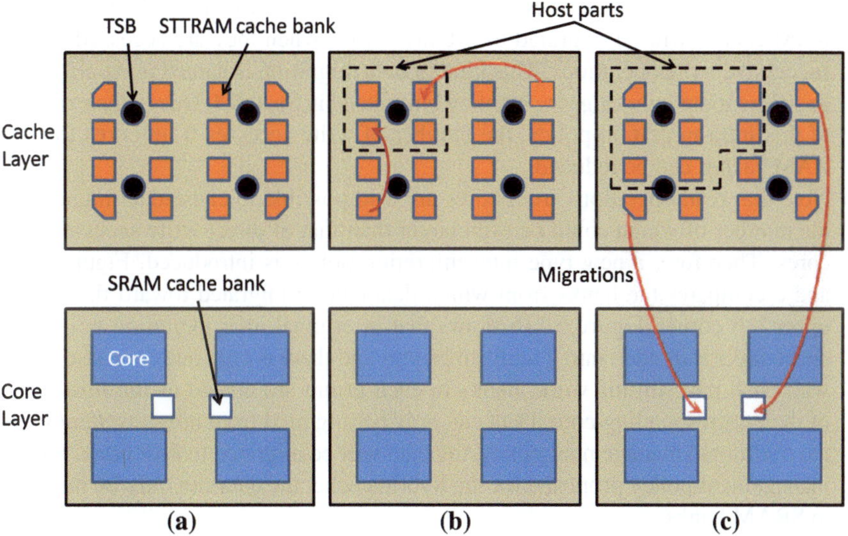

Fig. 6.11 SRAM–STT-RAM hybrid cache implementation **a** one placement method of SRAM and STT-RAM cache banks. **b** data migrations in original STT-RAM caches. **c** data migrations in hybrid SRAM–STT-RAM caches

STT-RAM banks), compensate this cache line loss with SRAM ones, and collect all the SRAM cache lines together to build several entire SRAM banks on the core layer. As shown in Fig. 6.11a, SRAM cache banks are placed in the center of the core layer instead of being distributed. In this method, SRAM and STT-RAM cache banks have no difference from the architectural point of view.

Note that after placing one way of SRAM cache lines in the core layer, the area of the core layer will increase and the area of the cache layer will decrease. In this work, the total size of all the SRAM cache lines is 256 KB, and the derived area overhead is about 12.5 %.

6.6.2.2 Hybrid Cache Management Policy

Another important issue is how to manage the hybrid L2 cache to improve the performance and reduce the power. Because the key point is to reduce the number of write operations to STT-RAM cache cells, we need to move as many write-intensive data in SRAM cache banks as possible. The management policy of the hybrid cache can be described as follows:

- The cache controller is aware of the locations of SRAM cache ways and STT-RAM cache ways. When there is a write miss, the cache controller first tries to place the data in the SRAM cache ways.
- Considering the high probability that a core write data to a specific group of cache lines repeatedly, data in STT-RAM caches should be migrated to SRAM caches if the some cache lines are frequently written to. In this work, data in STT-RAM caches will be migrated to SRAM caches when they are accessed by two successive write operations. This kind of data migration is named *intra-migration* to differentiate *inter-migration* policy introduced in Sect. 6.4. Due to the existence of this intra-migration policy, the number of write accesses from cores to STT-RAM caches can be reduced.
- Note that read operations from cores are also possible to cause data migrations, the number of which could be even larger than that of direct write accesses from cores. Therefore, a new type inter-migration policy is introduced. Figure 6.11b and c compares the banks from which data can be migrated toward the core in upper left corner. Figure 6.11b shows that in original inter-migration policy, the cache layer is divided into 4 uniform groups and there is only one core associative with each part. In this work, banks in each group are named as the *host banks* of their corresponding core. Data can only be migrated from *non-host banks*. For the traditional management policy, the data will be migrated to *host bank*. For the management policy proposed for the hybrid cache, the data can only be migrated to SRAM banks.

Two data migrations are illustrated in Fig. 6.11b for the traditional inter-migration. When using the hybrid SRAM–STT-RAM cache, the *host banks* for a core is redefined as shown in Fig. 6.11c. Two corresponding data migrations are also shown

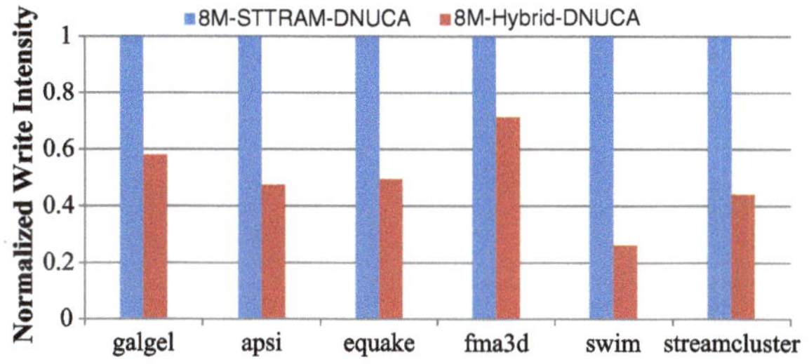

Fig. 6.12 The STT-RAM write intensity to STT-RAM before and after using hybrid SRAM–STT-RAM caches

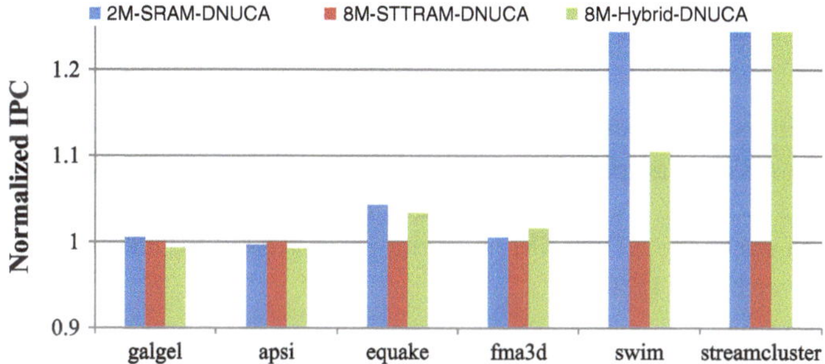

Fig. 6.13 The comparison of IPC among 2M SRAM cache, 8M STT-RAM pure cache, and 8M SRAM–STT-RAM hybrid cache (normalized by the IPC of 8M STT-RAM pure cache)

in Fig. 6.11c. Using this policy, there is no data migration between two STT-RAM cache lines, which reduces the number of write operations greatly. The drawback is that SRAM banks are shared by all cores so that their limited sizes may increase L2 miss rates. Considering we have 8M of total cache size, which is considerably large for most applications, our simulation results show that the increase in L2 miss rates is very small.

Figure 6.12 shows that the number of STT-RAM write operations per 1K instructions is reduced dramatically by using our hybrid SRAM–STT-RAM approach. As a result, the dynamic power associated with write operations to STT-RAM cells is also reduced and the performance is improved. Figure 6.13 shows the performance comparison. On average, the hybrid cache structure improves the performance by 5.65 %, which means that it mitigates the performance loss of STT-RAM caches from 8.48 to 2.61 % compared to their SRAM counterparts.

Figure 6.14 shows the power comparison. We observe that the total power is reduced except for "galgel." It is because both read and write intensities in "galgel" are so small that the dynamic power is very low. Consequently, the introduction of SRAM cache lines in the hybrid cache brings the leakage power back and eliminates the dynamic power reduction achieved by the hybrid structure. However, as the write

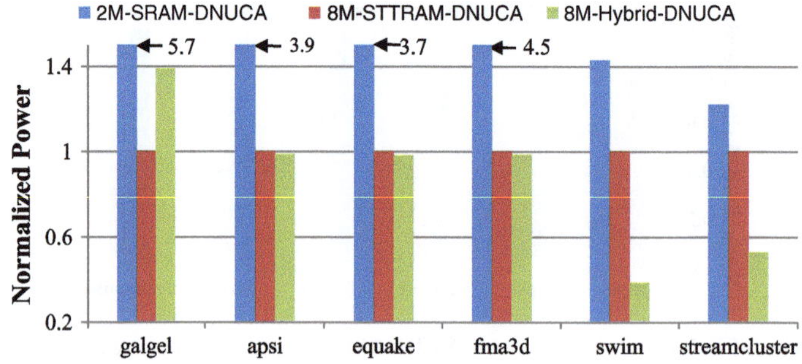

Fig. 6.14 The comparison of total power consumption among 2M SRAM cache, 8M STT-RAM pure cache, and 8M SRAM–STT-RAM hybrid cache (normalized by the total power consumption of 8M STT-RAM pure cache)

intensity increases, the SRAM–STT-RAM hybrid cache starts to save total power consumptions. For example, the total power consumption is cut by more than half for workloads such as "swim" and "streamcluster." On average, after the transition from SRAM caches to STT-RAM ones, our proposed hybrid cache further reduces the total power by 12.45 %.

6.6.3 Combination of Read-Preemptive Buffer and Hybrid Architecture

We combine the two techniques together as an optimized STT-RAM L2 cache architecture. In this architecture, we get more benefits from the advantages of the STT-RAM cache and, at the same time, mitigate the penalties caused by write operations. The performance and power comparisons are shown in Figs. 6.15 and 6.16, respectively. The average IPC is improved by 4.91 % compared to the SRAM SNUCA baseline, while power saving is 73.5 %. Table 6.6 gives an overview of the performance and power improvements.

6.7 Conclusion

STT-RAM is a promising candidate of on-chip memories, and the emerging 3D heterogeneous integration makes it feasible to stack STT-RAM as L2 caches for CMPs. In this work, we present a cache model for STT-RAM L2 cache stacking and evaluate its performance and power benefit. Even though replacing SRAM L2 cache with STT-RAM can result in significant power savings, the drawback comes from STT-RAM's long write latency and high write energy. As a result, for applications with high L2

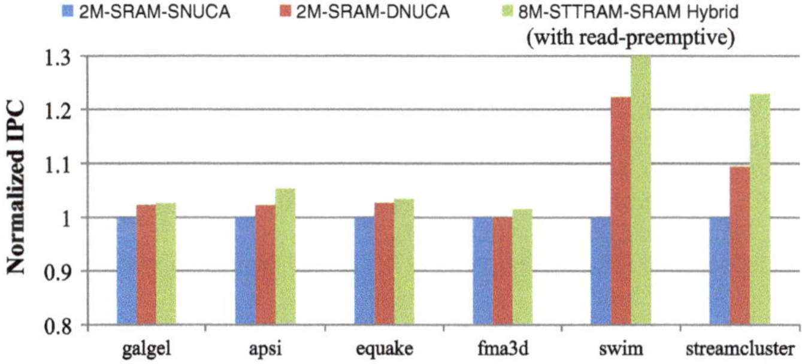

Fig. 6.15 The comparison of IPC among 2 MB SRAM SNUCA cache, 2 MB SRAM DNUCA cache, and 8 MB SRAM–STT-RAM hybrid cache with read-preemptive write buffer (normalized by the IPC of 2 MB SRAM SNUCA cache)

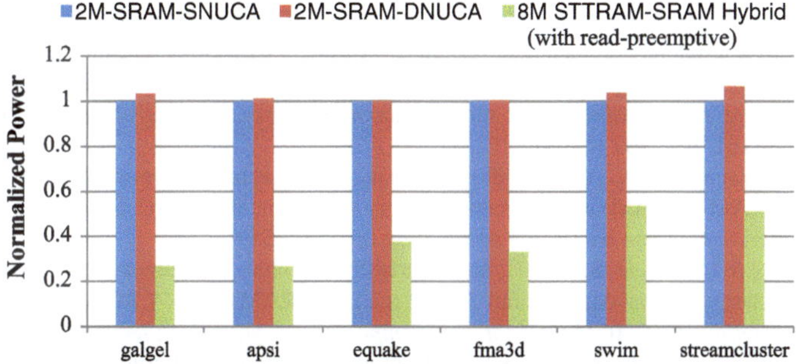

Fig. 6.16 The comparison of total power consumption among 2 MB SRAM cache, 2 MB SRAM cache, and 8 MB SRAM–STT-RAM hybrid cache with read-preemptive write buffer (normalized by the total power consumption of 2 MB SRAM cache)

Table 6.6 The performance and power improvement overview (use 2 MB SRAM L2 SNUCA cache as the baseline)

	Performance (%)	Total power (%)
Read-preemptive buffer	9.93	67.26
Hybrid cache	−2.61	85.45
Combined	4.91	73.5

cache write intensities, the performance can be degraded and the power saving can be reduced. Therefore, we propose two techniques: read-preemptive write buffer to mitigate the performance penalty caused by the long write latency and SRAM–STT-RAM hybrid L2 cache to reduce the number of STT-RAM write operations. Our result shows that these two techniques can make STT-RAM cache work effective for most workloads, regardless of their write intensities.

References

1. Black, B., Annavaram, M., Brekelbaum, N., et al. (2006). Die stacking (3D) microarchitecture. In *MICRO 39: Proceedings of the 39th Annual IEEE/ACM International Symposium on Microarchitecture* (pp. 469–479).
2. Borkar, S. (2008). 3D technology: A system perspective. In *Technical Digest of the International 3D System Integration Conference* (pp. 1–14).
3. Burger, D., Goodman, J. R., & Kagi, A. (1997). Limited bandwidth to affect processor design. *Micro, IEEE, 17*(6), 55–62.
4. Chishti, Z., Powell, M. D., & Vijaykumar, T. N. (2005). Optimizing replication, communication, and capacity allocation in CMPs. *SIGARCH Computer Architecture News, 33*(2), 357–368.
5. Davis, J. D., Laudon, J., & Olukotun, K. (2005). Maximizing CMP throughput with mediocre cores. In *PACT '05: Proceedings of the 14th International Conference on Parallel Architectures and Compilation, Techniques* (pp. 51–62).
6. Davis, W. R., Wilson, J., Mick, S., et al. (2005). Demystifying 3D ICs: The pros and cons of going vertical. *IEEE Design and Test of Computers, 22*(6), 498–510.
7. Desikan, R., Lefurgy, C. R., Keckler, S. W., & Burger, D. (2002). On-chip MRAM as a high-bandwidth low-latency replacement for DRAM physical memories. Technical report.
8. Diao, Z., Li, Z., Wang, S., et al. (2007). Spin-transfer torque switching in magnetic tunnel junctions and spin-transfer torque random access memory. *Journal of Physics: Condensed Matter, 19*(16), 165, 209 (13pp).
9. Dong, X., Wu, X., Sun, G., et al. (2008). Circuit and microarchitecture evaluation of 3D stacking magnetic RAM (MRAM) as a universal memory replacement. In *DAC '08: Proceedings of the 45th annual conference on Design automation* (pp. 554–559).
10. Ghosh, M., & Lee, H. H. S. (2007). Smart refresh: An enhanced memory controller design for reducing energy in conventional and 3D die-stacked DRAMs. In *MICRO '07: Proceedings of the 40th Annual IEEE/ACM International Symposium on Microarchitecture* (pp. 134–145).
11. Hosomi, M., Yamagishi, H., Yamamoto, T., et al. (2005). A novel non-volatile memory with spin torque transfer magnetization switching: Spin-RAM. In *International Electron Devices Meeting* (pp. 459–462).
12. http://parsec.cs.princeton.edu/
13. http://www.hpl.hp.com/research/cacti/
14. http://www.spec.org/
15. Jacob, P., Erdogan, O., Zia, A., et al. (2005). Predicting the performance of a 3D processor-memory chip stack. *IEEE Design and Test of Computers, 22*(6), 540–547.
16. Kahle, J. A., Day, M. N., Hofstee, H. P., et al. (2005). Introduction to the cell multiprocessor. *IBM Journal of Research and Development, 49*(4/5), 589–604.
17. Kgil, T., et al., D'Souza, S., Saidi, A., et al. (2006). PicoServer: Using 3D Stacking Technology to Enable a Compact Energy Efficient Chip Multiprocessor. *Proceedings of the 2006 ASPLOS Conference, 41*(11), 117–128.
18. Kim, C., Burger, D., & Keckler, S. (2002). An adaptive, non-uniform cache structure for wire-delay dominated on-chip caches. In *Proceedings the 10th International Conference on Architectural Support for Programming Languages and Operating Systems.*
19. Kim, J., Chung, S., Jang, T., et al. (2010). Vertical double gate Z-RAM technology with remarkable low voltage operation for DRAM application (pp. 163–164).
20. Kongetira, P., Aingaran, K., & Olukotun, K. (2005). Niagara: A 32-way multithreaded SPARC processor. *IEEE Micro, 25*(2), 21–29.
21. Lee, B. C., Ipek, E., Mutlu, O., & Burger, D. (2009). Architecting phase change memory as a scalable DRAM alternative. In *Proceedings of ISCA* (pp. 2–13).
22. Li, F., Nicopoulos, C., Richardson, T., et al. (2006). Design and management of 3D chip multiprocessors using network-in-memory. In *ISCA '06: Proceedings of the 33rd, Annual International Symposium on Computer Architecture* (pp. 130–141).
23. Liu, C. C., Ganusov, I., Burtscher, M., & Tiwari, S. (2005). Bridging the processor-memory performance gap with 3D IC technology. *IEEE Design and Test of Computers, 22*(6), 556–564.

24. Loh, G. H. (2008). 3D-stacked memory architectures for multi-core processors. In *ISCA '08: Proceedings of the 35th International Symposium on Computer, Architecture* (pp. 453–464).
25. Loh, G. H., & Hill, M. D. (2011). Efficiently enabling conventional block sizes for very large die-stacked dram caches. In *MICRO'11* (pp. 454–464).
26. Loh, G. H., & Hill, M. D. (2012). Supporting very large dram caches with compound-access scheduling and missmap. *IEEE Micro* (pp. 70–78).
27. Loi, G. L., Agrawal, B., Srivastava, N., et al. (2006). A thermally-aware performance analysis of vertically integrated (3-D) processor-memory hierarchy. In *DAC '06: Proceedings of the 43rd Annual Conference on Design automation* (pp. 991–996).
28. Lu, Z., Collaert, N., Aoulaiche, M., De Wachter, B., De Keersgieter, A., Schwarzenbach, W., et al. (2010). A novel low-voltage biasing scheme for double gate fbc achieving 5s retention and 10_{16} endurance at 85c. In *IEDM* (pp. 12.3.1–12.3.4). doi:10.1109/IEDM.2010.5703347.
29. Magnusson, P. S., Christensson, M., Eskilson, J., et al. (2002). Simics: A full system simulation platform. *Computer, 35*(2), 50–58.
30. Nigam, A., Smullen, C., Mohan, V., Chen, E., Gurumurthi, S., & Stan, M. (2011). Delivering on the promise of universal memory for spin-transfer torque ram (stt-ram). In *ISLPED 2011* (pp. 121–126). doi:10.1109/ISLPED.2011.5993623.
31. Qureshi, M., Franceschini, M., & Lastras-Montano, L. (2010). Improving read performance of phase change memories via write cancellation and write pausing. In *HPCA* (pp. 1–11). doi:10.1109/HPCA.2010.5416645.
32. Qureshi, M. K., Srinivasan, V., & Rivers, J. A. (2009). Scalable high performance main memory system using phase-change memory technology. In *Proceedings of ISCA* (pp. 24–33).
33. Smullen, C., Mohan, V., Nigam, A., Gurumurthi, S., & Stan, M. (2011). Relaxing non-volatility for fast and energy-efficient stt-ram caches. In *2011 IEEE 17th International Symposium on High Performance Computer Architecture (HPCA)* (pp. 50–61). doi:10.1109/HPCA.2011.5749716.
34. Tsai, Y. F., Xie, Y., Vijaykrishnan, N., & Irwin, M. J. (2005). Three-dimensional cache design exploration using 3DCacti. In *ICCD '05: Proceedings of the 2005 International Conference on, Computer Design* (pp. 519–524).
35. Xie, Y., Loh, G. H., Black, B., & Bernstein, K. (2006). Design space exploration for 3D architectures. *ACM Journal on Emerging Technologies in Computing Systems*, *2*(2), 65–103.
36. Zhao, W., Belhaire, E., Mistral, Q., et al. (2006). Macro-model of spin-transfer torque based magnetic unnel junction device for hybrid magnetic-CMOS design. In *IEEE International Behavioral Modeling and Simulation, Workshop* (pp. 40–43).
37. Zhou, P., Zhao, B., Yang, J., & Zhang, Y. (2009). A durable and energy efficient main memory using phase change memory technology. In *Proceedings of ISCA* (pp. 14–23).
38. Zhou, P., Zhao, B., Yang, J., & Zhang, Y. (2009). Energy reduction for stt-ram using early write termination. In *ICCAD* (pp. 264–268).

24. Loh, G. H. (2008). 3D-stacked memory architectures for multi-core processors. In ISCA '08: Proceedings of the 35th international symposium on Computer Architecture (pp. 453–464).
25. Loh, G. H., & Hill, M. D. (2011). Efficiently enabling conventional block sizes for very large die-stacked dram caches. In MICRO'11 (pp. 454–464).
26. Loh, G. H., & Hill, M. D. (2012). Supporting very large dram caches with compound access scheduling and missmap. IEEE Micro (pp. 70–78).
27. Loi, G. L., Agrawal, B., Srivastava, N., Lin, S.-C., Sherwood, T., & Banerjee, K. (2006). A thermally-aware performance analysis of vertically integrated (3-D) processor-memory hierarchy. In DAC '06: Proceedings of the 43rd annual Design Automation Conference (pp. 991–996).
28. Madan, N., Zhao, L., Muralimanohar, N., Udipi, A., Balasubramonian, R., Iyer, R., Makineni, S., & Newell, D. (2009). Optimizing communication and capacity in a 3D stacked reconfigurable cache hierarchy. In HPCA (pp. 262–274).
29. Muralimanohar, N., Balasubramonian, R., & Jouppi, N. P. [illegible] CACTI 6.0 [illegible]
30. [illegible]
31. Qureshi, M. K., Franceschini, M. M., & Lastras-Montaño, L. A. (2010). Improving read performance of phase change memories via write cancellation and write pausing. In HPCA (pp. 1–11). doi:10.1109/HPCA.2010.5416645.
32. Qureshi, M. K., Srinivasan, V., & Rivers, J. A. (2009). Scalable high performance main memory system using phase-change memory technology. In Proceedings of ISCA (pp. 24–33).
33. Smullen, C., Mohan, V., Nigam, A., Gurumurthi, S., & Stan, M. (2011). Relaxing non-volatility for fast and energy-efficient STT-RAM caches. In 2011 IEEE 17th International Symposium on High Performance Computer Architecture (HPCA) (pp. 50–61). doi:10.1109/HPCA.2011.5749716.
34. Tsai, Y. F., Xie, Y., Vijaykrishnan, N., & Irwin, M. J. (2005). Three-dimensional cache design exploration using 3DCacti. In ICCD '05: Proceedings of the 2005 International Conference on Computer Design (pp. 519–524).
35. Xie, Y., Loh, G. H., Black, B., & Bernstein, K. (2006). Design space exploration for 3D architectures. ACM Journal on Emerging Technologies in Computing Systems, 2(2), 65–103.
36. Zhao, W., Belhaire, E., Mistral, Q., et al. (2006). Macro-model of spin-transfer torque based magnetic tunnel junction device for hybrid magnetic-CMOS design. In IEEE International Behavioral Modeling and Simulation Workshop (pp. 40–43).
37. Zhou, P., Zhao, B., Yang, J., & Zhang, Y. (2009). A durable and energy efficient main memory using phase change memory technology. In Proceedings of ISCA (pp. 14–23).
38. Zhou, P., Zhao, B., Yang, J., & Zhang, Y. (2009). Energy reduction for stt-ram using early write termination. In ICCAD (pp. 264–268).

Chapter 7
STT-RAM Cache Hierarchy Design and Exploration with Emerging Magnetic Devices

Hai (Helen) Li, Zhenyu Sun, Xiuyuan Bi, Weng-Fai Wong, Xiaochun Zhu and Wenqing Wu

Abstract Spin-transfer torque random access memory (STT-RAM) is a promising new nonvolatile technology that has good scalability, zero standby power, and radiation hardness. The use of STT-RAM in last level on-chip caches has been proposed as it significantly reduced cache leakage power as technology scales down. Having a cell area only 1/9 to 1/3 that of SRAM, this will allow for a much larger cache with the same die footprint. This will significantly improve overall system performance, especially in this multicore era where locality is crucial. However, deploying STT-RAM technology in L1 caches is challenging because write operations on STT-RAM are slow and power-consuming. In this chapter, we propose a range of cache hierarchy designs implemented entirely using STT-RAM that delivers optimal power saving and performance. In particular, our designs use STT-RAM cells with various data retention times and write performances, made possible by novel magnetic tunneling junction (MTJ) designs. For L1 caches where speed is of the utmost importance, we propose a scheme that uses fast STT-RAM cells with reduced data retention time coupled with a dynamic refresh scheme. We will show that such a cache can achieve 9.2 % in performance improvement and saves up to 30 % of the total energy when compared to one that uses traditional SRAM. For lower-level caches with relatively

H. H. Li (✉) · Z. Sun · X. Bi
Department of Electrical and Computer Engineering, University of Pittsburgh,
3700 O'Hara Street, Pittsburgh, PA 15238, USA
e-mail: hai.helen.li@gmail.com

W. F. Wong
National University of Singapore,
13, Computing Drive, Singapore 117417, Republic of Singapore

X. Zhu
Qualcomm Inc.,
10185 Mckellar Ct., San Diego, CA 92109, USA

W. Wu
Qualcomm Inc.,
5665 Morehouse Dr., San Diego, CA 92121, USA

Y. Xie (ed.), *Emerging Memory Technologies*,
DOI: 10.1007/978-1-4419-9551-3_7,

larger cache capacities, we propose a design that has partitions of different retention characteristics and a data migration scheme that moves data between these partitions. The experiments show that on the average, our proposed multiretention-level STT-RAM cache reduces total energy by as much as 30–70 % compared to previous single retention-level STT-RAM cache, while improving IPC performance for both 2-level and 3-level cache hierarchies.

7.1 Introduction

Increasing capacity and cell leakage have caused the standby power of SRAM on-chip caches to dominate the overall power consumption of the latest microprocessors. Many circuit design and architectural solutions, such as V_{DD} scaling [14], power gating [17], and body biasing [13], have been proposed to reduce the standby power of caches. However, these techniques are becoming less effective as technology scaling has caused the transistor's leakage current to increase exponentially. Researchers have been prompted to look into the alternatives of SRAM technology. One possibility is the embedded DRAM (eDRAM) which is denser than SRAM. Unfortunately, it suffers from serious process variation issues [1]. Another alternative technology is the embedded phase change memory (PCM) [5], a new nonvolatile memory that can achieve very high density. However, its slow access speed makes PCM unsuitable as a replacement for SRAM.

Another substitute for SRAM, the spin-transfer torque RAM (STT-RAM), is receiving significant attention because it offers almost all the desirable features of a universal memory: the fast (read) access speed of SRAM, the high integration density of DRAM, and the nonvolatility of Flash memory. Also, the compatibility with the CMOS fabrication process and similarities in the peripheral circuitries makes the STT-RAM an easy replacement for SRAM.

However, there are two major obstacles to use STT-RAM for on-chip caches, namely its longer write latency and higher write energy. During an STT-RAM write operation in the sub-10 ns region, the magnetic tunnel junction (MTJ) resistance switching mechanism is dominated by *spin precession*. The required switching current rises exponentially as the MTJ switching time is reduced. As a consequence, the driving transistor's size must increase accordingly, leading to a larger memory cell area. The lifetime of memory cell also degrades exponentially as the voltage across the oxide barrier of the MTJ increases. As a result, a 10 ns programming time is widely accepted as the performance limit of STT-RAM designs and is adopted in mainstream STT-RAM research and development [6, 8, 12, 25, 30].

Several proposals have been made to address the write speed and energy limitations of the STT-RAM. For example, the early write termination scheme [32] mitigates the performance degradation and energy overhead by eliminating unnecessary writes to STT-RAM cells. The dual write speed scheme [30] improves the average access time of an STT-RAM cache by having a fast and a slow cache partition. A classic SRAM/STT-RAM hybrid cache hierarchy with 3D stacking structure was proposed in Ref. [25] to compensate the performance degradation caused by STT-RAM by migrate write-intensive data block into SRAM-based cache way.

In memory design, the *data retention time* indicates how long data can be retained in a nonvolatile memory cell after it has been written. In other words, it is the unit for measuring nonvolatility of a memory cell. Relaxing this nonvolatility can make the memory cells easier to be programmed and leads to a lower write current or faster switching speed. In Ref. [22], the volume (cell area) of the MTJ device is reduced to achieve better writability by sacrificing the retention time of the STT-RAM cache cells. A simple DRAM-style refresh scheme was also proposed to maintain the correctness of the data.

The key insight informing this paper is that the access patterns of L1 and lower-level caches in a multicore microprocessor are different. Based on this realization, we propose the use of STT-RAM designs with different nonvolatilities and write characteristics in different parts of the cache hierarchy so as to maximize power and performance benefits. A low-power dynamic refresh scheme is proposed to maintain the validity of the data. Compared to the existing works on STT-RAM cache designs, our work makes the following contributions:

- We present a detailed discussion on the trade-off between the MTJ's write performance and its nonvolatility. Using our macromagnetic model, we qualitatively analyze and optimize the device
- We propose a multiretention-level cache hierarchy implemented entirely with STT-RAM that delivers the optimal power saving and performance improvement based on the write access patterns at each level. Our design is easier to fabricate and has a lower die cost
- We present a novel refresh scheme that achieves the theoretically minimum refresh power consumption. A counter is used to track the life span of L1 cache data. The counter can be composed of SRAM cells or the spintronic memristor. As an embedded on-chip timer, spintronic memristor can save significant energy and on-chip area when compared to SRAM. Moreover, the cell size of the memristor is close to that of STT-RAM, making the layout easier
- We propose the use of a hybrid lower-level STT-RAM design for cache with large capacity that simultaneously offers fast average write latency and low standby power. It has two cache partitions with different write characteristics and nonvolatility. A data migration scheme to enhance the cache response time to write accesses is also described. The proposed hybrid cache structure has been evaluated in lower-level cache of both 2-level and 3-level cache hierarchies.

The rest of our chapter is organized as follows. Section 7.2 introduces the technical backgrounds of STT-RAM and spintronic memristor. Section 7.3 describes the trade-offs involved in MTJ nonvolatility relaxation and the memristor-based counter design. Section 7.4 proposes our multiretention STT-RAM L1 and L2 cache structures. Section 7.5 discusses our experimental results. Related works are summarized in Sect. 7.6, followed by our conclusion in Sect. 7.7.

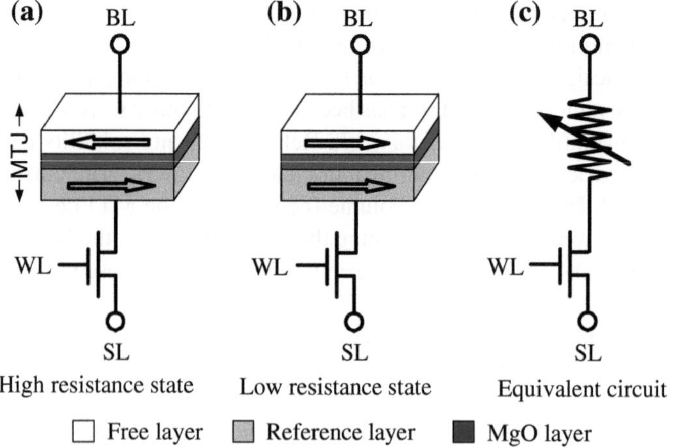

Fig. 7.1 1T1J STT-RAM design. **a** MTJ is in anti-parallel state; **b** MTJ is in parallel state; **c** the equivalent circuit

7.2 Background

7.2.1 STT-RAM

The data storage device in an STT-RAM cell is the magnetic tunnel junction (MTJ), as shown in Fig. 7.1a, b. A MTJ is composed of two ferromagnetic layers that are separated by an oxide barrier layer (e.g., MgO). The magnetization direction of one ferromagnetic layer (the *reference layer*) is fixed, while that of the other ferromagnetic layer (the *free layer*) can be changed by passing a current that is polarized by the magnetization of the reference layer. When the magnetization directions of the free layer and the reference layer are parallel (anti-parallel), the MTJ is in its low (high)-resistance state.

The most popular STT-RAM cell design is one-transistor-one-MTJ (or 1T1J) structure, where the MTJ is selected by turning on the word line (WL) that is connected to the gate of the NMOS transistor. The MTJ is usually modeled as a current-dependent resistor in the circuit schematic, as shown in Fig. 7.1c. When writing "1" (high-resistance state) into the STT-RAM cell, a positive voltage is applied between the source line (SL) and the bit line (BL). Conversely, when writing a "0" (low resistance state) into the STT-RAM cell, a negative voltage is applied between the SL and the BL. During a read operation, a sense current is injected to generate the corresponding BL voltage V_{BL}. The resistance state of the MTJ can be read out by comparing the V_{BL} to a reference voltage.

Table 7.1 Memristor design parameters

Length (L)	Width (w)	Thickness (h)	R_H (Ω)	R_L (Ω)	Γ_v ($nm^3 \times C^{-1}$)	J_{cr} ($A \times nm^{-2}$)
90 nm	45 nm	14 nm	7,500	2,500	2.01×10^{-14}	2×10^{-8}

7.2.2 Spintronic Memristor

As the fourth passive circuit element, the memristor has the natural property to record the historical profile of its electrical excitations [4]. In 2008, HP Lab reported the discovery of the memristor device, which was realized in a TiO_2 thin-film device [23]. In this work, we use the magnetic version of memristors, the tunneling magnetoresistance (TMR)-based spintronic memristor, as its device structure is similar to the MTJ, having a compatible manufacturing process.

Figure 7.2a illustrates the structure of a spintronic memristor [20, 27]. Like a MTJ, the free layer in a spintronic memristor is divided into two domains whose magnetization directions are, respectively, parallel or anti-parallel to the one of the reference layer. The domain wall can move along the length of the free layer when a polarized current is applied vertically.

As shown in Fig. 7.2b, the overall resistance of such a spintronic memristor can be modeled as two resistors connected in parallel with resistances R_L/α and $R_H/(1-\alpha)$, respectively [27]. Here, $0 \leq \alpha \leq 1$ represents the relative position of the domain wall which is the ratio of the domain wall position (x) over the total length of the free layer (L). The overall memristance can be expressed as

$$M(\alpha) = \frac{R_H \times R_L}{R_H \times \alpha + R_L \times (1-\alpha)}. \tag{7.1}$$

The domain wall moves only when the applied current density (J) is above the critical current density (J_{cr}) [15]. The domain wall velocity, $v(t)$, is determined by the spin-polarized current density J as [15]

$$v(t) = \frac{d\alpha(t)}{dt} = \frac{\Gamma_v}{L} \times J_{eff}(t), \quad J_{eff} = \begin{cases} J, J \geq J_{cr} \\ 0, J \geq J_{cr}. \end{cases} \tag{7.2}$$

where Γ_v is the domain wall velocity coefficient. It is determined by the device's structure and material. It is particularly noteworthy that the domain wall motion process has been successfully demonstrated in fabrication very recently [18].

Figure 7.3 shows the simulated programming property of a spintronic memristor when successive square waves are applied. The dashed gray line represents the pushing pulses applied across the spintronic memristor; the solid pink line is the corresponding current amplitude through the memristor. At the beginning of the simulated period, the pushing pulses raise up memristance, and therefore, the current amplitude decreases. Once the domain wall hits the device boundary, subsequent application of

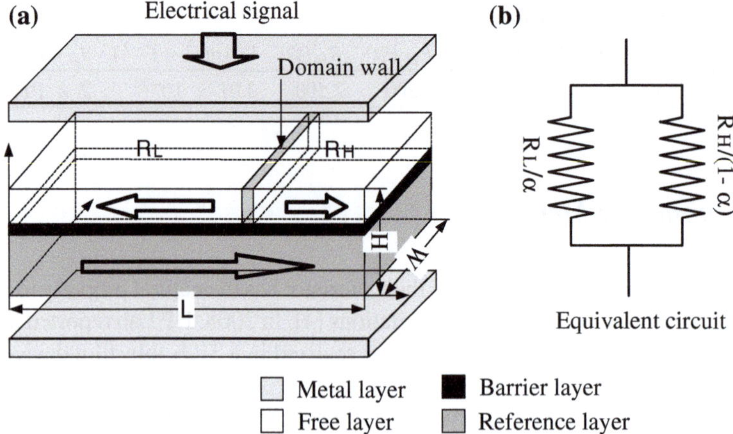

Fig. 7.2 TMR-based memristor. **a** the device view; **b** the equivalent circuit

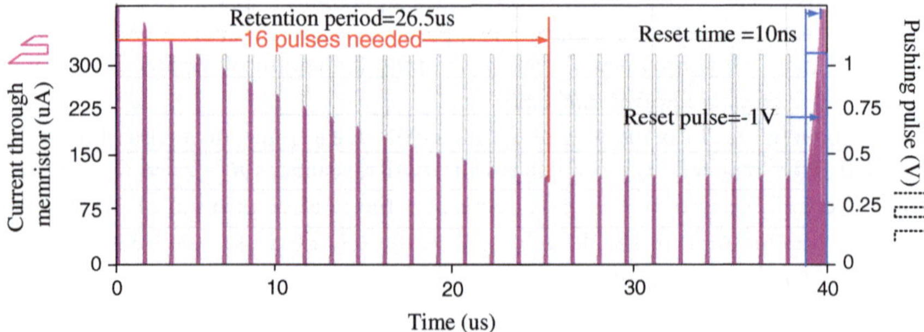

Fig. 7.3 The simulated programming property of a spintronic memristor

electrical excitations will not change the memristance. The larger the level number of memristance is, the more accurate the decision granularity will be. However, that also indicates a smaller sense margin. As we shall show in Sect. 7.4 that for our purpose, we partitioned the memristance of the spintronic memristor into a moderate number of 16 levels. The parameters of the spintronic memristor are summarized in Table 7.1.

7.3 Design

7.3.1 MTJ Write Performance Versus Nonvolatility

The *data retention time*, T_{store}, of a MTJ is determined by the *magnetization stability energy height*, Δ:

$$T_{\text{store}} = \frac{1}{f_0} e^{\Delta}. \tag{7.3}$$

f_0 is the thermal attempt frequency, which is of the order of 1 GHz for storage purposes [7]. Δ can be calculated by

$$\Delta = \left(\frac{K_u V}{k_\text{B} T}\right) = \left(\frac{M_s H_k V \cos^2(\theta)}{k_B T}\right), \tag{7.4}$$

where M_s is the saturation magnetization. H_k is the effective anisotropy field including magnetocrystalline anisotropy and shape anisotropy. θ is the initial angle between the magnetization vector and the easy axis. T is the working temperature. k_B is Boltzmann constant. V is the effective activation volume for the spin-transfer torque writing current. As Eqs. (7.3) and (7.4) show, the data retention time of an MTJ decreases exponentially when its working temperature, T, rises.

The required *switching current density*, J_C, of an MTJ operating in different working regions can be approximated as [21, 24]

$$J_C^{\text{THERM}}(T_{\text{sw}}) = J_{C0}\left(1 - \frac{1}{\Delta}\ln\left(\frac{T_{\text{sw}}}{\tau_0}\right)\right) \quad (T_{\text{sw}} > 10\,\text{ns}) \tag{7.5}$$

$$J_C^{\text{DYN}}(T_{\text{sw}}) = \frac{J_C^{\text{THERM}}(T_{\text{sw}}) + J_C^{\text{PREC}}(T_{\text{sw}})e^{(-A(T_{\text{sw}} - T_{\text{PIV}}))}}{1 + e^{(-A(T_{\text{sw}} - T_{\text{PIV}}))}} \quad (10\,\text{ns} > T_{\text{sw}} > 3\,\text{ns}) \tag{7.6}$$

$$J_C^{\text{PREC}}(T_{\text{sw}}) = J_{C0} + \frac{C\ln(\frac{\pi}{2\theta})}{T_{\text{sw}}} \quad (T_{\text{sw}} < 3\,\text{ns}). \tag{7.7}$$

Here, A, C, and T_{PIV} are the fitting parameters. T_{sw} is the switching time of MTJ resistance. $J_C = J_C^{\text{THERM}}(T_{\text{sw}})$, $J_C^{\text{DYN}}(T_{\text{sw}})$, or $J_C^{\text{PREC}}(T_{\text{sw}})$ are the required switching currents at T_{sw} in different working regions, respectively. The switching threshold current density J_{C0}, which causes a spin flip in the absence of any external magnetic field at 0 K, is given by

$$J_{C0} = (\frac{2e}{\hbar})\left(\frac{\alpha}{\eta}\right)(t_F M_s)(H_k \pm H_{\text{ext}} + 2\pi M_s). \tag{7.8}$$

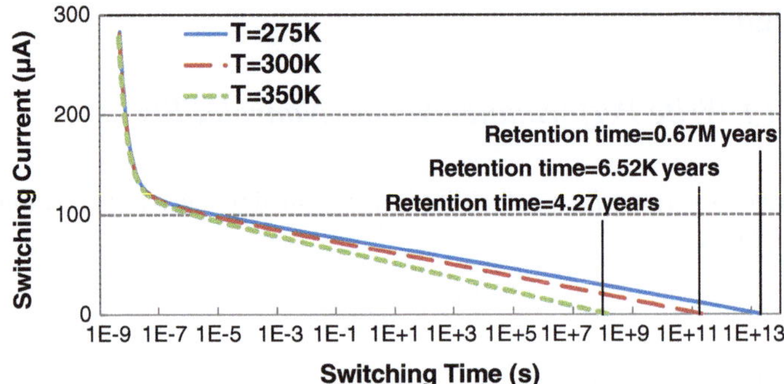

Fig. 7.4 The relationship between the switching current and the switching time of "Base" MTJ design

Here, e is the electron charge, α is the damping constant, τ_0 is the relaxation time, t_F is the free layer thickness, $\hbar$ is the reduced Planck's constant, H_{ext} is the external field, and η is the spin-transfer efficiency.

As proposed by [22], shrinking the cell surface area of the MTJ can reduce Δ and consequently decreases the required switching density J_C, as shown in Eq. (7.5). However, such a design becomes less efficient in the fast-switching region ($T_{\text{SW}} < 3\,\text{ns}$) because the coupling between Δ and J_C is less in this region, as shown in Eq. (7.7). Based on the MTJ switching behavior, we propose to change M_s, H_k, or t_F to reduce J_c. Such a technique can lower not only Δ but also J_{c0}, offering efficient performance improvement over the entire MTJ working range.

We simulated the switching current versus the switching time of a baseline $45 \times 90\,\text{nm}$ elliptical MTJ over the entire working range, as shown in Fig. 7.4. The simulation is conducted by solving the stochastic magnetization dynamics equation describing spin torque-induced magnetization motion at finite temperature [28]. The MTJ parameters are taken from [28], which are close to the measurement results recently reported in [31]. The MTJ data retention time is measured as the MTJ switching time when the switching current is zero. When the working temperature rises from 275 to 350 K, the MTJ's data retention time decreased from 6.7×10^6 to

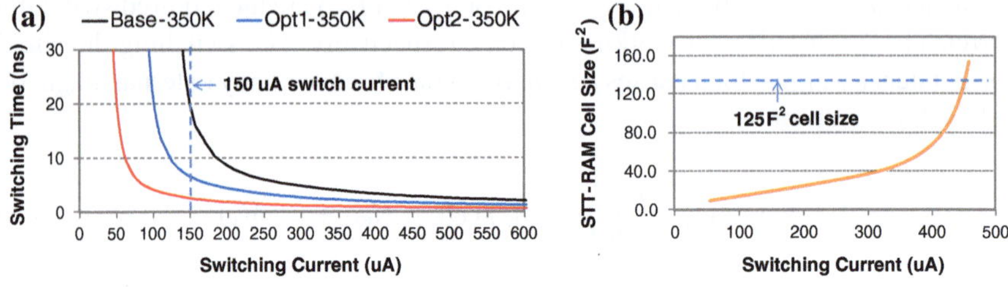

Fig. 7.5 **a** MTJ switching performances for different MTJ designs at 350 K. **b** The minimal required STT-RAM cell size at given switching current

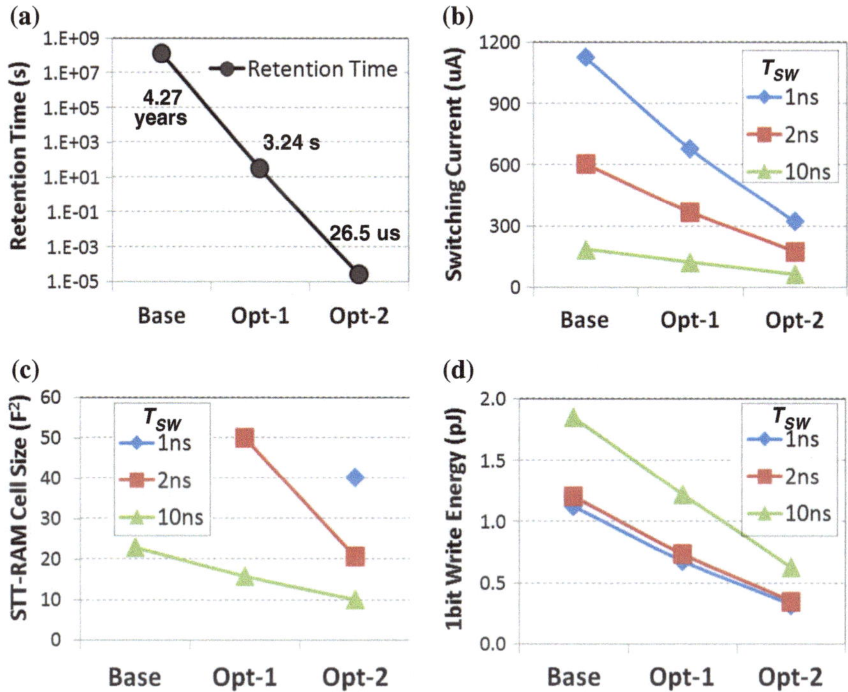

Fig. 7.6 Comparison of different MTJ designs at 350K: **a** the retention time. **b** the switching current. **c** STT-RAM cell size, and **d** the bit write energy

4.27 years. In the experiments reported in this work, we shall assume that the chip is working at a high temperature of 350 K.

7.3.2 STT-RAM Cell Design Optimization

To quantitatively study the trade-offs between the write performance and nonvolatility of an MTJ, we simulated the required switching current of three different MTJ designs with the same cell surface shapes. Besides the "Base" MTJ design shown in Fig. 7.4, two other designs ("Opt1" and "Opt2") that are optimized for better switching performance with degraded nonvolatility were studied. The corresponding MTJ switching performances of these three designs at 350 K are shown in Fig. 7.5a. The detailed comparisons of data retention times, the switching currents, the bit write energies, and the corresponding STT-RAM cell sizes of three MTJ designs at the given switching speed of 1, 2, and 10 ns are given in Fig. 7.6.

Significant write power saving is achieved if the MTJ's nonvolatility can be relaxed. For example, when the MTJ data retention time is scaled from 4.27 years

("Base") to 26.5 μs ("Opt2"), the required MTJ switching current decreases from 185.2 to 62.5 μA for a 10 ns switching time at 350 K. Or, at an MTJ switching current of 150 μA, the corresponding switching times of all three MTJ designs varied from 20 to 2.5 ns. A switching performance improvement of 8× can be obtained, as shown in Fig. 7.5a.

Since the switching current of an MTJ is proportional to its area, the MTJ is normally fabricated with the smallest possible dimension. The STT-RAM cell's area is mainly constrained by the NMOS transistor which needs to provide sufficient driving current to the MTJ. Figure 7.5b shows the minimal required NMOS transistor size when varying the switching current and the corresponding STT-RAM cell area at 45 nm technology node. The PTM model was used in the simulation [3], and the power supply V_{DD} is set to 1.0 V. Memory cell area is measured in F^2, where F is the feature size at a certain technology node.

According to the popular cache and memory modeling software CACTI [2], the typical cell area of SRAM is about 125 F^2. For an STT-RAM cell with the same area, the maximum current that can be supplied to the MTJ is 448.9 μA. A MTJ switching time of less than 1 ns can be obtained with the "Opt2" design under such as a switching current, while the corresponding switching time for the baseline design is longer than 4.5 ns. In this chapter, we will not consider designs that are larger than 125 F^2.

Since "Opt1" and "Opt2" require less switching current than the baseline design for the same write performance, they also consume less write energy. For instance, the write energies of "Base" and "Opt2" designs are 1.85 and 0.62 pJ, respectively, for a switching time of 10 ns. If the switching time is reduced to 1 ns, the write energy of "Opt2" design can be further reduced down to 0.32 pJ. The detailed comparisons on the write energies of different designs can be found in Fig. 7.6d.

7.4 Multiretention-level STT-RAM Cache Hierarchy

In this section, we will describe our multiretention-level STT-RAM-based cache hierarchy. Our multiretention-level STT-RAM cache hierarchy takes into account the difference in access patterns in L1 and the lower-level cache (LLC).

For L1, the overriding concern is access latency. Therefore, we propose the use of our "Opt2" nonvolatility-relaxed STT-RAM cell design as the basis of the L1 cache. In order to prevent data loss introduced by relaxing its nonvolatility, we propose a dynamic refresh scheme to monitor the life span of the data and refresh cells when needed. LLC caches are much larger than L1 cache. As such, a design built with only "Opt2" STT-RAM cells will consume too much refresh energy. Use of the longer retention "Base" or "Opt1" design is more practical. However, to recover the lost performance, we propose a hybrid LLC that has a regular and a nonvolatility-relaxed STT-RAM portions. Data will be migrated from one to other accordingly. The details of our proposed cache hierarchy will be given in the following subsections.

7.4.1 The Nonvolatility-Relaxed STT-RAM L1 Cache Design

As established earlier, using the "Opt2" STT-RAM cell design for L1 caches can significantly improve the write performance and energy. However, its data retention time of 26.5 μs may not be sufficient to retain the longest living data in L1. Therefore, a refresh scheme is needed. In Ref. [22], a simple DRAM-style refreshing scheme was used. This scheme refreshes all cache blocks in sequence, regardless of its data content. Read and write accesses to memory cells that are being refreshed must be stalled. As we shall show in Sect. 7.5.2, this simple scheme introduces many unnecessary refreshing operations whose elimination will significantly improve performance and save energy.

7.4.1.1 Dynamic Refresh Scheme

To eliminate unnecessary refresh, we propose the use counters to track the life span of cache data blocks. Refresh is performed only on cache blocks that have reached their full life span. In our refresh scheme, we assign one counter to each data block in the L1 cache to monitor its data retention status. Figure 7.7 illustrates our dynamic refresh scheme. The operation of the counter can be summarized as follows:

- *Reset*: On any write access to a data block, its corresponding counter is reset to '0'
- *Pushing*: We divide the STT-RAM cell's retention time into N_{mem} periods, each of which is T_{period} long. A global clock is used to maintain the countdown to T_{period}. At the end of every T_{period}, the level of every counter in the cache is increased by one
- *Checking*: The data block corresponding to a counter would have reached the maximum retention time when the counter reaches its highest level and hence needs

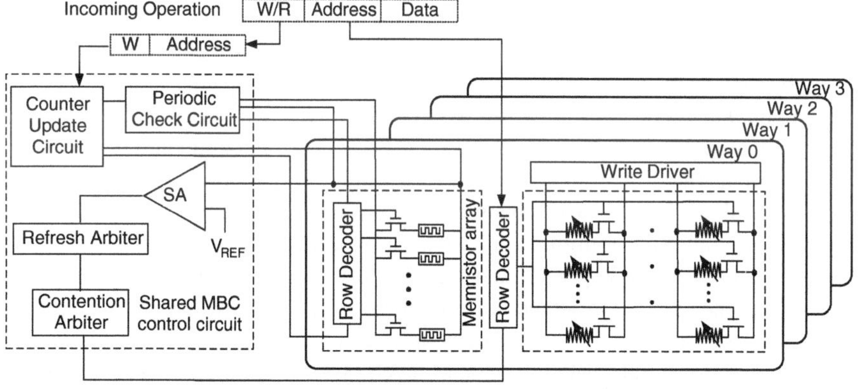

Fig. 7.7 Memristor counter-based refreshing scheme

to be refreshed. The overhead of such counter pushing scheme is very moderate. Take, for example, a 32 KB L1 cache built using the "Opt2" STT-RAM design and a counter can represent 16 values from 0 to 15. A pushing operation happens once every 3.23 ns, $=$ (26.5 μs/512/16) in the entire L1 cache. This is more than 6 cycles at a 2 GHz clock frequency. A larger cache may mean a higher pushing overhead.

The following is some design details of the proposed dynamic refresh scheme:

- *Cache access during refresh*: During a refresh operation, the block's data are read out into a buffer and then saved back to the same cache block. If a read request to the same cache block comes before the refresh finishes, the data are returned from this buffer directly. There is therefore no impact on the read response time of the cache. Should a write request comes, the refresh operation is terminated immediately, and the write request is executed. Again, no penalty is introduced
- *Reset threshold* N_{th}: However, we observe that during the life span of a cache block, updates happen more frequently within a short period of time after it has been written. Many resets of the cache block data occur far from their data retention time limits, giving us an optimization opportunity. We altered the reset scheme to eliminate counter resets that happen within a short time period after data have been written. We define a threshold level, N_{th}, that is much smaller than N_{mem}. The counter is reset only when its resistance is higher than N_{th}. The larger N_{th} is, the more resets are eliminated. On the other hand, the refresh interval of the data next written into the same cache block is shortened. However, our experiments in Sect. 7.5.2 shall show that such cases happen very rarely and the lifetimes of most data blocks in the L1 cache are much shorter than 26.5 μs.

7.4.1.2 Counter Design

In the proposed scheme, the counters are used in two ways: (1) to monitor the time duration for which the data have been written into the memory cells and (2) to monitor the read and write intensity of the memory cells. These counters can be implemented either by the traditional SRAM or by the recently discovered memristor device. The design detail of memristor as an on-chip analog counter will be introduced here. A Verilog-A model for spintronic memristor [27] was used in circuit simulations.

As demonstrated in Eqs. (7.1) and (7.2), when the magnitude of programming pulse is fixed, the memristance (resistance) of a spintronic memristor is determined by the accumulated programming pulse width. We utilize this characteristic to implement a high-density, low-power, and high-performance counter used in our cache refresh scheme: The memristance of a memristor can be partitioned into multiple levels, corresponding to the values the counter can record.

The maximum number of memristance levels is constrained by the minimal sense margin of the sense amplifier/comparator and the resolution of the programming pulse, i.e., the minimal pulse width. The difference between R_H and R_L of the spintronic memristor used in this work is 5,000 Ω (see Table 7.1), which is sufficiently

Table 7.2 Comparison between SRAM and memristor counter

A 4-bit counter	SRAM	Memristor
Area of a memory cell	$100 \sim 150\,F^2$	$33\,F^2$
Number of memory cells	4	1
Pushing and checking energy	0.7 pJ	0.45 pJ
Reset energy	0.46 pJ	7.2 pJ
Sense margin	$50 \sim 100$ mV at 45 nm tech.	46.875 mV

large to be partitioned into 16 levels. Moreover, we use the pushing current of 150 μA as the read current, further enlarging the sensing margin. The sense margin of the memristor-controlled counter $\Delta V = 46.875\,\text{mV}$ ($150\,\mu\text{A} \times 5{,}000\,\Omega/16$ levels) is at the same level as the sense margin in nowadays SRAM design.

The area of a memristor is only $2\,F^2$ (refer Table 7.1). The total size of a memristor counter including a memristor and a control transistor is below $33\,F^2$. For comparison, the area of a 6T SRAM cell is about $100 \sim 150\,F^2$ [33]. More importantly, the memristor counter has the same layout structure as STT-RAM and therefore can be easily integrated into STT-RAM array.

The memristance variation induced by process variations [10] is the major issue when utilizing memristors as data storage device. The counter design faces the same issue, but the impact is not that critical: As a timer, the memristance variation can be overcome by giving enough design margin to guarantee the on-time refresh.

Every *pushing and checking* operation of a SRAM counter should include two actions: increase the counter value by one and read it out. In the proposed memristor counter design, the injected current can obtain the two purposes simultaneously —pushing the domain wall to enable counter value increment and meanwhile serving as read current for data detection. The comparison between the two types of counter designs is summarized in Table 7.2. Note that the memristor counter has a larger energy consumption during a reset operation in which its domain wall moves from one end to the other.

7.4.2 Lower-level Cache with Mixed High- and Low-Retention STT-RAM Cells

The data retention time requirement in the mainstream STT-RAM development of $4 \sim 10$ years was inherited from Flash memory designs. Although such a long data retention time can save significant standby power of on-chip caches, it also entails a long write latency (~ 10 ns) and large write energy [25]. Relaxing the nonvolatility of the STT-RAM cells in the lower-level cache will improve write performance as well as save more energy. However, if further reducing retention time to μs scale, e.g., 26.5 μs of our "Opt2" cell design, the refresh energy dominates, and hence, any refresh scheme becomes impractical for the large lower-level cache.

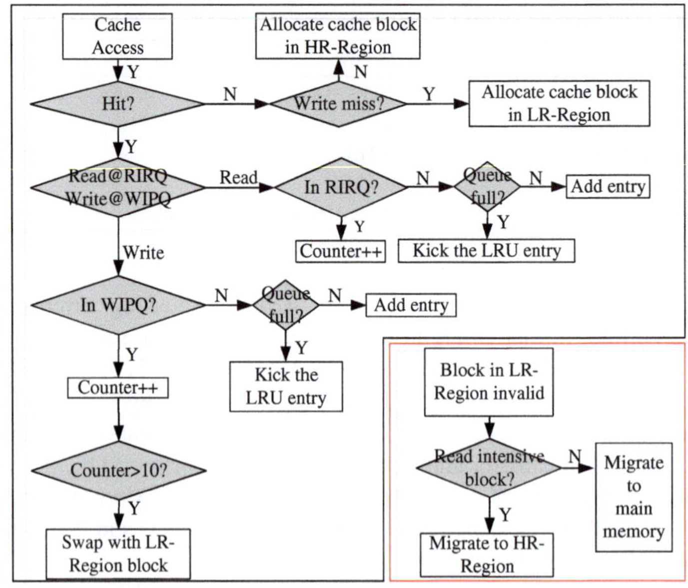

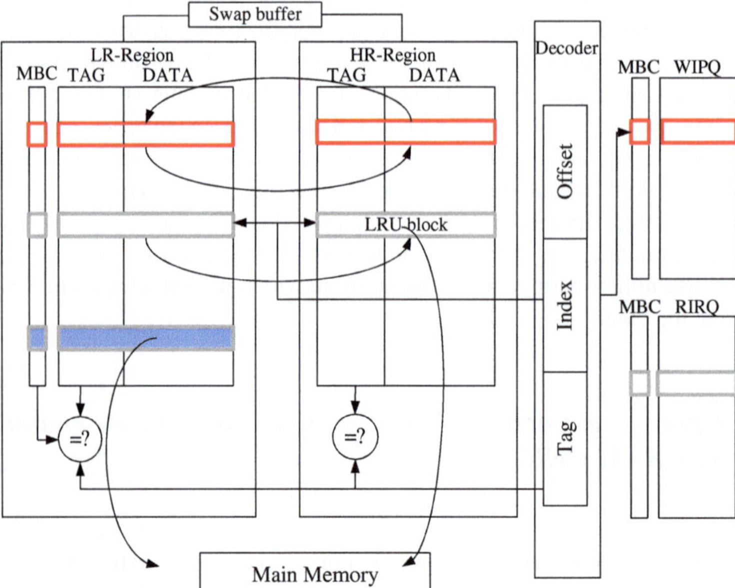

WIPQ: Write Intensive Predict Queue
RIRQ: Read Intensive Record Queue
LRU: Least Recent Used
LR-Region: Low-Retention Region
HR-Region: High-Retention Region
MBC: Memristor Bit Counter

Fig. 7.8 Hybrid lower-level cache migration policy: Flow graph (*top*). Diagram (*bottom*)

The second technique we proposed is a hybrid memory system that has both high- and low-retention STT-RAM portions to satisfy both the power and performance targets simultaneously. We take a L2 cache with 16 ways as a case study as shown in Fig. 7.8, way 0 of the 16-way cache is implemented with a low-retention STT-RAM design ("Opt2"), while ways 1–15 are implemented with the high-retention STT-RAM ("Base" or "Opt1"). Write-intensive blocks are primarily allocated from way 0 for a faster write response, while read-intensive blocks are maintained in the other ways.

Like our proposed L1 cache, counters are used in way 0 to monitor the blocks' data retention status. However, unlike in L1 where we perform a refresh when a memristor counter expires, here we move the data to the high-retention STT-RAM ways.

Figure 7.8 demonstrates the data migration scheme to move the data between the low and the high-retention cache ways based on their write access patterns. A write intensity prediction queue (WIPQ) of 16 entries is added to record the write access history of the cache. Every entry has two parts, namely the data address and an access counter.

During a read miss, the new cache block is loaded to the high-retention (HR) region (ways 1–15), following the regular LRU policy. On a write miss, the new cache block is allocated from the low-retention (LR) region (way 0), and its corresponding memristor counter is reset to '0'. On a write hit, we search the WIPQ first. If the address of the write hit is already in WIPQ, the corresponding access counter is incremented by one. Note that the block corresponding to this address may be in the HR- or the LR-region of the cache. Otherwise, the hit address will be added in to the queue if any empty entry available. If the queue is full, the LRU entry will be evicted and replaced by the current hit address. The access counters in the WIPQ are decremented periodically, for example, every 2,000 clock cycles, so that the entries that are in the queue for too long will be evicted. Once an access counter in a WIPQ entry reaches a preset value, $N_{HR \rightarrow LR}$, the data stored in the corresponding address will be swapped with a cache block in the LR-region. If the corresponding address is already in the LR-region, no further action is required. A read hit does not cause any changes to the WIPQ.

Likewise, a read intensity record queue (RIRQ) with the same structure and number of entries is used to record the read hit history of the LR-region. Whenever there is a read hit to the LR-region, a new entry is added into the RIRQ. Or if a corresponding entry already exists in the RIRQ, the value of the access counter is increased by one. When the memristor counter of a cache block B_i in the LR-region indicates the data are about to become unstable, we check to see whether this cache address is read intensive by searching the RIRQ. If B_i is read intensive, it will be moved to HR-region. The cache block being replaced by B_i in the HR-region will be selected using the LRU policy. The evicted cache block will be send to main memory. If B_i is not read intensive, it will be written back to main memory.

In a summary, our proposed scheme uses the WIRQ and RIRQ to dynamically classify cache blocks into three types:

1. *Write intensive*: The addresses of such cache blocks are kept in the WIRQ. They will be moved to the LR-region once their access counters in WIRQ reach $N_{HR\rightarrow LR}$;
2. *Read intensive but not write intensive*: The addresses of such cache blocks are found in the RIRQ but not in the WIRQ. As they approach to their data retention time limit, they will be moved to the HR-region
3. *Neither write nor read intensive*: Neither WIRQ nor RIRQ has their addresses. They are kept in HR-region or evicted from LR-region to main memory directly.

Identifying a *write intensive* cache blocks also appeared in some previous works. In Ref. [25], they check whether two successive write accesses go to the same cache block. It is highly possible that a cache block may be accessed several times within very short time and then becomes inactive. Our scheme is more accurate and effective as it monitors the read and write access histories of a cache block throughout its entire life span. The RIRQ ensures that *read intensive* cache blocks migrate from the LR-region to HR-region in a timely manner that, at the same time, also improves energy efficiency and performance.

7.5 Simulation Results and Discussion

7.5.1 Experimental Setup

We modeled a 2 GHz microprocessor with 4 out-of-order cores using MARSSx86 [16]. Assume a two-level or a three-level cache configuration and a fixed 200-cycle main memory latency. The MESI cache coherency protocol is utilized in the private L1 caches to ensure consistency, and the shared lower-level cache uses a write-back policy. The parameters of our simulator and cache hierarchy can be found in Tables 7.3 and 7.4.

Table 7.5 shows the performance and energy consumptions of various designs obtained by a modified NVSim simulator [19]. All the "*-hi*", "*-md*", and "*-lo*" configurations use the "Base", "Opt1", and "Opt2" MTJ designs, respectively. Note that as shown in Fig. 7.5, they scale differently. We simulated a subset of multithreaded workloads from the PARSEC 2.1, and the SPEC 2006 benchmark

Table 7.3 Simulation platform

Max issue width	4 insts	Fetch width	4 insts
Dispatch width	4 insts	Write-back width	4 insts
Commit width	4 insts	Fetch queue size	32 insts
Reorder buffer	64 entries	Max branch in pipeline	24
Load store queue size	32 entries	Functional units	2 ALU 2 FPU
Clock cycle period	0.5 ns	Main memory	200 cycle latency

suites so as to cover a wider spectrum of read/write and cache miss characteristics. We simulated 500 million instructions of each benchmark after their initialization.

SPICE simulations were conducted to characterize the performance and energy overheads of the memristor counter and its control circuit. The reset energy of a memristor counter is 7.2 pJ, and every pushing–checking operation consumes 0.45 pJ.

We compared the performance (in terms of instruction per cycle, IPC) and the energy consumption of different configurations for both 2-level and 3-level hybrid cache hierarchies. The conventional all-SRAM cache design is used as the baseline. The optimal STT-RAM cache configuration based on our simulations is summarized as follows. The detailed experimental results will be shown and discussed in Sects. 7.5.2, 7.5.3, and 7.5.4.

- An optimal 2-level STT-RAM cache hierarchy is the combination of (a) a L1 cache of the "L1-lo2" design and (b) a hybrid L2 cache of using the "L2-lo" in the LR-region and "L2-md2" in the HR-region;
- An optimal 3-level STT-RAM cache hierarchy is composed of (a) a L1 cache of the "L1-lo2" design, (b) a hybrid L2 cache of using the "L2-lo" in the LR-region and "L2-md1" in the HR-region and (c) a hybrid L3 cache of the "L3-lo" design in the LR-region and "L3-md2" in the HR-region.

7.5.2 Results for the Proposed L1 Cache Design

To evaluate the impacts of using STT-RAM in L1 cache design, we implemented the L1-cache with the different STT-RAM designs listed in the L1 cache portion of Table 7.5 while leaving the SRAM L2 cache unchanged. L1 I-cache are read operation, it is implemented by conventional STT-RAM. Compared to SRAM L1 I-cache, conventional STT-RAM has a faster read speed and large cache capacity that can reduce cache miss rate. Due to the smaller STT-RAM cell size, the overall area of L1 cache is significantly reduced. The delay components of interconnect and peripheral circuits also decrease accordingly. Even considering the relatively long sensing latency, the read latency of STT-RAM L1 cache is still similar or even slightly lower than that of a SRAM L1 cache. However, the write performance of STT-RAM L1 cache is always slower than that of the SRAM L1 cache for all the design configurations considered. The leakage power consumption of the STT-RAM

Table 7.4 Cache hierarchy configuration

Baseline 2-level cache hierarchy	Local L1 cache: 32 KB 4-way, 64 B cache block Shared L2 cache: 4 MB 16-way, 128 B cache block
3-level cache hierarchy	Local L1 cache: 32 KB 4-way, 64 B cache block Local L2 cache: 256 KB 8-way, 64 B cache block Shared L3 cache: 4 MB 16-way, 128 B cache block

Table 7.5 Cache configuration

	32 KB L1 Cache					
	SRAM	lo1	lo2	lo3	md	hi
Cell size (F^2)	125	20.7	27.3	40.3	22	23
MTJ switching time (ns)	/	2	1.5	1	5	10
Retention time	/	26.5 μs			3.24 s	4.27 year
Read latency (ns)	1.113	0.778	0.843	0.951	0.792	0.802
Read latency (cycles)	3	2	2	2	2	2
Write latency (ns)	1.082	2.359	1.912	1.500	5.370	10.378
Write latency (cycles)	3	5	4	4	11	21
Read dyn. energy (nJ)	0.075	0.031	0.035	0.043	0.032	0.083
Write dyn. energy (nJ)	0.059	0.174	0.187	0.198	0.466	0.958
Leakage power (mW)	57.7	1.73	1.98	2.41	1.78	1.82
	4 MB L2 or L3 Cache					
	SRAM	lo	md1	md2	md3	hi
Cell size (F^2)	125	20.7	22	15.9	14.4	23
MTJ switching time (ns)	/	2	5	10	20	10
Retention time	/	26.5 μs			3.24 s	4.27 year
Read latency (ns)	4.273	2.065	2.118	1.852	1.779	2.158
Read latency (cycles)	9	5	5	4	4	5
Write latency (ns)	3.603	3.373	6.415	11.203	21.144	11.447
Write latency (cycles)	8	7	13	23	43	23
Read dyn. energy (nJ)	0.197	0.081	0.083	0.070	0.067	0.085
Write dyn. energy (nJ)	0.119	0.347	0.932	1.264	2.103	1.916
Leakage power (mW)	4107	96.1	104	69.1	61.2	110

caches comes from the peripheral circuits only and is very low. The power supply to the memory cells that are not being accessed can be safely cut off without fear of data loss until the data retention limit is reached.

Figure 7.9 illustrates the ratio between read and write access numbers in L1 D-cache. Here, the read and write access numbers are normalized to the total L1 cache access number of `blackscholes`. The ratio reflects the sensitivity of the L1 cache in terms of performance, the dynamic energy toward per-read and per-write latency, and energy of the L1 cache.

Figure 7.10 shows the IPC performance of the simulated L1 cache designs normalized to the baseline all-SRAM cache. On average, implementing the L1 cache using the "Base" (used in "L1-hi") or "Opt1" (used in "L1-md") STT-RAM design incurs more than 32.5–42.5 % IPC degradation, respectively, due to the long write latency. However, the performance of the L1 caches with the low-retention STT-RAM design significantly improves compared to that of the SRAM L1 cache: the average normalized IPC's of 'L1-lo1', 'L1-lo2', and 'L1-lo3' are 0.998, 1.092, and 1.092, respectively. The performance improvement of 'L1-lo2' or 'L1-lo3' L1 cache w.r.t the baseline SRAM L1 cache comes from the shorter read latency even though its write latency is still longer (see Table 7.5). However, L1 read accesses are far more frequent than write access in most benchmarks as shown in Fig. 7.9. In some benchmarks whose read/write ratio is pretty high, for example, `swaptions`, the 'L1-lo2' or 'L1-lo3' design achieves a better than 20 % improvement in IPC.

The energy consumptions of the different L1 cache designs normalized to the baseline all-SRAM cache are summarized in Fig. 7.11a. The reported results include the energy overhead of the refresh scheme and the counters, where applicable. Not surprisingly, all three low-retention STT-RAM L1 cache designs achieved significant energy savings compared to the SRAM baseline. The "L1-lo3" design consumes more energy because of its larger memory cell size and larger peripheral circuit having more leakage and dynamic power, as shown in Table 7.5. Figure 7.11a also shows that implementing the L1 cache with the "Base" (used in "L1-hi") or "Opt1" (used in "L1-md") STT-RAM is much less energy efficient because (1) the MTJ switching time is longer, resulting in a higher write dynamic energy and (2) a longer operation time due to the low IPC.

Figure 7.11b presents the breakdowns of the read dynamic energy, the write dynamic energy, and the leakage energy in the baseline SRAM cache. First, the leakage occupies more than 30 % of overall energy, most of which can be eliminated in STT-RAM design. Second, when comparing to Fig. 7.9, we noticed that the dynamic read/write energy ratio is close to the read/write access ratio. The high read access ratio together with the lower per-bit read energy consumption of STT-RAM results in a much lower dynamic energy of STT-RAM L1 cache design. Therefore, "L1-lo1", "L1-lo2", and "L1-lo3" STT-RAM designs save up to 30 to 40 % of overall energy compared to the baseline SRAM L1 cache.

Figure 7.12a compares the refresh energy consumptions of the 'L1-lo2' L1 cache under different refresh schemes. In each group, the three bars from left to right

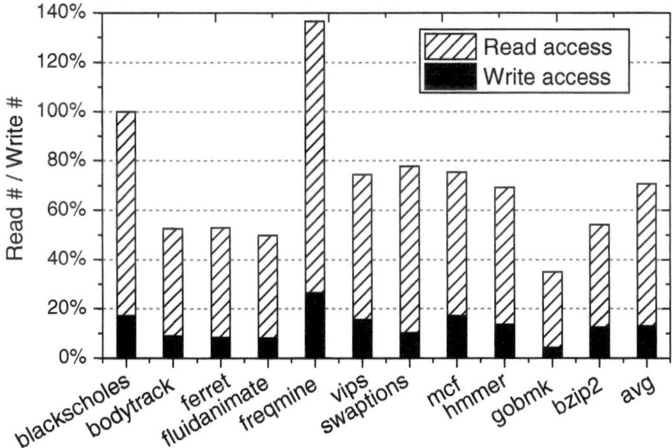

Fig. 7.9 The normalized L1 cache read and write access numbers. The access numbers are normalized to the total L1 access number of `blackscholes`

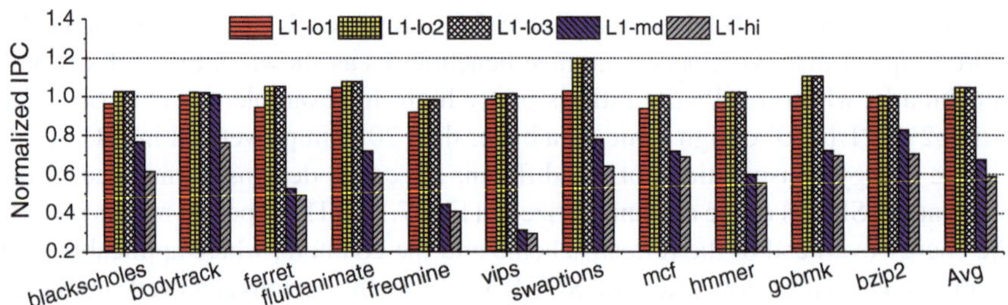

Fig. 7.10 IPC comparison of various L1 cache designs. The IPCs are normalized to all-SRAM baseline

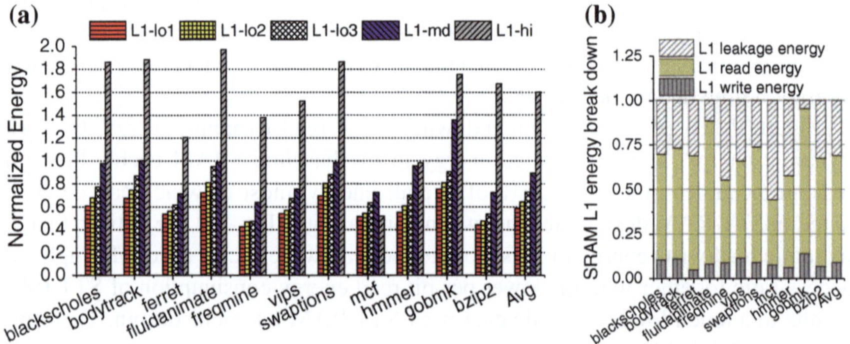

Fig. 7.11 **a** L1 cache overall energy comparison. The energy consumptions are normalized to SRAM baseline. **b** The breakdowns of energy consumption in SRAM-based L1 cache design

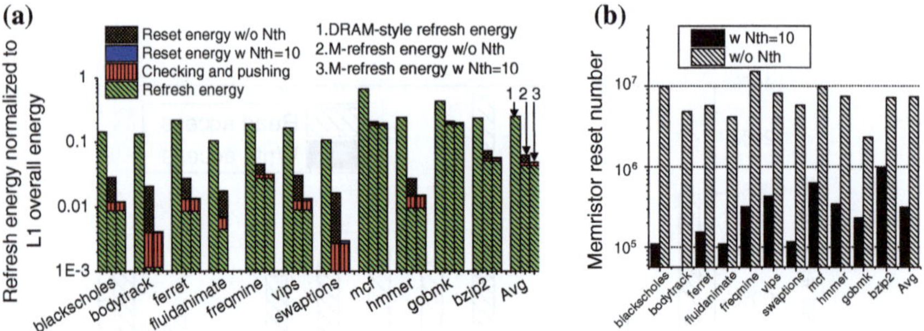

Fig. 7.12 **a** Refresh energy comparison of the different refresh schemes. **b** The number of counter reset operations in the refresh schemes without reset threshold N_{th} and with $N_{\text{th}} = 10$

represent the refresh energy consumptions of DRAM-style refresh scheme, refresh scheme without reset threshold N_{th} and with $N_{\text{th}} = 10$, respectively. The refresh energy consumptions are normalized to the overall L1 energy consumptions when

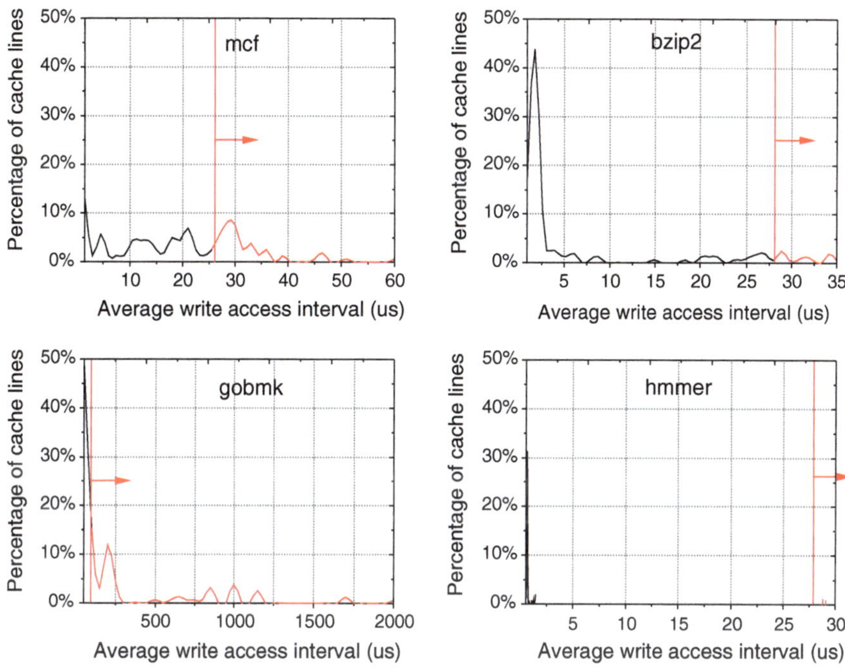

Fig. 7.13 Cache write access distributions of the selected benchmarks

implementing the refresh scheme with $N_{\text{th}} = 10$. Note that the y-axis is in logarithmic scale.

The energy consumption of the simple DRAM-style refresh scheme accounts for more than 20 % of the overall L1 cache energy consumption on average. In some extreme cases of low write access frequency, for example, `mcf`, this ratio is as high as 80 % because of the low dynamic cache energy consumption. The total energy consumption of our proposed refresh scheme consists of the checking and pushing, the reset, and the memory cell refresh.

As we discussed in Sect. 7.4.1, the introduction of the reset threshold N_{th} can further reduce the refresh energy consumption by reducing the number of counter resets. This is confirmed in Fig. 7.12a, b. The number of counter reset operations is reduced by more than 20× on average after setting a reset threshold N_{th} of 10, resulting in more than 95 % of the reset energy being saved. The energy consumption for the refresh scheme is very marginal, accounting for only 4.35 % of the overall L1 cache energy consumption. By accurately monitoring the life span of the cache line data, our refresh scheme significantly reduced the refresh energy in all the benchmarks.

The refresh energy saving by utilizing the dynamic refresh scheme is determined by the cache write access distribution and intensity. Figure 7.13 demonstrates the distribution of average write access intervals obtained from four selected benchmarks. In each subfigure, the STT-RAM retention time is represented by the red vertical line.

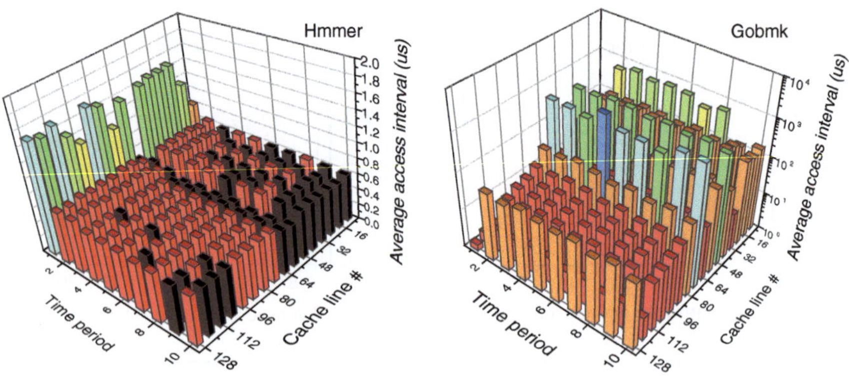

Fig. 7.14 Cache write access intensities of different cache lines

Therefore, the data stored in those cache lines on the right side of the red line need refreshment to maintain correctness.

We also collected the cache write access intensities with time. The results of two selected benchmarks are shown in Fig. 7.14. For illustration purpose, we divide the overall simulation time into ten periods and partition cache lines into eight groups. Figure 7.14 exhibits the average write access intervals for all the cache line groups in each time period. Benchmark `hammer` has a relatively uniform cache write intensity. Its average write access interval is less than 2 μs, which is much shorter than the STT-RAM data retention time 26.5 μs. Often the cache lines are updated by regular write access without refreshed. Therefore, the dynamic refresh scheme can reduce the refresh energy of `hammer` significantly—from 30 to 1 % of the total energy consumption when DRAM-style refresh is utilized. On the contrary, benchmark `gobmk` demonstrates a completely uneven write access intensities among different cache lines. Moreover, the access intervals of many cache lines are longer than the data retention time, making refresh necessary. The dynamic refresh scheme does not benefit too much in such a type of programs.

7.5.3 Evaluating the Hybrid Cache Design in 2-Level Cache Hierarchy

First, we evaluate the proposed hybrid cache design within L2 cache in 2-level cache hierarchies. In comparing the different L2 cache designs, we fixed the L1 cache to the 'L1-lo2' design. In our proposed hybrid L2 cache, way 0 assumes the 'L2-lo' design for the best read latency and the smallest leakage power among all three low-retention STT-RAM designs. Ways 1 to 15 are implemented using the 'L2-md1', 'L2-md2', or 'L2-md3' (all "Opt1" MTJ designs) because a 3.24s retention time is good enough for most applications, and they have the minimal refresh over-

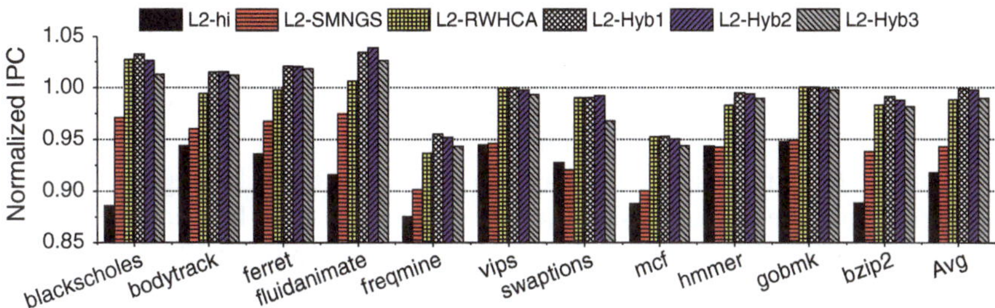

Fig. 7.15 Performance comparison of different 2-level cache designs. The IPCs are normalized to all-SRAM baseline

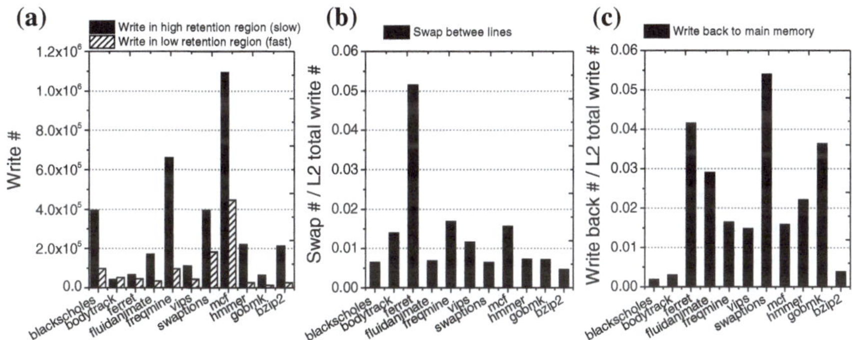

Fig. 7.16 The hybrid L2 cache statistics. **a** The write access numbers in HR- and LR-regions. **b** The ratio of data swaps between HR- and LR-regions among all the L2 accesses. **c** The ratio of data writing back to main memory among all the L2 accesses

head. The three resultant configurations are labeled as 'L2-Hyb1', 'L2-Hyb2', and 'L2-Hyb3', respectively. We compare our hybrid L2 cache with the single retention-level STT-RAM design of [22] and the *read/write aware high-performance architecture* (RWHCA) of [29] and label them as 'L2-SMNGS' and 'L2-RWHCA', respectively. For 'L2-SMNGS', we assumed that the L2 cache uses 'L2-md1' because its cell area of 22 F^2 is compatible with the 19 F^2 one reported in [22]. Instead of using 'L2-hi' in ways 1–15, 'L2-RWHCA' uses 'L2-md2' as it has an access latency that is similar to the one assumed in [29] but a much lower energy consumption. Except for hybrid, all other L2 STT-RAM schemes use the simple DRAM refresh when refresh is needed. To be consistent with the previous section, we normalize the simulation results to the all-SRAM design.

Figure 7.15 compares the normalized IPC results of the different L2 cache designs. As expected, the regular STT-RAM L2 cache with 'L2-hi' design shows the worst performance among all the configurations, especially for benchmarks with high L1 miss rates and L2 write frequencies (such as `mcf` and `swaptions`). Using relaxed retention STT-RAM design, 'L2-SMNGS' improves performance, but on the average, it still suffers 6 % degradation compared to the all-SRAM baseline due to its longer write latency. Among the three hybrid schemes we proposed, 'L2-Hyb1' is comparable in performance (99.8 % on average) to the all-SRAM cache design. As we prolong the MTJ switching time by reducing STT-RAM cell size in 'L2-Hyb2' and 'L2-Hyb3', IPC performance suffers.

Figure 7.16a compares the write access numbers in HR- and LR-regions in hybrid L2 cache. Some benchmarks, such as `mcf` and `freqmine`, have a large amount of write accesses falling into HR-region, resulting in significant IPC performance degradation. In the contrast, other programs such as `bodytrack` and `ferret` obtain IPC improvement compared to all-SRAM baseline, which mainly benefits from the less L2 write accesses. Although `blackscholes` sends more data to HR-region than `bodytrack` and `ferret`, it has low chances to swap data between HR- and LR-regions and to write data back to main memory, as shown in Fig. 7.16b, c, respectively. So the performance of `blackscholes` also improves. In summary, all our hybrid L2 caches outperform both 'L2-SMNGS' and 'L2-RWHCA' due to their lower read latencies.

Since the savings in leakage energy by using STT-RAM designs in the L2 cache are well established, we compared the *dynamic* energy consumptions of different L2 cache designs. The energy overheads of the data refresh in LR-region and the data migration between LR- and HR-regions in our hybrid L2 caches are included in the dynamic energy. Due to the lower write energy in the LR-region, 'L2-Hyb1' has the lowest dynamic energy consumption, as shown in Fig. 7.19a. As the STT-RAM cell size is reduced, the write latency and write energy consumption increased. Thus, the corresponding dynamic energy of 'L2-Hyb2' and 'L2-Hyb3' grows rapidly. Figure 7.19b shows the leakage energy comparison. Compared to 'L2-RWHCA' which is a combination of SRAM/STT-RAM [29], all the other configurations have much lower leakage energy consumptions. 'L2-hi', 'L2-SMNGS', and 'L2-Hyb1' have similar leakage energies because their memory array sizes are quite close to each other. However, 'L2-Hyb2' and 'L2-Hyb3' benefit from their much smaller memory cell size.

The overall cache energy consumptions of all the simulated cache configurations are summarized in Fig. 7.17. On the average, 'L2-Hyb2' and 'L2-Hyb3' consume about 70 % of the energy of 'L2-SMNGS' and 26.2 % of 'L2-RWHCA'. In summary, our proposed hybrid scheme outperforms the previous techniques in [22] and [29] both in terms of performance and in terms of total energy (by an even bigger margin).

7.5.4 Deployment in 3-level Cache Hierarchies

We also evaluate four 3-level cache designs whose parameters were given in Table 7.5. These designs are

1. The all-SRAM cache hierarchy;
2. '3L-SMNGS' that uses the "md1" STT-RAM design in all the three level of caches, just like 'L2-SMNGS' [22];

3. '3L-MultiR'—a multiretention 3-level STT-RAM cache hierarchy with "L1-lo2" , "L2-md2", and "L3-hi";
4. '3L-MultiR-Hyb'—a multiretention 3-level STT-RAM cache hierarchy with "L1-lo2", as well as the proposed hybrid cache design used in both L2 and L3 caches. Here, 'Hyb1' is used in L2 cache for the performance purpose, while 'Hyb2' is used in L3 cache to minimize the leakage energy.

In [22], the IPC performance degradations for using the single retention STT-RAM ('md1') were from 1 to 9 % when compared to an all-SRAM design. Our simulation result of '3L-SMNGS' (8 % performance degradation on average) matches this well. Comparatively, the average IPC performance degradation of '3L-MultiR' is only 1.4 % on average, as shown in Fig. 7.18. The performance gain of '3L-MultiR' over '3L-SMNGS' comes mainly from "L1-lo2". '3L-MultiR-Hyb' has the best performance which is on average 8.8 and 2.1 % better than '3L-SMNGS' and '3L-MultiR', respectively. Most of the write accesses in L2 and L3 caches of '3L-MultiR-Hyb' are allocated into the fast region, boosting up the system performance. Under the joint effort of "L1-lo2" and hybrid lower-level cache, '3L-MultiR-Hyb' can even achieve a slightly higher IPC can all-SRAM design.

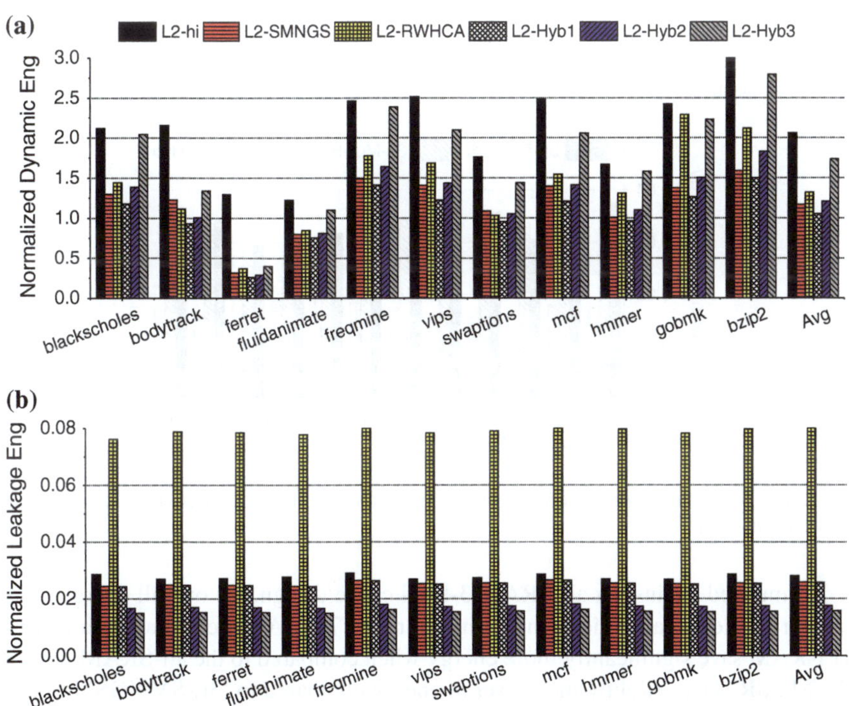

Fig. 7.17 Dynamic and leakage energy comparison of L2 cache (normalized to SRAM baseline)

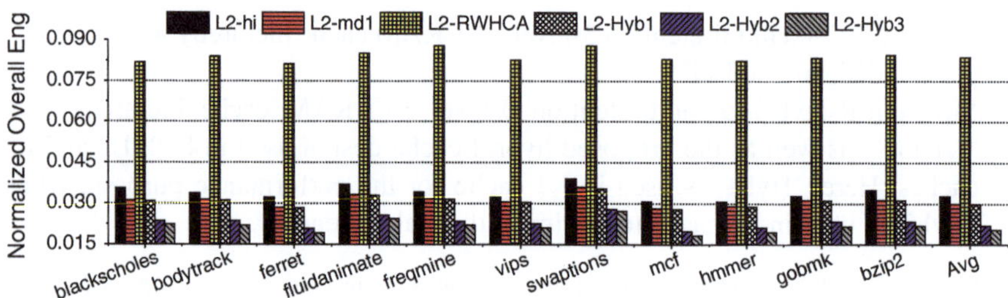

Fig. 7.18 Overall cache energy consumption comparison of 2-level cache designs (normalized to the all-SRAM design)

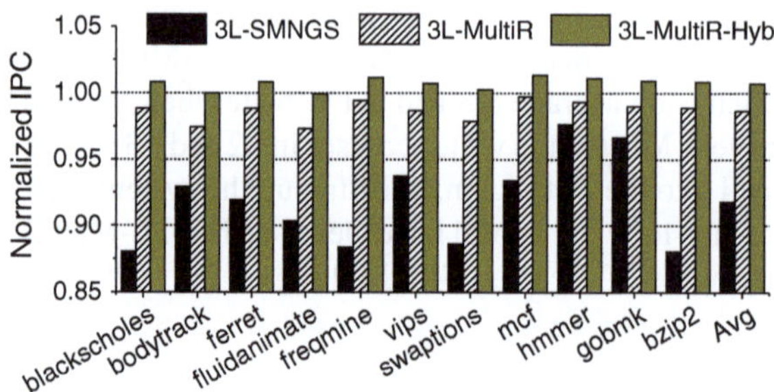

Fig. 7.19 Performance comparison of different 3-level cache designs. The IPCs are normalized to all-SRAM baseline

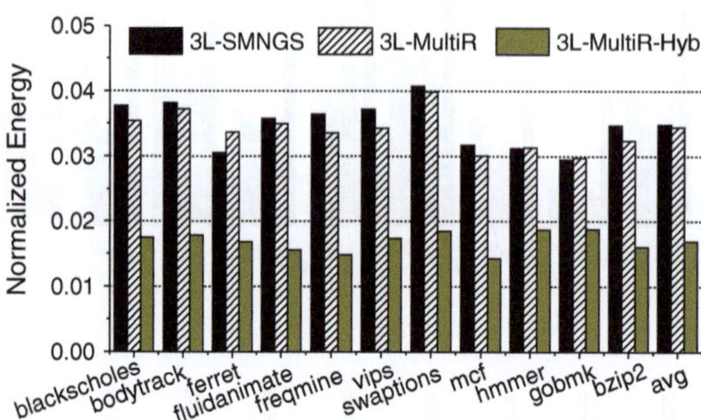

Fig. 7.20 Overall cache energy consumption comparison 3-level cache designs (Normalized to the all-SRAM design)

Normalized against an all-SRAM 3-level cache design, the overall energy comparison of 3-level cache hierarchy is shown in Fig. 7.20. All three combinations with STT-RAM save significantly more energy when compared to the all-SRAM design. '3L-MultiR' saves slightly more overall energy compared to '3L-SMNGS' because the 'Lo" STT-RAM cell design has a lower per-bit access dynamic energy than the "md" design. In '3L-MultiR-Hyb', shared L3 cache which embedded "md2" is much larger than local L2 cache which uses "md1". Thereby, the leakage of L3 dominates

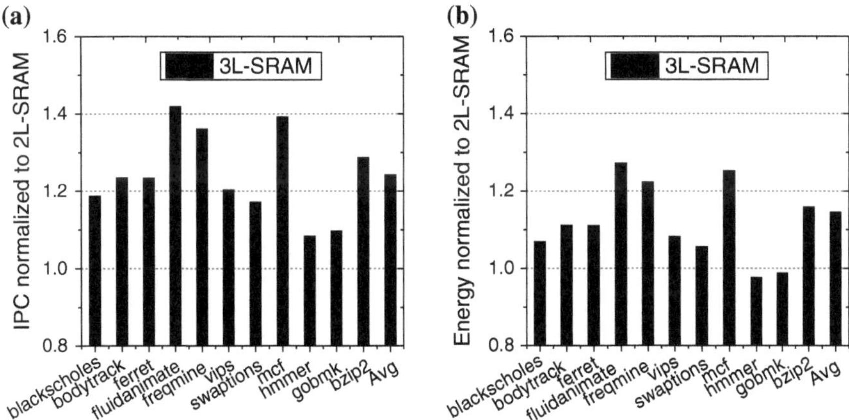

Fig. 7.21 IPC and overall cache energy comparison between 2-level and 3-level SRAM cache designs (normalized to the 2-level SRAM design)

the overall energy consumption. The leakage power ratio between "md2" and "hi" is 69.1/110 (see Table 7.5). This is why the overall energy of '3L-MultiR-Hyb' is only 60 % of '3L-MultiR' whose L3 is "hi".

7.5.5 Comparison Between 2-Level and 3-Level Cache Hierarchies

First, we directly compare 2-level and 3-level caches both implemented by SRAM. Figure 7.21a shows the IPC comparison. The 3-level SRAM cache outperforms the 2-level SRAM cache by 24.2 % in IPC performance because the 3-level cache hierarchy includes 256 KB private L2 cache within each core, enlarging the cache capacity by 32 %. Accordingly, the leakage energy increases. Figure 7.21b compares their overall energy consumptions. The total energy of 3-level SRAM cache is 14.4 % greater than that of 2-level SRAM cache.

The 2-level cache hierarchy with hybrid LR- and HR-regions ('2L-Hybrid') is compared with the 3-level multiretention STT-RAM cache hierarchy ('3L-MuliR'). With regard to IPC performance, '2L-Hybrid' is 14.36 % worse than '3L-MuliR', as shown in Fig. 7.22a. Compared to SRAM-based cache, that is to say the hybrid design actually shrinks the performance degradation between 2-level and 3-level cache hierarchies. On the one hand, since the leakage energy of STT-RAM cell is very small, the leakage energy increasing has a much smaller scalar than the growth of cache capacity. On the other hand, the access to L3 cache is filtered by L2 cache, which induces a smaller dynamic energy in '3L-MuliR' than that of '2L-Hybrid'. So the overall energy of '3L-MuliR' is not increased as 3L SRAM does. The overall energy comparison between '2L-Hybrid' and '3L-MuliR' is shown in Fig. 7.22b.

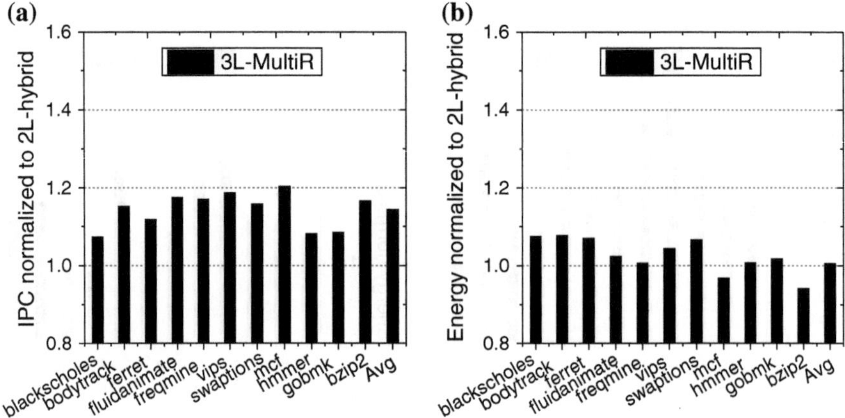

Fig. 7.22 IPC and overall cache energy comparison between 2-level and 3-level STT cache designs (normalized to the 2-level hybrid STT design)

7.6 Related Work

STT-RAM has many attractive features such as the nanosecond access time, CMOS process compatibility, and nonvolatility. The unique programming mechanism of STT-RAM—changing the MTJ resistance by passing a spin-polarized current [9]—ensures good scalability down to the 22nm technology node with a programming speed that is below 10 ns [26]. Early this year, Zhao, et al. [31] reported a subnanosecond switching at the 45 nm technology node for the in-plane MTJ devices.

Dong, et al. [8] gave a comparison between the SRAM cache and STT-RAM cache in a single-core microprocessor. Desikan, et al. [6] conducted an architectural evaluation of replacing on-chip DRAM with STT-RAM. Sun, et. al. [25] extended the application of STT-RAM cache to chip multiprocessor (CMP) and studied the impact of the costly write operation in STT-RAM on power and performance.

Many proposals have been made to address the slow write speed and high write energy of STT-RAM. Zhou et al. [32] proposed an early write termination scheme to eliminate the unnecessary writes to STT-RAM cells and save write energy. A dual write speed scheme was used to improve the average access time of STT-RAM cache that distinguishes between the fast and slow cache portions [30]. A SRAM/STT-RAM hybrid cache hierarchy and some enhancements, such as write buffering and data migration, were also proposed in [25, 29]. The SRAM and STT-RAM cache ways are fabricated on the different layers in the proposed 3D integration. The hardware and communication overheads are relatively high. None of these works considered using STT-RAM in L1 due to its long write latency.

In early 2011, Smullen et al. [22] proposed trading-off the nonvolatility of STT-RAM for write performance and power improvement. The corresponding DRAM-style refresh scheme to assure the data validity is not scalable for a large cache capacity. However, the single retention-level cache design is lack of optimization space to maximize the benefits of STT-RAM writability and nonvolatility trade-offs. Also, the MTJ optimization technique they proposed, namely shrinking the cell surface area of the MTJ, is not efficient in the fast-switching region (< 10 ns), as discussed in Sect. 7.3.

The macromagnetic model used in our work was verified by a leading magnetic recording company and calibrated with the latest in-plane MTJ measurement results [31]. However, we note that our model was not able to reproduce the MTJ parameters given in [22], which are overly optimistic in the fast-switching region (<3 ns) in terms of write energy and performance, as well as data retention time.

7.7 Conclusion

In this chapter, we proposed a multiretention-level STT-RAM cache hierarchy that trades off the STT-RAM cell's nonvolatility for energy saving and performance improvement. Taking into consideration the differences in data access behavior, we proposed a low-retention L1 cache with a counter-controlled refresh scheme and a hybrid structure for lower-level cache with both low- and high-retention portions. A memristor-controlled refresh scheme was proposed for the STT-RAM L1 cache to ensure data validity with the minimized hardware cost. For L2, a data migration scheme between the low- and the high-retention portions of the cache yielded fast average write latency and low standby power. Compared to the classic SRAM or a SRAM/STT-RAM hybrid cache hierarchy, our proposal uses only STT-RAM. This can save significant die cost and energy consumption. Moreover, compared to the previous STT-RAM-relaxed retention design that only has a single retention level, our design utilizes multiple retention levels, resulting in an architecture that is optimized for the data access patterns of the different cache levels.

Our experimental results show that our proposed multiretention-level STT-RAM hierarchy achieves on average a 73.8 % energy reduction over the SRAM/STT-RAM mixed design, while maintaining a nearly identical IPC performance. Compared with the previous single-level relaxed retention STT-RAM design, we obtained a 5.5 % performance improvement and a 30 % overall energy reduction by having multiple retention levels in 2-level hierarchy. The multiretention STT-RAM cache with proposed hybrid STT-RAM lower-level cache achieves on average of 6.2 % performance improvement and 40 % energy saving compared to the previous single-level relaxed retention STT-RAM design for a 3-level cache hierarchy. Compared to traditional SRAM L1 cache, the L1 cache with a ultralow-retention STT-RAM augmented by the proposed refresh scheme can achieve a 9.2 % performance improvement and a 30 % energy saving.

With technology scaling, and the increasing complexity of fabrication, we believe that our proposed cache hierarchy will become even more attractive because of its performance, low energy consumption, and CMOS compatibility.

References

1. Barth, J., Plass, D., Nelson, E., Hwang, C., Fredeman, G., Sperling, M., Mathews, A., Reohr, W., Nair, K., Cao, N. (2010). A 45 nm SOI embedded DRAM macro for POWER7TM 32 MB on-chip L3 cache. *IEEE international solid-state circuits conference digest of technical papers* (ISSCC), 342–343.
2. CACTI. http://www.hpl.hp.com/research/cacti/
3. Cao, Y., Sato, T., Orshansky, M., Sylvester, D., Hu, C. (2000). New paradigm of predictive MOSFET and interconnect modeling for early circuit design. *IEEE custom integrated circuit conference*, 201–204.
4. Chua, L. (2002). Memristor-the missing circuit element. *IEEE transastions on circuit theory*, *18*, 507–519.
5. De Sandre, G., Bettini, L., Pirola, A., Marmonier, L., Pasotti, M., Borghi, M., Mattavelli, P., Zuliani, P., Scotti, L., Mastracchio, G. (2010). A 90 nm 4 Mb embedded phase-change memory with .2 V $12 nsreadaccesstimeand$ MB/s write throughput. *IEEE international solid-state circuits conference digest of technical papers* (ISSCC), 268–269.
6. Desikan, R., Lefurgy, C. R., Keckler, S. W., Burger, D. (2008). On-chip MRAM as a high-bandwidth low-latency replacement for DRAM physical memories. http://www.cs.utexas.edu/ftp/pub/techreports/tr02-47.pdf
7. Diao, Z., Li, Z., Wang, S., Ding, Y., Panchula, A., Chen, E., Wang, L. C., Huai, Y. (2007) Spin-transfer torque switching in magnetic tunnel junctions and spin-transfer torque random access, memory (Vol. 19). p. 165209.
8. Dong, X., Wu, X., Sun, G., Xie, Y., Li, H., Chen, Y. (:2008). Circuit and microarchitecture evaluation of 3D stacking magnetic RAM (MRAM) as a Universal Memory Replacement. *ACM/IEEE desgin automation conference* (DAC), 554–559.
9. Hosomi, M., et al. (2005). A novel nonvolatile memory with spin torque transfer magnetization switching: Spin-RAM. *IEEE international electron devices meeting*, 459–462.
10. Hu, M., Li, H., Chen, Y., Wang, X., Pino, R. E. (2011). Geometry variations analysis of TiO 2 thin-film and spintronic memristors. *Proceedings of the 16th Asia and South Pacific design automation conference*, 25–30.
11. IntelQ8200. http://ark.intel.com/Product.aspx?id=36547
12. Kawahara, T., Takemura, R., Miura, K., Hayakawa, J., Ikeda, S., Lee, Y. M., et al. (2008). 2 Mb SPRAM (SPin-Transfer Torque RAM) With Bit-by-Bit bi-directional current write and parallelizing-direction current read. *IEEE Journal of solid-state circuit*, *43*, 109–120.
13. Kim, C. H., Kim, J. J., Mukhopadhyay, S., & Roy, K. (2005). A forward body-biased low-leakage SRAM cache: device, circuit and architecture considerations. *IEEE transations on very large scale integration* (VLSI) *system*, *13*, 349–357.
14. Kirolos, S., Massoud, Y. (2007). Adaptive SRAM design for dynamic voltage scaling VLSI systems. *IEEE international midwest symposium on circuits and systems* (MWSCAS), 1297–1300.
15. Li, Z., Zhang, S. (2004). Domain-wall dynamics driven by adiabatic spin-transfer torques. *Physics Review B*, *70*, 024417.
16. Marss86. http://www.marss86.org/
17. Nair, P., Eratne, S., John, E. (2007). A quasi-power-gated low-leakage stable SRAM cell. *IEEE international midwest symposium on circuits and systems* (MWSCAS), 761–764.
18. Nebashi, R., Sakimura, N., Tsuji, Y., Fukami, S., Honjo, H., Saito, S., Miura, S., Ishiwata, N., Kinoshita, K., Hanyu, T. (2011). A content addressable memory using mMagnetic domain wall motion cells. *Symposium on VLSI circuits*, 300–301.
19. NVSim. http://www.rioshering.com/nvsimwiki/index.php
20. Parkin, S. (2009). Racetrack memory: A storage class memory based on current controlled magnetic domain wall motion. *Device research conference*, 3–6.
21. Raychowdhury, A., Somasekhar, D., Karnik, T., De, V. (2009). Design space and scalability exploration of 1T–1STT MTJ memory arrays in the presence of variability and disturbances. *IEEE international electron devices meeting*, 1–4.

22. Smullen, C. W., Mohan, V., Nigam, A., & Gurumurthi, S., Stan, M. R. (2011). Relaxing non-volatility for fast and energy-efficient STT-RAM caches.
23. Strukov, D. B., Snider, G. S., Stewart, D. R., & Williams, R. S. (2008). The missing memristor found. *Nature*, *453*, 80–83.
24. Sun, J. Z. (2000). Spin-current interaction with a monodomain magnetic body: A model study. *Physics Review B*, 62.
25. Sun, G., Dong, X., Xie, Y., Li, J., Chen, Y. (2009). A novel architecture of the 3D stacked MRAM L2 cache for CMPs. *IEEE symposium on high-performance computer architecture* (HPCA), 239–249.
26. Wang, X., Chen, Y., Li, H., Dimitrov, D., Liu, H. (2008). Spin torque random access memory down to 22 nm technology (Vol. 44). *IEEE transactions on magnetics*, 2479–2482.
27. Wang, X., et al. (2009). Spintronic Memristor through Spin Torque Induced Magnetization Motion. *IEEE electron device letters*, *30*, 293–297.
28. Wang, X., Zhu, W., Xi, H., & Dimitrov, D. (2008). Relationship between symmetry and scaling of spin torque thermal switching barrier. *IEEE transactions on magnetics*, *44*, 2479–2482.
29. Wu, X., Li, J., Zhang, L., Speight, E., & Xie, Y. (2009). Power and performance of read-write aware hybrid caches with non-volatile memories. *Design, automation and test in Europe conference and exhibition* (pp. 737–742).
30. Xu, W., et al. (2011). Design of last-level on-chip cache using spin-torque transfer RAM (STT RAM). *IEEE transations on very large scale integration* (VLSI) *system*, *483–493*, 2011.
31. Zhao, H. and Lyle, A. and Zhang, Y. and Amiri, PK and Rowlands, G. and Zeng, Z. and Katine, J. and Jiang, H. and Galatsis, K. and Wang, KL.: Low writing energy and sub nanosecond spin torque transfer switching of in-plane magnetic tunnel junction for spin torque transfer RAM, *Journal of Applied Physics*, *109*, 07C720.
32. Zhou, P., Zhao, B., Yang, J., Zhang, Y. (2009). Energy reduction for STT-RAM using early write termination. *IEEE/ACM international conferernce on computer-aided design* (ICCAD), 264–268.
33. ITRS (2011). The International Technology Roadmap for Semiconductors. http://www.itrs.net.

22. Smullen, C. W., Mohan, V., Nigam, A., Gurumurthi, S., Stan, M. R. (2011). Relaxing non-volatility for fast and energy-efficient STT-RAM caches.
23. Strukov, D. B., Snider, G. S., Stewart, D. R., Williams, R. S. (2008). The missing memristor found. *Nature* 453, 80–83.
24. S[illegible] (1989) [illegible] *Phys Rev B* [illegible]
25. Sun, G., Dong, X., Xie, Y., Li, J., Chen, Y. (2009). A novel architecture of the 3D stacked MRAM L2 cache for CMPs. [illegible] *High-Performance Computer Architecture*, HPCA [illegible]
26. [illegible]
27. [illegible]
28. [illegible]
29. [illegible]
30. [illegible]
31. Zhao, [illegible] Wang, X. [illegible] writing energy and [illegible] torque transfer switching of in plane magnetic tunnel junction for spin torque transfer RAM. *Journal of Applied Physics* [illegible]
32. Zhou, P., Zhao, B., Yang, J., Zhang, Y. (2009). Energy reduction for STT-RAM using early write termination. *IEEE/ACM International Conference on Computer-Aided Design (ICCAD)*, 264–268.
33. ITRS (2011). The International Technology Roadmap for Semiconductors. http://www.itrs.net.

Chapter 8
Resistive Memories in Associative Computing

Engin Ipek, Qing Guo, Xiaochen Guo and Yuxin Bai

Abstract As CMOS scaling continues into the billion transistor era, power dissipation and off-chip bandwidth limitations are threatening to bring an end to microprocessor performance growth. Data-intensive applications such as data mining, information retrieval, video processing, and image coding demand significant computational power and generate substantial memory traffic, which places a heavy strain on both off-chip bandwidth and overall system power. Associative computing using ternary content-addressable memories (TCAM) is an attractive solution to curb both power dissipation and off-chip bandwidth demand in a wide range of data-intensive applications. When associative lookups are implemented using TCAM, data are processed directly on the TCAM chip, which decreases off-chip traffic and lowers bandwidth demand. Often, a TCAM-based system also improves energy efficiency by eliminating instruction processing and data movement overheads that are present in a purely RAM-based system. This chapter reviews TCAM designs using CMOS and resistive memory technologies and presents a case study on a novel memory system that aims at cost-effective, modular integration of a high-capacity TCAM within a general-purpose computing platform.

E. Ipek (✉)
Department of Computer Science, Department of Electrical and Computer Engineering, University of Rochester, Rochester, NY14627, USA
e-mail: engin.ipek@rochester.edu

Q. Guo
Department of Computer Science, University of Rochester, Rochester, NY14627, USA
e-mail: qing.guo@rochester.edu

X. Guo · Y. Bai
Department of Electrical and Computer Engineering, University of Rochester, Rochester, NY14627, USA
e-mail: xiaochen.guo@rochester.edu

Y. Bai
e-mail: yuxin.bai@rochester.edu

Y. Xie (ed.), *Emerging Memory Technologies*,
DOI: 10.1007/978-1-4419-9551-3_8,

8.1 Location-addressed Computing

Conventional von Neumann architectures store and retrieve data by explicitly addressing memory locations. This elegant model simplifies computer system design and implementation from both hardware and software perspectives; consequently, location-addressed computing has been a commercial success for the past 60 years. Unfortunately, despite this ease of implementation and long-lived success, location-addressed computing also introduces significant sources of inefficiency that degrade system performance. Since processing units and data storage are physically apart, programs are responsible for indexing the memory locations, fetching operands, and storing results, which often results in significant data movement across the memory hierarchy. Data-intensive applications such as data mining, information retrieval, video processing, and image coding demand significant computational power and generate substantial memory traffic, which places a heavy strain on both off-chip bandwidth and overall system power in a location-addressed computer system. Device, circuit, and architecture innovations are needed to surmount this problem.

8.2 Associative Computing

An effective way of addressing power and bandwidth limitations on many data-intensive workloads is to use memories that are accessed based on content (also named association [27]), rather than index. Associative computing, which leverages memory systems that store and retrieve data by content, has been broadly applied to both software and hardware designs. The best known software solution is a hash table, whereby data are located by an address computed through a hash function. A hash table has O(1) lookup time and is proven more efficient than other data structures in handling sparse data (i.e., when the number of keys in use is far less than the total number of possible keys). On the hardware side, a simple example of associative computing is the use of content-addressable memory (CAM), which has seen wide use since 1970s [5]. Nowadays, CAMs are commonly used in highly associative caches, translation lookaside buffers (TLBs), and microarchitectural queues (e.g., issue queues). Compared to a hash table, a CAM has no collision problems and offers better storage utilization and shorter search time; moreover, a CAM avoids the software overhead of rehashing and chaining.

8.2.1 CAM Basics

A CAM conceptually implements the inverse function of a RAM. Figure 8.1 shows an example CAM structure. A CAM consists of a data array, peripheral circuitry, and result generation network (e.g., an encoder). The data array comprises a mesh of cells for storage, which are connected by horizontal matchlines and vertical searchlines. Each row of the data array holds a searchable entry. On a CAM lookup, a search key

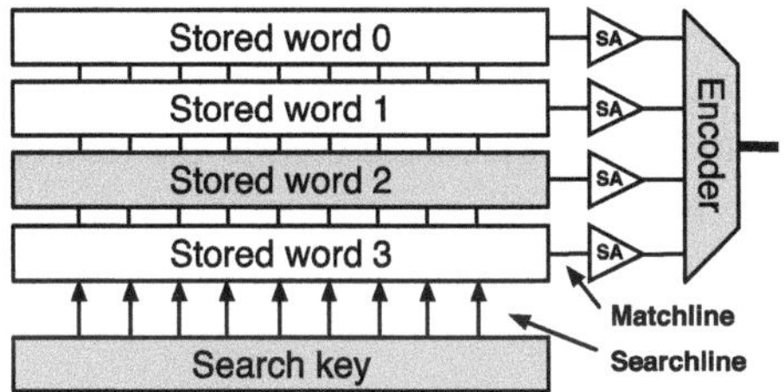

Fig. 8.1 A conceptual view of a CAM. A CAM consists of a data array (where the stored words reside), peripheral circuitry (sense amplifiers and search key registers), and result generation network (encoder)

is applied to the search key register (shaded block under the data array in Fig. 8.1); subsequently, the searchlines and matchlines are driven to compare the search key to all stored entries simultaneously. The search result on each matchline is sensed by a sense amplifier. A result generation network then operates on the output of the sense amplifiers and returns the final result. In this example, the encoder will return the index of the highest-priority matching row.

8.2.2 Ternary CAMs

A ternary CAM (TCAM) is a special type of associative memory that allows for both storing and searching with a wildcard (X) in addition to a logic 0 or 1. A wildcard, when part of either the search key or the stored data word, matches against both binary states as well as another wildcard. (Table 8.1 shows the truth table of a TCAM.) This flexibility can be exploited by a surprisingly large number of applications.

TCAM has been widely used in networking for decades. The primary commercial application of earlier generation TCAM was in routers [39], where it was used to store the routing table and to perform fast lookups through longest prefix matching [59]. With technology scaling over the last 20 years, TCAM capacity has increased from

Table 8.1 Truth table of a TCAM

Stored content	Search key	Search Result
0	0	Match
0	1	Mismatch
1	0	Mismatch
1	1	Match
X[a]	—	Match
—[b]	X	Match

[a] X is a wildcard
[b]—can be a 0, 1, or wildcard

64 Kb [64] to 9 Mb [25], while typical array width has grown from 72 to 576 bits. Numerous networking applications have emerged to leverage the benefits of TCAM, including packet classification [28], access control list filtering [40], and network intrusion detection [7].

8.3 TCAM Design Issues

The main design challenge of content-addressable memories is achieving high search throughput without unduly sacrificing power, or area efficiency.[1] Throughput can be improved by increasing the number of cells that are searched in parallel (which often comes with an increase in power consumption), and by speeding up the sensing and result generation circuitry (which often results in a reduction in area efficiency). This trade-off between system performance, power consumption, and area efficiency must be carefully balanced in TCAM design. This section presents existing TCAM proposals, with a particular emphasis on the cell structure. Circuit- and architecture-level design issues in existing TCAMs are discussed thoroughly in other work [47].

8.3.1 CMOS-based TCAM Cells

A TCAM cell serves two functions: bit storage and bit comparison. Figure 8.2 shows a *NOR-style* TCAM cell: the two pairs of cross-coupled inverters act as bistable storage elements that hold the value, and the two access transistors $M1$ and $M2$ are used to read and write the cell contents. Transistors $M3 - M6$ form two pull-down paths. On a search, the matchline ML is driven $high$, and searchlines supply the complementary search values SL and $\overline{SL}$. The cross-coupled inverters supply the data stored in the cell, and the bottom four NMOS transistors ($M3 - M6$) compare the search key to the data. On a mismatch, one of the pull-down paths is activated,

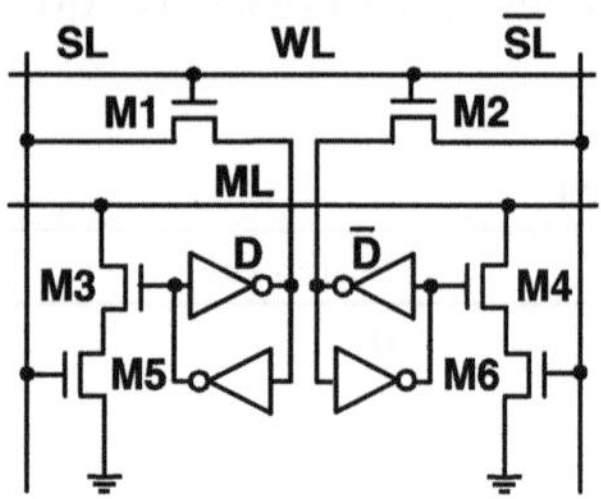

Fig. 8.2 An example *NOR-style* TCAM cell. Two pairs of cross-coupled inverters store complementary data bits (D and $\overline{D}$). $M1andM2$ are access transistors, and transistors $M3 - M6$ compare the search key against the data stored in the cell

[1] Area efficiency is defined as the area of the data arrays divided by the total area.

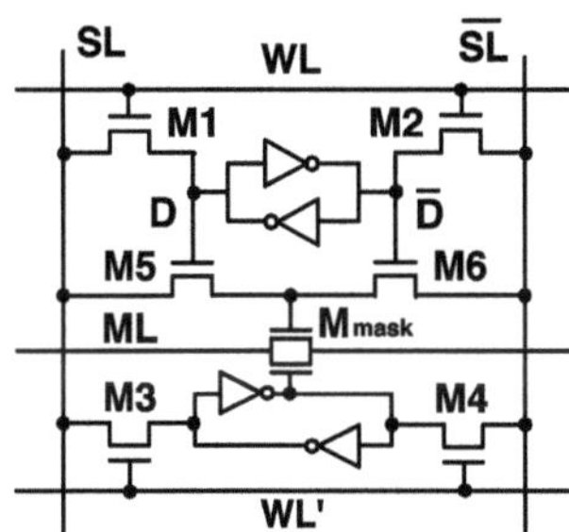

Fig. 8.3 An example *NAND-style* TCAM cell. The cross-coupled inverters (*top*) store complementary data bits, while the cross-coupled inverters (*bottom*) can be programmed to store a wildcard. $M1 - M4$ are access transistors. Transistors $M5$, $M6$, and M_{mask} compare the search key against the data stored in the cell

connecting the matchline to ground; on a match, all pull-down paths are inactive, leaving the matchline in the *high* state. To store a wildcard, both D and $\overline{D}$ are programmed to 1; to search with a wildcard, both SL and $\overline{SL}$ are driven *low*. A wildcard always results in a match when it is part of the search key or the stored word.

An example *NAND-style* TCAM cell is shown in Fig. 8.3, where a pair of cross-coupled inverters stores the data bit (D) and its complement ($\overline{D}$). Transistors $M5$, $M6$, and M_{mask} work as switches. On a match, M_{mask} turns on, transmitting the matchline (ML) state from the previous stage to the next stage; on a mismatch, M_{mask} electrically disconnects the matchline from the following stage. To store a wildcard, another pair of cross-coupled inverters (at the bottom of the cell) is programmed to keep M_{mask} *on*, thereby allowing the matchline current to flow through the cell regardless of the stored data. To search with a wildcard, both searchlines (SL and $\overline{SL}$) are driven *high*.

On the one hand, the *NOR* cell provides full rail voltage at the gates of all transistors, leading to faster matchline evaluation and a larger noise margin than the $NAND$ cell; on the other hand, the $NAND$ cell has no active *VDD-GND* path and thus offers a power advantage. However, an $NAND$ TCAM row has to propagate the matchline state in serial, resulting in a sluggish matchline evaluation.

A CMOS-based TCAM cell is relatively large. Even with an area-efficient implementation that uses half-latches instead of cross-coupled inverters [4], its area is $541F^2$ (where F is the feature size), which is 3.8× as large as an SRAM cell, and over 90× as large as a DRAM cell. Table 8.2 shows a comparison among CMOS-

Table 8.2 Comparison of CMOS-based memory technologies [16]

Memory	Chip capacity (MB)	$ / chip	$ / MB	Access speed (ns)	Watts / chip	Watts / MB
DRAM	128	10–20	0.08–0.16	40–80	1–2	0.008–0.016
SRAM	9	50–70	5.5–7.8	3–5	1.5–3	0.17–0.33
TCAM	4.5	200–300	44.5–66.7	4–5	15–20	3.33–4.44

based TCAM, SRAM, and DRAM chips. State-of-the-art TCAM devices are as fast as SRAM, but the cost per bit is 8× higher due to the lower density of TCAM. Furthermore, TCAM is 10× more power hungry than SRAM and DRAM. These and other limitations restrict the commercial use of TCAM to niche networking applications.

8.3.2 Resistive TCAM Designs

As CMOS scales to 22-nm and beyond, charge-based memory technologies such as DRAM, Flash, and SRAM are starting to experience scalability problems [23]. In response, the industry is exploring resistive memory technologies that can serve as scalable, long-term alternatives to charge-based memories. Resistive memories, which include phase-change memory (PCM) and spin-torque transfer magnetoresistive RAM (STT-MRAM), store information by modulating the resistance of nanoscale storage elements and are expected to scale to much smaller geometries than charge-based memories. Resistive memories are non-volatile, which provides near-zero leakage power and immunity to radiation induced transient faults; however, because resistive memories need to change material states to update stored data, they generally suffer from high write energy, long write latency, and limited write endurance.

8.3.2.1 Phase-Change Memory

Phase-change memory (PCM) [29] is argurably the most mature among all resistive memory technologies, as evidenced by 128Mb parts that are currently in production [43], as well as gigabit array prototypes [9, 58]. A PCM cell is formed by sandwiching a chalcogenide phase-change material such as $Ge_2Sb_2Te_5$ (GST) between two electrodes. PCM resistance is determined by the atomic ordering of this chalcogenide storage element and can be changed from less than 10 KΩ in crystalline state to greater than 1 MΩ in amorphous state [23, 58]. On a write, a high-amplitude current pulse is applied to the cell to induce Joule heating. A slow reduction in write current causes the device to undergo a fast annealing process, whereby the material reverts to a crystalline state. Conversely, an abrupt reduction in current causes the device to retain its amorphous state. On a read, a sensing current lower than the write current passes through the cell, and the resulting voltage is sensed to infer the cell's content. Since the ratio of the high (R_{HI}) and low (R_{LO}) resistances is as high as 100, a large sensing margin is possible; however, the absolute resistance is in the mega-ohm range, which leads to large RC delays, and hence, slow reads [23]. PCM also suffers from finite write endurance; the typical number of writes that can be performed before a cell wears out is 10^6–10^8 [23]. Consequently, several architectural techniques have been proposed to alleviate PCM's wear-out problem [8, 22, 52, 57].

8.3.2.2 Spin-Torque Transfer Magnetoresistive RAM

As a universal embedded memory candidate, Spin-torque Transfer Magnetoresistive RAM (STT-MRAM) has a read speed as fast as SRAM [69], practically unlimited write endurance [13], and favorable delay and energy scaling characteristics [23]. Multimegabit array prototypes at competitive technology nodes (e.g., 45 and 65 nm) have already been demonstrated [32, 63], and the ITRS projects STT-MRAM to be in production by 2013 [23]. In STT-MRAM, information is stored by modulating the magnetoresistance of a thin film stack called a magnetic tunnel junction (MTJ). An MTJ is typically implemented using two ferromagnetic $Co_{40}Fe_{40}B_{20}$ layers, and an MgO tunnel barrier that separates the two layers. One of the ferromagnetic layers, the *pinned layer*, has a fixed magnetic spin, while the magnetic spin of the *free layer* can be altered by applying a high-amplitude current pulse through the MTJ. Depending on the direction of the current, the magnetic polarity of the free layer can be made either parallel or antiparallel to that of the pinned layer. In the case of anti-parallel alignment, the MTJ exhibits high resistance (12.5 KΩ at 22 nm [23]); in the case of parallel alignment, a low resistance (5 KΩ) is observed [23]. Although the typical $\frac{R_{HI}}{R_{LO}}$ ratio (2.5) is lower than that of PCM (100), it is still relatively easy to sense the state of a single bit [61].

8.3.2.3 Resistive TCAM Cell Designs

Resistive memories offer high density and near-zero leakage power, making them promising to build area- and power-efficient TCAMs. Matsunaga et al. [38] propose a 2T-2R TCAM cell (Fig. 8.4). The proposed cell consists of two resistor–transistor pairs, where each resistor is implemented using a resistive memory element (e.g., an MTJ in STT-MRAM, or a GST stack in PCM) that can be programmed to high- or low-resistance states. To store a logic 1 or 0, the resistor on the left is programmed to store the data bit (D), while the resistor on the right is programmed to store the complement of the data ($\overline{D}$). For example, when storing a logic 1, the resistor on the left is programmed to R_{HI}, and the resistor on the right is programmed to R_{LO}. To store a wildcard, both resistors are programmed to R_{HI}.

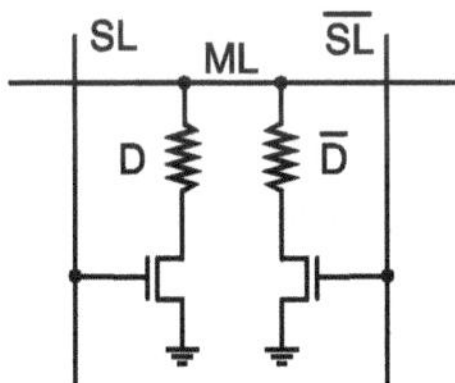

Fig. 8.4 A 2T-2R resistive TCAM cell [38]. A resistor is a programmable memory element, e.g., an MTJ in STT-MRAM, or a GST stack in PCM. The transistors compare the search key against the data stored in the resistors

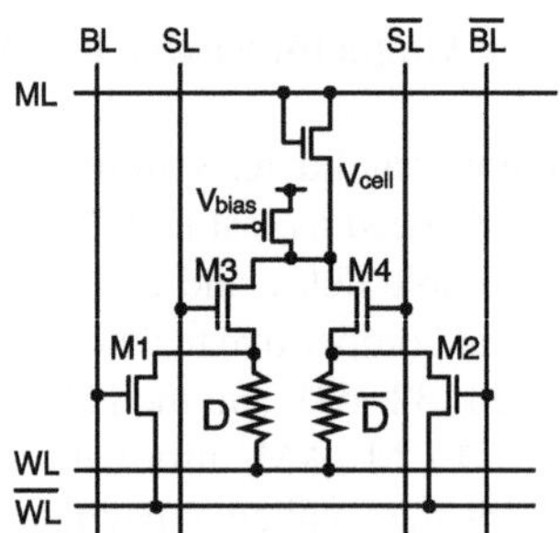

Fig. 8.5 A 6T-2R resistive TCAM cell [36]. A resistor is a programmable memory element, e.g., an MTJ in STT-MRAM, or a GST stack in PCM. $M1andM2$ are access transistors. Transistors $M3andM4$ compare the search key against the data stored in the resistors. A diode-connected NMOS transistor serves as self matchline discharge control

To search for a logic 0 or 1, SL and $\overline{SL}$ are driven with the search bit and its complement, respectively, turning one of the access transistors on, and the other off. To search with a wildcard, both SL and $\overline{SL}$ are driven low. A match is decided based on the effective resistance between the matchline and ground. If a resistor in its high-resistance state is in series with an *on* transistor—adding a resistance of $R_{HI}+R_{ON}$ (where R_{ON} represents the on-resistance of an access transistor) between the matchline and ground—the search results in a match; conversely, a resistance of $R_{LO} + R_{ON}$ connected to the matchline indicates a mismatch. Searching with a wildcard causes both SL and $\overline{SL}$ to be driven low, making the matchline-to-ground resistance virtually infinite, which still falls within the matching region.

The 2T-2R cell shows a significant density advantage over existing CMOS-based TCAM designs (Sect. 8.3.1); however, it lacks the ability to scale up to longer search widths. Specifically, when searching a TCAM row (with a length of N bits), the effective matchline-to-ground resistance (the parallel resistance of all the T-R paths connecting the matchline to ground) decides the search result. In the worst case, a row mismatches with one mismatching bit and $(N - 1)$ wildcards as part of the search key results in an effective resistance of $R_{LO} + R_{ON}$; since a match can have a matchline-to-ground resistance of $\frac{R_{HI}+R_{ON}}{N}$, and this quantity has to be greater than $R_{LO} + R_{ON}$ to distinguish a match from a mismatch, there is a limit to how large N can be.

An alternative design with a 6T-2R TCAM cell structure using STT-MRAM [36] is shown in Fig. 8.5. A diode-connected NMOS transistor is added to the cell for matchline (ML) discharge control. The diode nMOS serves as a ML voltage keeper. Upon a search operation, ML is initialized to V_{dd}. On a match, each cell's V_{cell} becomes $V_{\text{cell-high}}$ and the ML voltage keeper is cut off; on a mismatch, V_{cell}s of the mismatched cells become $V_{\text{cell-low}}$, and the ML is discharged via those ML voltage keepers. As such, the matchline output voltage swing is $(V_{\text{cell-high}} - V_{\text{cell-low}})$. The self-adjusted NMOS increases the number of bits that can be simultaneously searched to 144 bits, albeit, with a loss of density.

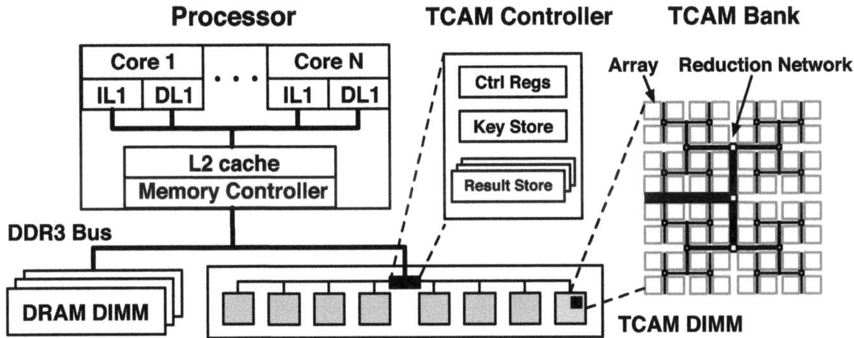

Fig. 8.6 Illustrative example of a computer system with the resistive TCAM DIMM

8.4 Case Study: A Resistive TCAM Accelerator

This section presents a resistive TCAM design, which aims at cost-effective, modular integration of a high-capacity TCAM system within a general-purpose computing platform [18]. The key idea is to build a resistive TCAM chip which can be integrated on a DDR3-compatible DIMM and selectively placed on the memory bus. Figure 8.6 presents an example computer system with the TCAM DIMM. A multicore processor connects to main memory through an on-chip memory controller. The TCAM DIMM sits side-by-side with DRAM on the DDR3 bus. An on-DIMM TCAM controller serves as the interface to DDR3 and is in charge of DIMM control. The processor communicates with the controller through a set of memory-mapped control registers (for configuring functionality) and a memory-mapped key store that resides with the controller (for buffering the search key). A 2 KB result store on the controller die buffers search results for multiple processes. All TCAM chips share the on-DIMM command, address, and data buses; however, a search operation is restricted to be on a single chip due to power constrains. Each TCAM chip has eight banks; a bank comprises a set of arrays that are searched against the search key, as well as a hierarchical reduction network for counting the number of matches and picking the highest-priority matching row.

8.4.1 Cell Structure

The area of a TCAM cell not only affects the cost per bit in a multimegabit array, but also has a profound effect on speed and energy since it determines the length (and thus, the capacitance) of the matchlines and searchlines that need to be charged and discharged on every access. Figure 8.7 demonstrates an area-efficient resistive TCAM cell, which consists of three resistor–transistor pairs. (For comparison, refer to the 2T-2R resistive TCAM cell discussed in Sect. 8.3.2.) The first two resistors

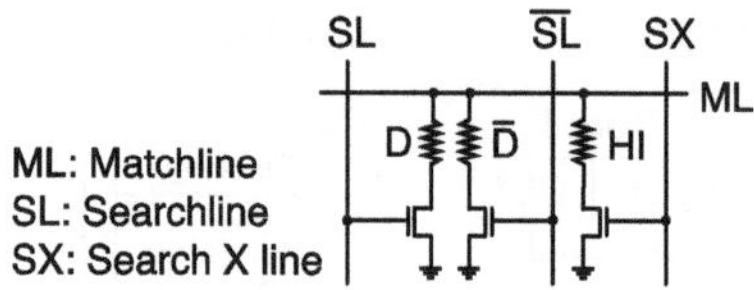

Fig. 8.7 A 3T-3R resistive TCAM cell. A third transistor-resistor pair is added to the cell to support searching with a wildcard

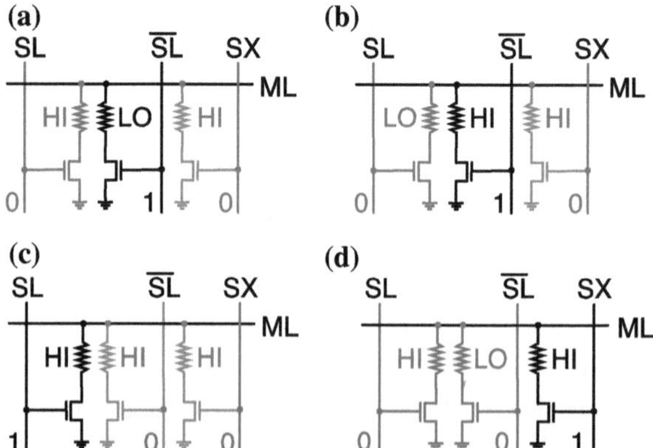

Fig. 8.8 Illustrative example of a resistive TCAM cell on a match (**b**, **c**, **d**) and on a mismatch (**a**). Inactive circuit paths are shown in *light gray*

store the data bit and its complement, whereas the third resistor is permanently programmed to R_{HI} to support searching with a wildcard. When searching with a wildcard, SL and $\overline{SL}$ are disabled and SX is driven $high$; hence, a resistor in its R_{HI} state is connected to the matchline regardless of the data stored in the cell. Examples are shown in Fig. 8.8: (a) demonstrates a mismatch case when the search bit is 0 and stored bit is 1; (b) presents a match scenario which searches for a 0 when a 0 is stored; (c) shows a case where a wildcard is stored in the cell, in which searching for either a 0 or a 1 will result in a match; and (d) illustrates a search for a wildcard.

The resistive TCAM cell shows a significant density advantage over its CMOS-based counterparts. Even with an area-efficient implementation that uses half-latches instead of cross-coupled inverters [4], CMOS-based TCAM cell has an area of $541F^2$ (where F is the feature size). The resistive TCAM cell consists of 3 1T-1R cells sharing matchline and GND contacts. At 22 nm, this amounts to an area of $27F^2$ ($3\times$ the 1T-1R PCM cell size projected by ITRS [23]), which is $\frac{1}{20}$ of a CMOS TCAM cell's area.

Compared to the 2T-2R cell (Sect. 8.3.2), the 3T-3R TCAM cell adds a third resistor–transistor pair to reduce worse-case mismatch resistance, which significantly increases the number of bits that can be simultaneously searched (Sect. 8.4.2). Compared to the 6T-2R cell (Sect. 8.3.2), which adds assist circuitry to each cell to increase the parallel-search width, the 3T-3R cell design has much better density.

8.4.2 Row Organization

TCAM cells cascade in parallel on a matchline to form a row. An example TCAM row is shown in Fig. 8.9. The key idea is that, on a match, each cell will connect the matchline to ground through an R_{HI} path, whereas at least one cell will provide an R_{LO} path to ground on a mismatch. Hence, the effective parallel resistance to ground (and hence, the matchline voltage) will always be higher in the case of a match.

A *precharge-low* sensing scheme is employed on the matchline, where the matchline is discharged in the precharge phase and actively driven in the evaluation phase. The input of the inverter is also charged *high* in the precharge phase. On a match, the gate voltage of M_{sense} is higher than it is on a mismatch. M_{sense} and the keeper PMOS M_k are sized so that on a match, the gate-to-source voltage of M_{sense} is large enough to win against M_k, pulling the inverter input low and in turn driving the matchline output ($MLso$) *high*. On a mismatch, the input to the inverter stays *high* and $MLso$ outputs a logic zero.

8.4.2.1 Searching

Searching a TCAM row involves distinguishing the effective matchline-to-ground resistance on a word match, where all bits stored in a row match the corresponding search bits, from the effective resistance on a word mismatch, where one or more bits mismatch the search key. In a TCAM row, each matching bit contributes a resistance of $R_{HI} + R_{ON}$, and each mismatching bit contributes an $R_{LO} + R_{ON}$ to the parallel resistance of the row. On a word match, the total resistance between the matchline and ground is

$$R_{\text{match}} = \frac{R_{HI} + R_{ON}}{N} \tag{8.1}$$

where N is the number of bits in a word. On a worst-case mismatch ($N - 1$ matching bits plus one mismatching bit), the matchline-to-ground resistance is

$$R_{\text{mismatch}} = \frac{(R_{LO} + R_{ON})(R_{HI} + R_{ON})}{(N-1)(R_{LO} + R_{ON}) + (R_{HI} + R_{ON})} \tag{8.2}$$

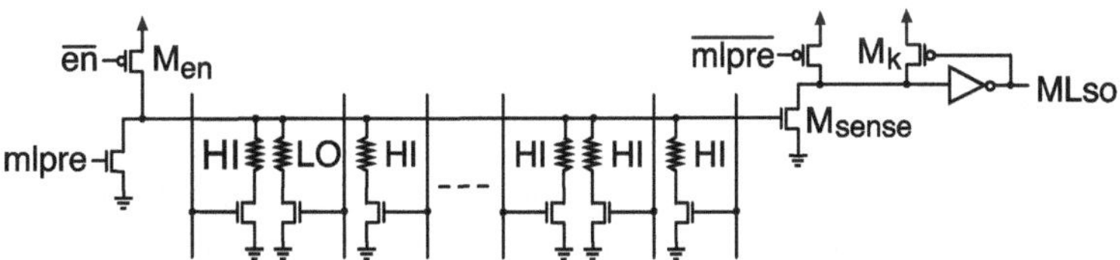

Fig. 8.9 A resistive TCAM row structure. TCAM cells cascade in parallel on a matchline

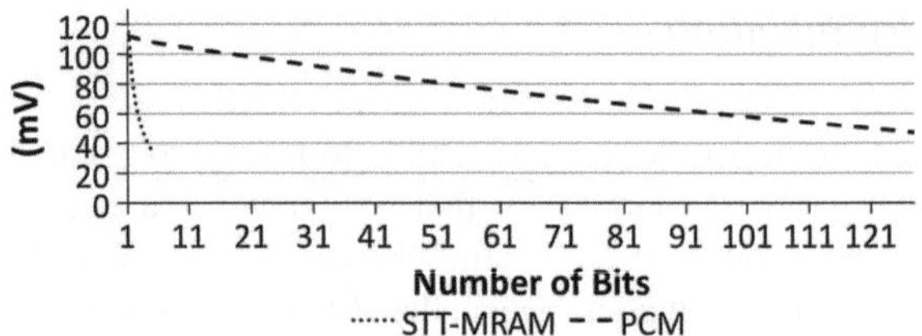

Fig. 8.10 Sensing margin as a function of search width. Simulation results are generated by Cadence (see Sect. 8.4.7 for the experimental setup)

The ratio between match and worst-case mismatch resistances is

$$\frac{R_{\text{match}}}{R_{\text{mismatch}}} = 1 + \frac{R_{HI} - R_{LO}}{N(R_{LO} + R_{ON})} \tag{8.3}$$

This $\frac{R_{\text{match}}}{R_{\text{mismatch}}}$ ratio must be high enough to tolerate process-voltage-temperature (PVT) variations, which affect both the transistors and the resistive devices. Therefore, a resistive memory technology with a greater difference between its high and low resistances (e.g., PCM) allows a larger number of bits to be sensed in parallel, whereas a technology with a lower $\frac{R_{HI}}{R_{LO}}$ ratio (e.g., STT-MRAM) has a tighter constraint on the number of bits that can be searched at the same time.

At the evaluation stage, current is supplied through the enabling PMOS transistor (M_{en}), and the voltage drop on the TCAM cells is sensed by a matchline sense amplifier. M_{en} is sized appropriately to limit the current through the resistors so that a search will not accidentally change cell states.

Figure 8.10 shows Cadence (Spectre) simulation results on PCM and STT-MRAM. As the number of simultaneously searched bits increases, both technologies encounter a significant drop on the voltage difference between a match and a mismatch, but the drop is much steeper in the case of STT-MRAM. Taking PVT variations into account, a smaller voltage difference (i.e., sensing margin) results in lower yield; hence, to tolerate variations and to improve yield, the number of bits that can be searched simultaneously must be limited. With a 128-bit search width, the PCM-based TCAM array functions correctly even in the face of extreme variations in R_{HI} (500 KΩ-1 MΩ) and R_{LO} (15–30 KΩ). These resistance ranges cover 98 % of all devices fabricated in a recent 45-nm PCM prototype [58]; nevertheless, if the variation was to be higher, the search width could be reduced to improve the margin.

8.4.2.2 Writing

In order to maintain the density advantage of resistive memories in TCAM design, it is important not to unduly increase the number of transistors in a cell. Writing would normally require a write port, which would need additional access transistors and wires. Instead, a novel *bulk sequential write* mechanism leverages the matchlines for writing in order to eliminate the need for extra write ports.

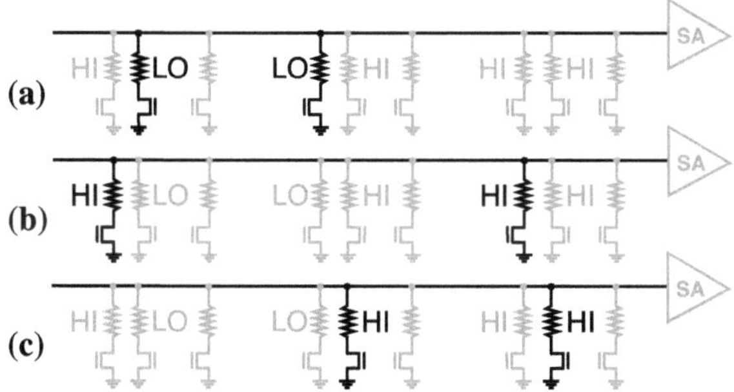

Fig. 8.11 Illustrative example of writing the three bit pattern "10X" to a row by (**a**) programming all *low* resistors, (**b**) programming all leftmost *high* resistors, and (**c**) programming all *middle high* resistors. Inactive circuit paths are shown in *light gray*

Figure 8.11 shows an example of how a TCAM row gets written under bulk-sequential writes. During a write, the sense amplifier and all search-X lines are disabled. Writing takes place in two phases. In the first phase, all resistors to be updated to R_{LO} are connected to the matchline by enabling the relevant searchlines and are programmed by applying the appropriate switching current through the matchline. Next, resistors to be updated to R_{HI} are written in two steps. In the first step, the leftmost resistor in each cell to be updated with a 1 or X is programmed; in the second step, the middle resistor in every cell to be updated with a 0 or X is written. This two-step procedure when writing R_{HI} limits the maximum number of resistors to be programmed simultaneously to 128. Minimum-pitch local wires at 22 nm can carry enough current to write these 128 bits in parallel, and a $200F^2$ write driver supplies adequate switching current.

8.4.2.3 Reading

Since the TCAM accelerator is ultimately integrated on a discrete, DDR3-compatible DIMM, it can be selectively included in systems which run workloads that can leverage associative search. Nevertheless, even for such systems, it is desirable to use the TCAM address space as ordinary RAM when needed. Fortunately, this is straightforward to accomplish with the TCAM array.

When the TCAM chip is configured as a RAM module, data are stored and retrieved column-wise in the TCAM array by the TCAM controller, and the searchlines SL, $\overline{SL}$, and SX serve as wordlines. The key observation that makes such configurability possible is that reading a cell is equivalent to searching the corresponding row with a "1" at the selected column position, and with wildcards at all other bit positions. An illustrative figure of the array and additional details on the required array-level support to enable configurability is outlined in Sect. 8.4.3.

8.4.3 Array Architecture

Figure 8.12 shows an example 1K×1K TCAM array. Cells (C) are arranged in a 2D organization with horizontal matchlines (ML) and vertical searchlines (SL). Searchlines are driven by search key registers and drivers, while matchlines connect to sense amplifiers (SA). A hierarchical reduction network facilitates counting the number of matching lines and selecting one of the multiple matching rows (Fig. 8.13).

An important design decision that affects area efficiency, power, speed, and reliability is the size of an array. Larger arrays improve area efficiency by amortizing the area overhead of the peripheral circuitry over more cells; however, the sensing margin deteriorates with the number of bits that are searched in parallel (Fig. 8.10). Table 8.3 compares absolute area and area efficiency for 1Gb TCAM chips constructed from arrays of different sizes (see Sect. 8.4.7 for the experimental setup). As array size increases from 128 × 128 to 1K × 1 K, overall chip area reduces by more than 2×.

Interestingly, it is possible to enjoy the high area efficiency of a large array while also delivering the large sensing margin of a smaller array through *matchline segmentation*. The key idea is to build a wide row, only a part (segment) of which can be searched simultaneously, providing additional peripheral circuitry to support iterative searching for a key larger than the size of a segment. Figure 8.12 shows an example 1K × 1 K array, where each row is partitioned into 128-bit segments. To perform a full 1-Kb-wide search across a row, the array is accessed eight times, once per each segment. On each access, searchlines connected to the relevant segment are

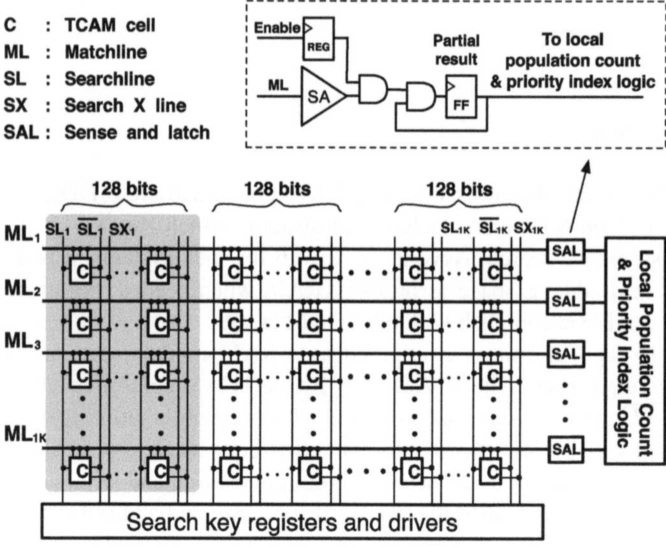

Fig. 8.12 Illustrative example of a 1K×1K TCAM array with 128-bit-wide matchline segmentation

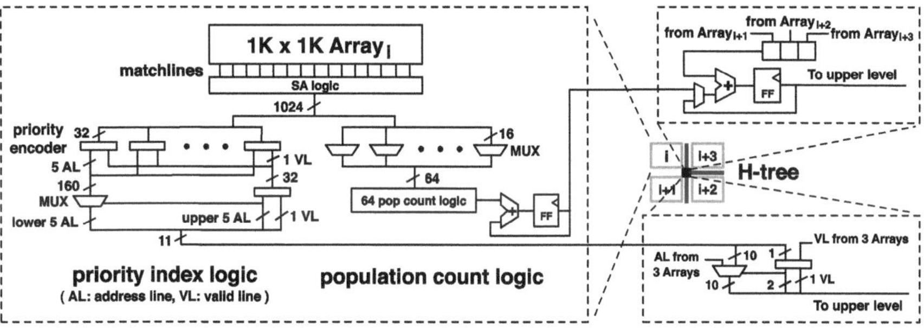

Fig. 8.13 Architecture of the population count logic, priority index logic, and the reduction network

Table 8.3 Total chip area and area efficiency with different array sizes

Array size	Total chip area (mm^2)	Area efficiency (%)
1K × 1K	36.00	38.98
512 × 512	41.04	34.19
256 × 256	51.49	27.25
128 × 128	77.51	18.10

driven with the corresponding bits from the search key, while all other searchlines are driven *low*. Each row is augmented with a helper flip-flop that stores the partial result of the ongoing search operation as different segments are accessed. This helper flop is initially set *high*; at the end of each cycle, its content is *ANDed* with the sense amplifier's output, and the result overwrites the old content of the helper flop. Hence, at the end of eight cycles, the helper flop contains a logic 1 if and only if all segments within the row match. To permanently disable a row, an additional "enable" flop is added to the design.

8.4.4 Response Solver

The resistive TCAM is capable of providing two different results on a search operation: (1) a population count indicating the total number of matching rows in the system, and (2) a priority index indicating the address of the matching row with the lowest index. Once the search is complete, the array's local population count and priority logic start operating on the results (Fig. 8.13).

Priority Index Logic. The highest-priority row among all matching rows is selected in a hierarchical fashion. 32 priority encoders are used in the first level of the hierarchy, where each encoder selects one valid matchline out of 32 helper flops; also selected are the 5-bit addresses of the corresponding matchlines. Then, 32 valid bits with corresponding addresses enter the second level priority encoder.

Table 8.4 Energy and delay comparison of different array sizes

Array size	Energy of search (nJ)	Energy with priority index (nJ)	Energy with pop. count (nJ)	Delay of local search (ns)	Delay with priority index (ns)	Delay with pop. count (ns)
1 × 1 K	244.97	248.59	249.64	2.50	21.57	60.28
512 × 512	176.16	180.22	181.30	1.75	21.91	61.48
256 × 256	96.21	100.85	102.36	1.00	23.78	64.63
128 × 128	55.57	61.78	63.33	0.50	28.29	67.67

Finally, an 11-bit result, containing 1 valid bit and a 10-bit address, is selected and forwarded to the reduction network.

Population Count Logic. The local population count for the array is computed iteratively, accumulating the result of a 64-bit population count each cycle (itself computed using 16 four-input lookup tables and a carry save adder tree). It takes a total of 16 cycles to walk the array and accumulate the population count, after which the result is sent to the reduction network for further processing.

Reduction Network. The reduction network is a quad tree, whose leaves are the results of different arrays' population count and priority logic. Each internal node of the quad tree takes the outputs of its four children, processes them by counting or priority encoding, and then forwards the result to its parent (Fig. 8.13). Once the results of a search propagate to the root of the quad tree, final results are obtained and placed in an SRAM-based result store that resides with the TCAM controller. For a fixed-capacity chip, the size of an array affects the size, and hence, the latency and energy of the reduction network. Table 8.4 shows the energy and delay when searching for a 128-bit key in an eighth of a 1-Gbit chip constructed from different size arrays. As array size increases, delay and energy within an array increase due to longer searchlines and matchlines; however, delay and energy consumed in the reduction network decrease because there are fewer levels of priority encoding and population count computation in the hierarchy. Since search energy dominates total energy and reduction network delay dominates total delay, enlarging the array size results in higher energy and lower delay.

Result Store. Search throughput can be improved significantly by pipelining the reduction network, the local population count, and priority computations; however, this requires the ability to buffer the results of in-flight search operations and to retrieve them at a later time. To facilitate such buffering, the on-DIMM TCAM controller provides a 2-KB SRAM mapped to the system's physical address space. As part of the search operation, the application sets up a number of memory-mapped control registers, one of which indicates the location where the search results should be placed. In this way, the application overlaps multiple search operations in a pipelined fashion and retrieves the results through indexed reads from the result store.

8.4.5 System Interface

The level of the memory hierarchy at which the TCAM is placed can have a significant impact on overall system cost, performance, and flexibility. On the one hand, an integrated solution that places the TCAM on the processor die would result in the shortest possible communication latency, at the expense of much desired modularity. On the other hand, treating the TCAM as an I/O device and placing it on the PCI or PCI Express bus would result in a modular system architecture (in which the TCAM can be selectively included), but the high latency of I/O buses would limit performance. Instead, this section explores an intermediate solution that places the TCAM on a DDR3-compatible DIMM with an on-DIMM TCAM controller. The resulting design requires no changes to existing processors, memory controllers, or motherboards, while providing modularity to enable selective inclusion of TCAM in systems that benefit from it. Moreover, with the ability to configure the TCAM as a regular RAM, users can also leverage the TCAM DIMM as a byte-addressable, persistent memory module.

8.4.5.1 Processor Interface

Software is given access to a TCAM accelerator by mapping the TCAM address range to the system's physical address space. In addition, the on-DIMM TCAM controller maintains a 2-KB RAM array (Sect. 8.4.4) to implement memory-mapped control registers, search key buffer, and the result store. All accesses issued to TCAM are uncacheable and subject to strong ordering, which prevents requests to the memory-mapped TCAM address range from being reordered. (This can be accomplished, for example, by marking the corresponding physical pages strong-uncacheable in the page attribute table of any Intel X86 processor since the Intel 386 [21].)

Communication between the processor and TCAM takes place in four ways:

- `Device Configuration.` The processor configures the TCAM system by writing to memory-mapped control registers. Configurable system attributes include key length, required result type (population count, priority index, or both), and whether the module should operate in RAM or TCAM mode.
- `Content Update.` The processor stores data into TCAM control registers, after which the TCAM controller updates the TCAM array.
- `Search.` The processor stores the search key into the memory-mapped TCAM key buffer, which resides with the TCAM controller. As soon as the last word of the key is received, the TCAM controller initiates a search operation whose results are written to the appropriate words within the memory-mapped result store.
- `Read.` After a search, the processor loads the outcome from the result store.

8.4.5.2 TCAM Controller

To plug into an existing DIMM socket with no modifications to the memory controller, the TCAM's memory bus interface should be fully compatible with DDRx and its timing constraints. Here, we discuss the TCAM interface in the context of a modern DDR3 protocol and an FR-FCFS [55]-based memory controller, but the memory controller's scheduling policy is orthogonal to the TCAM DIMM, since a DDR3 compatible TCAM DIMM is compatible with any DDR3-compatible memory controller by definition. In DDR3, a complete read (write) transaction includes a row activation (activate a row in a bank and move data to a row buffer), a column read (write) involving data transfer, and a precharge (write data back to the row and precharge the bitlines). When consecutive requests hit an open row, no activate or precharge operation is needed. The minimum number of cycles between a write and a consecutive read in DDR3 is less than the time needed to perform a search; hence, without additional support, it is possible for the memory controller to issue a read from the result store before the corresponding search operation is complete.

To avert this problem, the software library used for accessing TCAM inserts an appropriate number of dummy writes between the search and the subsequent read from the result store, generating the required number of idle cycles on the DDR3 bus for the search operation to complete. (Recall that all requests to TCAM control registers are processed in order, since the corresponding physical pages are marked strong-uncacheable.) This is accomplished by issuing writes to a known location in physical memory, which allows the TCAM controller to detect the dummy write and silently drop it upon receipt.

Although injecting dummy writes (i.e., bubbles) into the DDR3 bus guarantees correctness, it also lowers the utilization of the bus and wastes precious bandwidth. To improve search throughput, the TCAM chip is pipelined (Sect. 8.4.3) so that shortly after an array search is complete, the next search is allowed to begin, largely overlapping the array search operation of a later request with the reduction network and result store update of an earlier request. Consequently, when performing multiple search operations, it becomes possible to pipeline search and read operations, thereby reducing the number of required dummy writes. Figure 8.14 shows an example of pipelined search with a 128-bit key, where pipelining improves performance by almost 3×.

8.4.6 Software Support

A user-level library serves as the API to the programmer and implements four functions.

- `Create`. This function is called at the beginning of an application to map a virtual address range to a portion of the TCAM's physical address space. Physical pages in TCAM are marked strong-uncacheable (Sect. 8.4.5), and the OS is signaled to

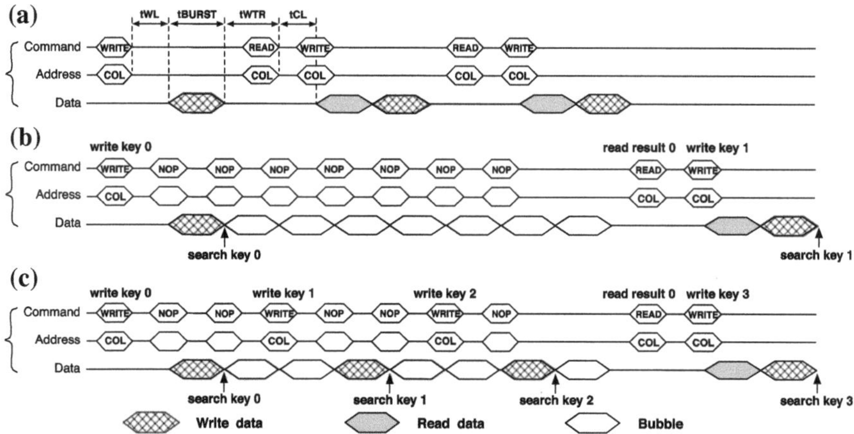

Fig. 8.14 A consecutive write–read–write command sequence in DDR3 (**a**), a consecutive search–read command sequence in TCAM without pipelining (**b**), and a consecutive search–read command sequence in TCAM with pipelining (**c**). Dummy writes are labeled "NOP"

allocate content-addressable physical pages by leveraging one of the reserved flags of the mmap system call [62]. Upon successful termination, the function returns a pointer to the newly allocated TCAM space, which helps distinguish independent TCAM regions.

- `Destroy`. This function releases the allocated TCAM space.
- `Store`. This function updates the TCAM array. The function accepts a mask indicating the bit positions that should be set to wildcards, as well as the data to be stored in the remaining positions. Internally, the function communicates to the TCAM controller by writing two memory-mapped control registers, and the TCAM controller writes the data into the TCAM array. To hide the long latency of writing to TCAM, the library function distributes consecutive writes to different TCAM banks. In the case of a bank conflict, an appropriate number of dummy writes are inserted by the library to cover the busy period. (In the evaluated benchmarks, the database is stored in a sequential order; thus, no bank conflicts were observed.)
- `Search`. This function performs a search operation in three steps, by (1) storing the search key into the memory-mapped TCAM search key register, (2) issuing enough dummy writes to pad the waiting period between the search operation and the subsequent read from the result store, and (3) reading the results from the memory-mapped result store. Two different flavors of the search function are provided: a "single" search call performs a search with a single key, whereas a "batch" search call searches for multiple independent keys in a pipelined fashion to improve throughput (keys are stored in memory and are passed to the batch search function through a pointer). A pointer to the TCAM region to be searched is given as an argument to the search call, and the library ensures that only results from the specified region are returned by storing a (searchable) unique region ID with each TCAM row.

8.4.6.1 Multiprogramming

Supporting multiprogramming with the proposed TCAM system requires mechanisms to prevent the search results of one application from being affected by data that belongs to another application. Hence, although conventional virtual memory solves protection problems in the case of TCAM reads and writes, search operations need additional support.

To enable multiprogramming, each process is assigned an address space identifier (ASID); when the process allocates space in the TCAM address range, the OS records the ASID in a memory-mapped control register within the same physical page that contains the key buffer and the result store for that process. On a write, the TCAM controller stores the corresponding ASID along with each word in a given row. On a search, after the process writes the key buffer, the TCAM controller appends the ASID to the search key; as a result, all rows that belong to other processes result in mismatches, and do not affect search results.

8.4.6.2 Handling Misfits

Although the OS naturally supports virtualization of the physical address space, searching a data structure larger than the allocated TCAM space requires extra care. Specifically, two types of misfits are possible: (1) a horizontal misfit, in which the key is larger than the width of a TCAM array (e.g., 2 KB key for a $1\text{K} \times 1\text{K}$ array), and (2) a vertical misfit, in which the number of rows to be searched exceeds the capacity of the allocated TCAM region.

To solve the horizontal misfit problem, the word is broken into a series of 1-Kb-wide subwords, each of which is stored in a consecutive row of an array (if the final subword is narrower than 1 Kb, it is padded with enough wildcards to cover the row). On a search, the TCAM controller partitions the key in units of 1-Kb subwords and searches the array one row after another. At the end of a search operation, the content of the helper flip-flop that connects to a row is shifted into the helper flip-flop of the next row and is $ANDed$ with the outcome of the next row's search operation. Hence, the helper flop of the final row contains the complete search result.

On a vertical misfit, the search is staged over multiple local search operations, in which the missing pages are transferred from DRAM to TCAM. This process is transparent to the user and is handled by the TCAM library. Since data transfer between TCAM and DRAM can be expensive, the search is optimized for minimizing data movement. For example, if the search region is larger than the capacity of the TCAM, the library function partitions the search region to fit the TCAM space and does batch search in each subregion. The final results are calculated by merging all of the partial results. This is obviously cheaper than doing each single search in the entire region, which would generate constant data movement.

8.4.7 Experimental Setup

The TCAM accelerator is evaluated on seven data-intensive applications from existing benchmark suites, running on a model of an eight-core processor with a DDR3-1066 memory system.

Circuits. The modeling of the TCAM array and sensing circuitry uses BSIM-4 predictive technology models (PTM) [68] of NMOS and PMOS transistors at 22 nm, and circuit simulations use Cadence (Spectre). Resistances and capacitances on searchlines, matchlines, and the H-tree are modeled based on interconnect projections from ITRS [23]. All peripheral circuits (decoders, population count logic, priority index logic, and reduction network nodes) are synthesized using Cadence Encounter RTL Compiler with FreePDK at 45 nm, and results are scaled to 22 nm (relevant parameters are shown in Table 8.5). Resistive memory parameters based on ITRS projections are listed in Table 8.6.

Architecture. An eight-core system with a 1-GB TCAM DIMM is modeled using (modified) SESC simulator [54]. Energy results for the cores and the DRAM subsystem are evaluated using MCPAT [31]. Details of the experiments are shown in Table 8.7.

Applications. Evaluated applications are extracted from MiBench [19], Phoenix [67], NU-MineBench [45], and SPEC2000 [26] benchmark suites. Table 8.8 presents a summary of each benchmark, as well as a description of how it is adapted to TCAM. Aside from BitCount and Vortex—which are sequential applications—all baselines are simulated with eight threads. TCAM-accelerated versions of all codes use one thread.

8.4.8 Evaluation

This section first evaluates the contribution of different hardware structures to search energy, search delay, and overall area. Second, performance and energy improvements of a single-threaded, TCAM-accelerated version of each application over a baseline parallel execution with eight threads are presented.

8.4.8.1 TCAM Delay, Energy, and Area: Where are the Bottlenecks?

Figure 8.15 shows the breakdown of search delay, search energy, and die area over the reduction network, local population count/priority logic, and array search operations

Table 8.5 Technology parameters [23, 68]

Technology (nm)	Voltage (V)	FO4 delay (ps)
45	1.1	20.25
22	0.83	11.75

Table 8.6 Resistive memory parameters [23]

Technology	R_{HI} (KΩ)	R_{LO} (KΩ)	Cell size (F^2)	Write current (μA)
PCM	1024	15	8.4	48
STT-MRAM	12.5	5	8	35

Table 8.7 Core parameters

Core	8 cores, 4.0 GHz, 4-issue
Functional units	Int/FP/Ld/St/Br units 2/2/2/2/2, Int/FP Mult 1/1
IQ, LSQ, ROB size	IssueQ 32, LoadQ/StoreQ 24/24, ROB 96
Physical registers	Int/FP 96/96
Branch predictor	Hybrid, local/global/meta 2 K/2 K/8 K, 512-entry direct-mapped BTB, 32-entry RAS
IL1 cache (per core)	32 KB, direct-mapped, 32 B block, 2-cycle hit time
DL1 cache (per core)	32 KB, 4-way, LRU, 32 B block, hit/miss delay 3/3, MESI protocol
L2 cache (shared)	4 MB, 8-way, LRU, 64B block, 24-cycle hit time
Memory controller	4-channel, 64-entry queue, FR-FCFS, page interleaving
DRAM Subsystem	DDR3-1066 MHz
Timing (DRAM cycles) [42]	tRCD: 7, tCL: 7, tWL: 6, tCCD: 4, tWTR: 4, tWR: 8, tRTP: 4, tRP: 7, tRRD: 4, tRAS: 20, tRC: 27, tBURST: 4, tFAW: 20

Table 8.8 Evaluated applications

Benchmarks	Input	Description	TCAM content	TCAM search keys
Apriori [1]	95,554 transactions, 1000 items, 2.7 MB	Association rule mining	Transaction database	Candidate itemsets
BitCount [19]	75,000 array size, 64 bits per element	Non-recursive bit count by bytes	Integer data array	N-bit vectors with a single one and N-1 wildcards
Histogram [67]	34,843,392 pixels, 104 MB	Pixel value distribution in bitmap image	Pixel values	Distinct RGB values
ReverseIndex	78,355 files, 14,025 folders, 1.01GB	Extract links and compile an index from links to files	URLs	URLs
StringMatch	50 MB non-encrypted file and non-encrypted keys	String search on encrypted file	Encrypted strings	Encrypted keys
Vortex [26]	3 inter-related databases, 200 MB	Insert, delete, and lookup operations	Unique item ID and valid bit	Unique item ID
WordCount	10 MB text file	Count frequencies of distinct words	English words	Distinct keywords

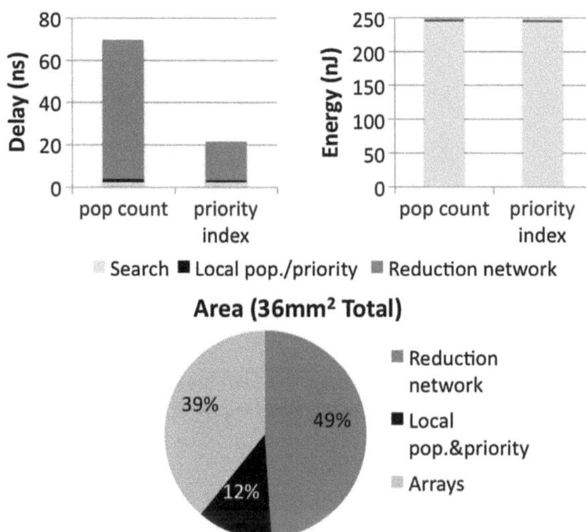

Fig. 8.15 Delay, energy, and area breakdown over main components of a 1-Gb TCAM chip

for a 1-Gb TCAM chip. Delay and energy results correspond to a chip-wide search with a 128-bit key; in each case, results are reported for both population count and priority index configurations of the system.

Overall delay is dominated by the delay of the reduction network in both priority index and population count configurations. This is because each node of the reduction network, which is implemented as a quad tree, depends on its children. To improve area efficiency, a reduction network node implements population count using an accumulator that iteratively sums the counts sent from four lower-level nodes; this adds three clock cycles per network node. As a result, obtaining the global population count takes over 3× longer than the priority index.

Although searching the array contributes a small fraction of the overall delay, it almost entirely dominates energy consumption. This is because all matchlines are activated with every search operation. Since the array search—which is needed for priority and population count configurations—dominates energy, both configurations of the system are equally energy-hungry.

The area efficiency of TCAM is 39 %, which is competitive but less than the projected DRAM area efficiency (50 % [23]) at 22 nm. Nearly half the area is devoted to the reduction network, which is responsible for distributing the 128-bit key to all arrays, and aggregating results.

8.4.8.2 System Performance and Energy

Figure 8.16 shows performance and energy evaluations on seven data-intensive benchmarks. The TCAM accelerator achieves significant performance improve-

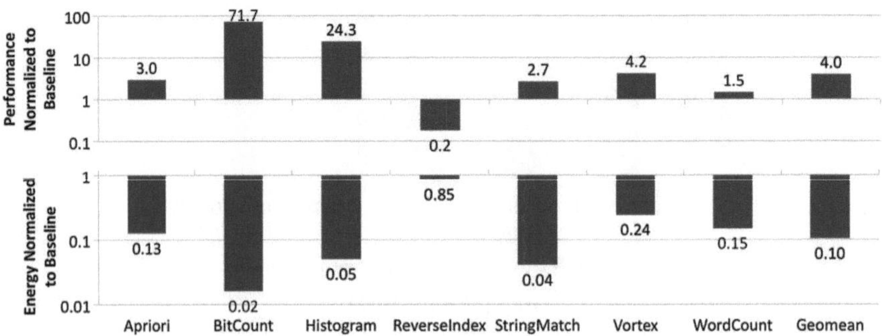

Fig. 8.16 TCAM performance (*top*) and energy (*bottom*) normalized to the RAM-based baseline. All baseline applications except BitCount and Vortex (*sequential codes*) execute with eight threads

ment over the baseline multicore system on six of the seven benchmarks (except ReverseIndex), with an average speedup of 4×. Highest speedups are obtained on BitCount (71.7×) and Histogram (24.3×), where over 99 % of the baseline runtime is in search and comparisons—operations amenable to TCAM acceleration. ReverseIndex performs considerable preprocessing, whereby a set of HTML files are parsed to extract links to others files. This portion of the application cannot leverage the TCAM accelerator; we have measured the maximum theoretical (i.e., Amdahl's Law) speedup on a single-threaded version of this benchmark, finding it to be only 1.06×. The TCAM accelerator achieves a 1.04× speedup over this sequential version, but is nearly 5× slower than the parallel version of the benchmark which uses eight threads (recall that the TCAM-enabled version of the benchmarks are single-threaded).

Figure 8.16 also shows overall system energy with the TCAM accelerator, normalized to the eight-core baseline. Six of the seven benchmarks achieve significant energy reduction, with an overall average of 10×. The efficiency gains are due to two factors. First, TCAM eliminates off-chip data movement and instruction processing overheads by processing data directly on the chip; second, the faster execution time leads to lower leakage energy. As one would expect, energy savings on ReverseIndex are lower than other applications due to the limited applicability of TCAM in this benchmark.

8.5 Related Work

This section presents related work on resistive memories, TCAM, processing in memory, and programming languages.

Resistive Memories. Recent research on TCAM is seeking opportunities to utilize the low leakage power and high scalability of resistive memories to improve power efficiency. PCM-based 2T–2R (two pairs of transistors and resistors in parallel) TCAM cells have been demonstrated in prior work [12, 53]. In a patent [50], a 2T-2R PCM TCAM is proposed with two additional bit lines. Matsunaga et al. [37] propose a bit-serial 2T-2R TCAM design using MTJ devices. Xu et al. [66] propose an STT-MRAM CAM design with high sensing and search speed. Matsunaga et al. [35] later present a 6T-2R cell, which adds assist circuitry to each cell to increase the sensing margin and search width. Alibart et al. [3] propose a TCAM cell with a pair of memristors and demonstrate how layout can be done in such a way that it takes full advantage of the potential for memristor densities. Eshraghian et al. [15] evaluate four types of memristor-based CAM designs. Other proposals for improving TCAM power efficiency include a stacked 3D-TCAM design [33] and a scheme for virtualizing TCAM [6].

TCAM. TCAMs are commonly used in networking [28, 49]. Recent work applies TCAM to a wider range of applications. Panigrahy et al. use TCAM in sorting and searching [59], Goel and Gupta solve set query problems [16], Shinde et al. study similarity search and locality sensitive hashing [60]. Hashimi and Lipasti [20] propose a TCAM accelerator as a functional unit. Other applications of TCAM include decision tree training [24], search engines [11], spell checking [19], sequential pattern mining [30], packet classification [28], IP routing [49], parametric curve extraction [41], Hough transform [44], Huffman encoding [34], Lempel-Ziv compression [65], image coding [48], and logic minimization [2].

Processing in memory. Processing in memory has been proposed to reduce memory bandwidth demand in prior work. Elliott et al. [14] build a computational RAM, which adds processing elements (PE) of a SIMD machine directly to the sense amplifiers of a 4 Mb DRAM chip. Gokhale et al. [17] propose the processor-in-memory (PIM) chip. PIM can be configured in conventional memory mode or in SIMD mode to speedup massively parallel SIMD applications. Oskin et al. [46] propose Active Pages, which adds reconfigurable logic elements to each DRAM subarray to process data.

Programming Language Support. Potter [51] proposes ASC—an associative programming paradigm that virtually maps any RAM-based algorithm to an associative computing framework. Structured data such as matrix, stack, queue, tree, and graph are implemented with two-dimension tables. ASC has been applied to the STARAN computer system [5, 10].

8.6 Conclusion

Associative computing with TCAMs is arguably an effective solution to curb both power dissipation and off-chip bandwidth challenges in modern microprocessor performance scaling. This chapter reviews existing TCAM designs using CMOS and resistive memory technologies and further focuses on a case study on recent work which aims at cost-effective, modular integration of a high-capacity TCAM system within a general-purpose computing platform. High-capacity resistive TCAM chips are placed on a DDR3-compatible DIMM and are accessed through a user-level software library with zero modifications to the processor or the motherboard. The modularity of the resulting memory system allows TCAM to be selectively included in systems running workloads that are amenable to TCAM-based acceleration; moreover, when executing an application or a program phase that does not benefit from associative search capability, the TCAM DIMM can be configured to provide ordinary RAM functionality. On a set of seven data-intensive applications that span such diverse application domains as data mining, relational database management, text parsing, and image processing, the proposed system delivers an average performance improvement of 4× and an average energy reduction of 10× over a general-purpose, eight-core system.

We believe that the resistive TCAM is part of a larger trend toward leveraging resistive memory technologies in designing systems with qualitatively new capabilities. With technology scaling, power and bandwidth restrictions will likely become more stringent while resistive memories move to mainstream, making such systems increasingly appealing and viable.

References

1. Agrawal, R.& Srikant, R. (1994). Fast algorithmsfor mining association rules. In *Proceedings of the 20thvery large databases conference*, Santioago de Chile, Chile, Sept1994.
2. Ahmad, S., & Mahapatra, R. (2007). An efficientapproach to on-chip logic minimization. *IEEE Transactions onVery Large Scale Integration (VLSI) Systems, 15*(9),1040–1050.
3. Alibart, F., Sherwood, T., & Strukov,D. (2011) Hybrid CMOS/nanodevice circuits for high throughputpattern matching applications. In *NASA/ESA Conference onAdaptive Hardware and Systems (AHS)*, June 2011.
4. Arsovski, I., Chandler, T., & Sheikholeslami, A. (2003). A ternary content-addressable memory TCAM based on 4T static storage and including a current-race sensing scheme. *Journal of Solid-State Circuits, 38*(1), 155–158.
5. Batcher, K. E. (1974). Staran parallel processor system hardware. AFIPS: In *Proceedings of the national computer conference and exposition.*
6. Bhattacharya, S., & Gopinath, K.(2011). Virtually cool ternary content addressable memory. In*Proceedings of the 13th USENIX conference on hot topics in OS*, May 2011.
7. Chang, Y. -K., Tsai, M. -L., & Su, C. -C.(2008). Improved TCAM-based pre-filtering for network intrusion detection systems. In *22nd International Conference on Advanced Information Networking and Applications*, Mar 2008.
8. Cho, S., & Lee, H. (2009).Flip-N-Write: A simple deterministic technique to improve PRAM write performance, energy and endurance. In *International Symposium on Microarchitecture*, New York.
9. Chung, H., et al. (2011). A 58nm 1.8V 1Gb PRAM with 6.4MB/s program BW. In *IEEE International solid-state circuits conference digest of technical papers*, Feb 2011.
10. Davis, E. W. (1974). Staran parallel processor systemsoftware. In *Proceedings of the May 6–10. (1974). nationalcomputer conference and exposition*. AFIPS. Vol 74.
11. de Kretser, O., & Moffat, A. (2000) Needles andHaystacks: a search engine for personal information collections. In*23rd Australasian Computer Science Conference*, 2000.
12. Derharcobian, N.,& Murphy, C. N. (2010).Phase-change memory (PCM) based universal content-addressable memory(CAM) configured as binary/ternary CAM. United States Patent US7,675,765 B2, Agate Logic, Inc., Mar 2010.

13. Driskill-Smith, A. (2010). (Grandis, Inc). Latest advances and future prospects of STT-MRAM. In Non-Volatile Memories Workshop, University of California, San Diego, Apr 2010.
14. Elliott, D., Snelgrove, W., & Stumm, M.(1992). Computational ram: A memory-SIMD hybrid and its applicationto DSP. In *Proceedings of the IEEE custom integratedcircuits conference*, pp.30.6.1–30.6.4, May 1992.
15. Eshraghian, K., et al. (2011). Memristor MOScontent addressable memory (MCAM): Hybrid architecture for futurehigh performance search engines. *IEEE Transactions on VeryLarge Scale Integration (VLSI) Systems, 19*(8), 1407–1417.
16. Goel, A., & Gupta, P. (2010).Small subset queries and bloom filters using ternary associativememories, with applications. SIGMETRICS: In *Proceedings ofthe ACM SIGMETRICS international conference on measurement andmodeling of computer systems.*
17. Gokhale, M., et al. (1995). Processing in memory: the terasys massively parallel PIM array. *Computer, 28*(4), 23–31.
18. Guo, Q., Guo, X., Bai, Y., & İpek, E. (2011). Aresistive tcam accelerator for data-intensive computing. In*Proceedings of the 44th Annual IEEE/ACM InternationalSymposium on Microarchitecture, MICRO-44 '11*, ACM, pp.339–350.
19. Guthaus, M. R., et al. (2001). MiBench: A free, commerciallyrepresentative embedded benchmark suite. In *IEEE 4th annualworkshop on workload characterization*, Austin, TX, Dec 2001.
20. Hashmi, A., & Lipasti, M. (2008).Accelerating search and recognition with a TCAM functional unit. In*IEEE International conference on computer design*. Oct 2008.
21. Intel Corporation. (2011). Intel 64 and IA-32 achitectures software developer's manual volume 3A: System programming guide, Part 1, May 2011. http://www.intel.com/products/processor/manuals
22. Ipek, E., et al. (2010). Dynamicallyreplicated memory: Building reliable systems from nanoscaleresistive memories. In *International conference onarchitectural support for programming languages and operatingsystems*. Mar 2010
23. ITRS. International Technology Roadmap for Semiconductors: 2010 Update. (2010). http://www.itrs.net/links/2010itrs/home2010.htm
24. Joshi, M., Karypis, G., & Kumar, V. (1998).ScalParC: A new scalable and efficient parallel classificationalgorithm for mining large datasets. In *Symposium:Proceedings of the international parallel processing*. pp. 573–579.
25. Kasai, G., et al. (2003). 200MHz/200MSPS 3.2W at 1.5VVdd, 9.4Mbits ternary CAM with new charge injection match detectcircuits and bank selection scheme. In *Proceedings of theIEEE custom integrated circuits conference*, pp. 387–390, Sept2003.
26. KleinOsowski, A., & Lilja, D. J. (2002). MinneSPEC: A new SPEC benchmark workload for simulation-based computer architecture research. *Computer Architecture Letters, 1*(1), 7–7.
27. Kohonen, T. (1980). *Content-addressable memories*. Berlin: Springer.
28. Lakshminarayanan, K.,Rangarajan, A., & Venkatachary, S. (2005). Algorithms for advanced-packet classification with ternary CAMs. In *Proceedings ofthe conference on applications, technologies, architectures, andprotocols for computer communications, SIGCOMM*, pp. 193–204.
29. Lee, B., et al. (2009). Architectingphase-change memory as a scalable DRAM alternative. In*International symposium on computer architecture*, Austin,TX, June 2009.
30. Leleu, M., Rigotti, C., & FrançoisBoulicaut, J. (2003). GO-SPADE: Mining sequential patterns overdatasets with consecutive repetitions. In *Proceedings ofinternational conference on machine learning and data mining inpattern recognition (MLDM)*, pp. 293–306, 2003.
31. Li, S., et al. (2009). McPAT: An integratedpower, area, and timing modeling framework for multicore andmanycore architectures. Architecture: In *Internationalsymposium on computer*, pp. 469–480.
32. Lin, C. J., et al. (2009). 45nm lowpower CMOS logic compatible embedded STT-MRAM utilizing areverse-connection 1T/1MTJ cell. In *IEEE electron devicesmeeting*, Dec 2009.
33. Lin, M., Luo, J., & Ma, Y. (2008). A low-powermonolithically stacked 3D-TCAM. In *IEEE internationalsymposium on circuits and systems*, pp. 3318–3321, May 2008.

34. Liu, L. -Y., et al. (1994).CAM-based VLSI architectures for dynamic huffman coding. Digest ofTechnical Papers. In *IEEE international conference onconsumer electronics*.
35. Matsunaga, S., et al. (2011). Fully parallel6T–2MTJ nonvolatile TCAM with single-transistor-based selfmatch-line discharge control. In *2011 Symposium on VLSICircuits (VLSIC)*, June 2011.
36. Matsunaga, S., Katsumata, A., Natsui, M.,Fukami, S., Endoh, T., Ohno, H., & Hanyu, T. (2011). Fully parallel6t–2mtj nonvolatile tcam with single-transistor-based selfmatch-line discharge control. In *VLSI Circuits (VLSIC), 2011Symposium on*, pp. 298–299, June 2011.
37. Matsunaga, S., et al. (2009). Standby-power-free compact ternary content-addressable memory cell chip using magnetic tunnel junction devices. *Applied Physics Express*, *2*(2), 23004.
38. Matsunaga, S., Hiyama, K., Matsumoto, A., Ikeda, S., Hasegawa, H., Miura, K., et al. (2009). Standby-power-free compact ternary content-addressable memory cell chip using magnetic tunnel junction devices. *Appl Physics Express*, *2*(2), 23004.
39. Mcauley, A. J., & Francis, P. (1993).Fast routing table lookup using CAMs. In *Twelfth annualjoint conference of the IEEE computer and communications societiesProceedings*.
40. McGeer, R., & Yalagandula, P. (2009). Minimizingrulesets for TCAM implementation. In *Proceedings. The 28thconference on, computer communications*, Apr 2009.
41. Meribout, M., Ogura, T., & Nakanishi, M. (2000). On using the CAM concept for parametric curve extraction. *IEEE Transactions on Image Processing*, *9*(12), 2126–2130.
42. Micron Technology, (2006). Inc., http://www.micron.com//get-document/?documentId=425. 1Gb DDR3 SDRAM, 2006.
43. Micron Technology, Inc. http://www.micron.com/products/pcm
44. Nakanishi, M., & Ogura, T. (1996). A real-timeCAM-based hough transform algorithm and its performance evaluation.In *Proceedings of the 13th International Conference on,Pattern Recognition*, Aug 1996.
45. Narayanan, R., et al. (2006).Minebench: A benchmark suite for data mining workloads. In*IEEE international symposium on workload characterization*.Oct 2006.
46. Oskin, M., Chong, F., & Sherwood, T. (1998). Activepages: a computation model for intelligent memory. In *The25th, annual international symposium on computer architecture*,1998.
47. Pagiamtzis, K., & Sheikholeslami, A. (2006). Content addressable memory (CAM) circuits and architectures: A tutorial and survey. *IEEE Journal of Solid State Circuits*, *41*(3), 712–727.
48. Panchanathan, S., & Goldberg, M. A.(1989). Content-addressable memory architecture for image codingusing vector quantization. *IEEE Transactions on SignalProcessing*, *39*(9), 2066–2078.
49. Pei, T. -B., & Zukowski, C. (1991).VLSI implementation of routing tables: tries and CAMs. In*Tenth annual joint conference of the IEEE computer andcommunications societies proceedings*.
50. Phase-change based TCAM architectures and cell. http://ip.com/IPCOM/000187746.
51. Potter, J., Baker, J., Scott, S., Bansal, A., Leangsuksun, C., & Asthagiri, C. (1994). Asc: an associative-computing paradigm. *Computer*, *27*(11), 19–25.
52. Qureshi, M. K., et al. (2009). Enhancing lifetimeand security of PCM-based main memory with start-gap wear leveling.In *Proceedings of the 42nd annual IEEE/ACM internationalsymposium on microarchitecture*, 2009.
53. Rajendran, B., et al. (2011). Demonstration of CAMand TCAM using phase change devices. In *3rd IEEEinternational memory, workshop*, May 2011.
54. Renau, J., et al. (2005). SESC simulator, Jan 2005. http://sesc.sourceforge.net.
55. Rixner, S., et al. (2000). Memory access scheduling.In *Proceedings of the 27th, annual international symposiumon computer architecture*, May 2000.
56. Rudolph, J. A. (1972). A production implementation ofan associative array processor: Staran. In *Proceedings ofthe December 5–7, 1972, fall joint computer conference, part I,AFIPS '72 (Fall, part I)*, (pp. 229–241), ACM, New York, USA, 1972.
57. Schechter, S., Loh, G. H., Straus, K., &Burger, D. (2010). Use ecp, not ecc, for hard failures in resistivememories. In *Proceedings of the 37th, annual internationalsymposium on computer architecture*, 2010.

58. Servalli, G. (2009). A 45 nm generation phasechange memory technology. In *2009 IEEE internationalelectron devices meeting*, 2009.
59. Sharma, S., & Panigrahy, R. (2002). Sorting andsearching using ternary CAMs. In *Proceedings 10th symposiumon high performance interconnects*, Dec 2002.
60. Shinde, R., et al. (2010). Similarity search andlocality sensitive hashing using ternary content addressablememories. In *Proceedings of the 2010 internationalconference on management of data*, Jun 2010.
61. Slaughter, J. M. (2009). Materials for magnetoresistive random access memory. *Annual Review of Materials Research*, *39*, 277–296.
62. Stevens, R. W., & Rago, S. A. (2005). *Advanced programming in the UNIX(R) environment* (2nd ed.). Boston: Addison-Wesley Professional.
63. Tsuchida, K., et al. (2010). A 64Mb MRAM withclamped-reference and adequate-reference schemes. In*Proceedings of the IEEE international solid-state circuitsconference*, 2010.
64. Wade, J., & Sodini, C. (1985). Dynamic cross-coupledbitline content addressable memory cell for high density arrays. In*International electron devices meeting*, 1985.
65. Wei, B., et al. (1993). A single chip lempel-ziv datacompressor. In *IEEE international symposium on circuits andsystems*, 1993.
66. Xu, W., Zhang, T., & Chen, Y. (2010). Design of spin-torque transfer magnetoresistive RAM and CAM/TCAM with high sensing and search speed. *IEEE Transactions on Very Large Scale Integration Systems*, *18*(1), 66–74.
67. Yoo, R. M., Romano, A., & Kozyrakis, C.(2009). Phoenix rebirth: Scalable MapReduce on a large-scaleshared-memory system. In *Proceedings of the 2009 IEEEinternational symposium on workload characterization*, 2009.
68. Zhao, W., & Cao, Y. (2006). New generation of predictivetechnology model for sub-45 nm design exploration. In*International symposium on quality, electronic design*,2006.
69. Zhao, W., Chappert, C., & Mazoyer, P. (2009). Spin transfer torque (STT) MRAM-based runtime reconfiguration FPGA circuit. *ACM Transactions on Embedded Computing Systems*, *9*(2), 14:1–14:16.

58. Servalli, G. (2009). A 45nm generation phase change memory technology. In: IEEE International Electron Devices Meeting, 2009.
59. Sharma, S., & Panigrahy, R. (2002). Sorting and searching using ternary CAMs. In: Proc. [illegible] High performance interconnects, Dec 2002.
60. Shinde, R., et al. (2010). Similarity search and locality sensitive hashing using ternary content addressable memories. In: Proceedings of the 2010 international conference on Management of data. ACM, 2010.
61. [illegible]
62. [illegible]
63. [illegible]
64. [illegible]
65. [illegible] design exploration. In: International symposium on quality electronic design, 2008.
66. Zhao, W., Chappert, C., & Mazoyer, P. (2009). Spin transfer torque (STT)-MRAM-based [illegible] In: IEEE Transactions on Computer-Aided Design of Integrated Circuits and Systems, 28(11) [illegible]

Chapter 9
Wear-Leveling Techniques for Nonvolatile Memories

Jue Wang, Xiangyu Dong, Yuan Xie and Norman P. Jouppi

Abstract Nonvolatile memories (NVMs) are promising technologies for replacing SRAM or eDRAM in low-level on-chip caches and main memories because they can save standby power and provide high cache capacity. However, limited write endurance is a common problem for NVM technologies. The current memory management policies are not write variation aware and result in significant nonuniformity in terms of writing to memory blocks, which would cause heavily written nonvolatile cache blocks to fail much earlier than most other blocks. Thus, wear-leveling techniques are important for NVM-based memory system to balance write traffic and extend the system lifetime. Some wear-leveling techniques have been proposed for NVM-based main memories and on-chip caches. In this chapter, we use inter-/intra set write variation aware cache policy (i^{2}WAP) as an example to show how to design an endurance-aware management policy for nonvolatile caches. i^{2}WAP has two features: first, Swap-Shift (SwS), an enhancement based on previous main memory wear leveling to reduce cache inter-set write variations; second, probabilistic set line flush (PoLF), a novel technique to reduce cache intra-set write variations. Implementing i^{2}WAP only needs two global counters and two global registers. By adopting i^{2}WAP, we can improve the lifetime of on-chip nonvolatile caches by 75 % on average and up to 224 %.

J. Wang · Y. Xie (✉)
Pennsylvania State University, Pennsylvania, US
e-mail: yuanxie@cse.psu.edu

J. Wang
e-mail: jzw175@cse.psu.edu

X. Dong
Qualcomm Technology, Inc, San Diego, US
e-mail: xiangyud@qualcomm.com

N. P. Jouppi
Google, Inc., CA, USA
e-mail: jouppi@acm.org

Y. Xie (ed.), *Emerging Memory Technologies*,
DOI: 10.1007/978-1-4419-9551-3_9,

9.1 Overview of Wear-Leveling Techniques

The scaling of SRAM and DRAM is increasingly constrained by technology limitations such as leakage power and cell density. Recently, some emerging nonvolatile memory (NVM) technologies, such as phase-change RAM (PCM or PCRAM), spin-torque transfer RAM (STTRAM or MRAM), and resistive RAM (ReRAM), have been explored as alternative memory technologies. Compared to SRAM and eDRAM, these nonvolatile memory technologies have advantages in high density, low standby power, better scalability, and nonvolatility.

However, adoption of NVM at the main memory or on-chip cache level has a write endurance issue. For example, PCM is only expected to sustain 10^8 writes before experiencing frequent stuck-at-1 or stuck-at-0 errors [6, 12, 14–16, 21]. For ReRAM, the write endurance bar is much improved but is still around 10^{11} [9]. For STTRAM, although a prediction of up to 10^{15} write cycles is often cited, the best endurance test result for STTRAM devices so far is less than 4×10^{12} cycles [4]. Although the absolute write endurance values for NVMs seem sufficiently high for use, the actual problem is that the current cache management policies are not write variation aware. These polices were originally designed for SRAM/DRAM and result in significant nonuniformity in terms of writing to memory blocks, which would cause heavily written NVM blocks to fail much earlier than most other blocks.

There are various previous architectural proposals to extend NVM lifetimes. These work can be classified by two basic types of techniques: The first category is wear-leveling technique which focuses on evenly distributing unbalanced write frequencies to all memory lines. Zhou et al. [22] proposed a row shifting and a segment swapping policy for PCM main memory. Their key idea is to periodically shift the memory row and swap the memory segments to even out the write accesses. Qureshi et al. [13] proposed fine-grain wear leveling (FGWL) and start-gap wear leveling [12] to shift cache lines within a page to achieve uniform wear out of all lines in the page. Seong et al. [15] addressed potential attacks by applying security refresh. They used a dynamic randomized address mapping scheme that swaps data using random keys upon each refresh due. There are also some other work for NVM caches. Joo et al. [7] extended wear-leveling techniques for main memory and proposed the bit-line shifting and word-line remapping techniques for nonvolatile caches. Wang et al. [20] proposed i^2WAP to minimize both inter- and intra-set write variations, and in this chapter we will give detail description of the method proposed in [20]

The second category is about error corrections. The conventional error correction code (ECC) is the most common technique in this category. Dynamically replicated memory [6] reuses memory pages containing hard faults by dynamically forming pairs of pages that act as a single one. Error correction pointer (ECP) [14] corrects failed bits in a memory line by recording the position and its correct value. Seong et al. [16] propose SAFER to efficiently partition a faulty line into groups and correct the error in the group. FREE-p [21] proposed an efficient means of implementing line sparing. These architectural techniques add different types of data redundancy to solve the access errors caused by limited write endurance.

It should be noted that all the techniques from the second category are orthogonal to the wear-leveling techniques from the first category. And the total system lifetime could be extended further after combining them together.

9.2 Background

Write endurance is defined as the number of times a memory cell can be overwritten, and emerging NVMs commonly have a limited write endurance. In this section, the failure mechanisms of different NVMs are introduced.

9.2.1 PCM Failure Mechanism

PCM devices have two major failure modes: *stuck-RESET* and *stuck-SET* [8]. Stuck-RESET failures are caused by void formation or delamination that catastrophically disconnects the electrical path between GST (i.e., the storage element of PCM) and access device. Instead, stuck-SET failures are caused by the aging of GST material that becomes more reluctant to create an amorphous (RESET) phase after continuously experiencing write cycles, resulting in a degradation of the PCM RESET-to-SET resistance ratio. Both of these failure modes are commonly observed in PCM devices. ITRS [5] projects that the average PCM write endurance is in a range between 10^7 and 10^8.

9.2.2 ReRAM Failure Mechanism

There are several different types of endurance failure mechanisms are observed in ReRAM devices. One of the them is anode oxidation-induced interface reaction [3]. High temperature, large current/power process and oxygen ions produced during forming/SET process cause the oxidation at the anode–electrode interface. Another endurance failure mechanism is extra vacancy attributed reset failure effect [3]. Electric field-induced extra oxygen vacancy generation during switching may increase the filament size or make the filament rougher, accompanying with the reduced resistance in high-resistance states (R_{HRS}) and resistance in low-resistance states (R_{LRS}) as well as the increased reset voltage. Recent ReRAM prototypes demonstrate the best write endurance ranging from 10^{10} [11] to 10^{11} [9].

9.2.3 STTRAM Failure Mechanism

STTRAM uses magnetic tunnel junction (MTJ) as the storage element, and there are two challenges in controlling the MTJ reliability: time-dependent dielectric breakdown (TDDB) and resistance drift [18]. TDDB is detected as an abrupt increase in junction current owing to a short forming through the tunneling barrier. Resistance drift is a gradual reduction in the junction resistance over time that can eventually lead to reduced read margin. The best endurance test of STTRAM so far is less than 4×10^{12} [4].

9.3 Improving Nonvolatile Cache Lifetime by Reducing Inter- and Intra-set Write Variations

The write variation brought by normal memory management policy would cause heavily written memory blocks to fail much earlier than their expected lifetime. Thus, eliminating memory write variation is one of the most critical problems that must be addressed before NVMs can be used to build practical and reliable memory systems.

There are many wear-leveling techniques to extend the lifetime of nonvolatile main memories and caches. In this chapter, we use inter-/intra-set Write variation-aware cache policy (i^2WAP) [20] as an example to show how to improve nonvolatile cache lifetime by reducing write variations.

The difference between cache and main memory operational mechanisms makes the wear-leveling techniques for NVM main memories inadequate for NVM caches. This is because writes to caches have *intra-set variations* in addition to *inter-set variations* while writes to main memories only have *inter-set variations*. According to our analysis, intra-set variations can be comparable to inter-set variations for some workloads. This presents a new challenge in designing wear-leveling techniques for NVM caches. In order to demonstrate how severe the problem is for NVM caches, we first do a quick experiment.

9.3.1 Definition

The objective of cache wear leveling is to reduce write variations and make write traffic uniform. To quantify the cache write variation, we first define the *coefficient of inter-set variations* (InterV) and the *coefficient of intra-set variations* (IntraV) as follows:

$$\text{InterV} = \frac{1}{W_{\text{aver}}}\sqrt{\frac{\sum_{i=1}^{N}\left(\sum_{j=1}^{M} w_{i,j}/M - W_{\text{aver}}\right)^2}{N-1}} \tag{9.1}$$

$$\text{IntraV} = \frac{1}{W_{\text{aver}} \times N} \sum_{i=1}^{N} \sqrt{\frac{\sum_{j=1}^{M} \left(w_{i,j} - \sum_{j=1}^{M} w_{i,j}/M \right)^2}{M-1}} \tag{9.2}$$

where $w_{i,j}$ is the write count of the cache line located at set i and way j, W_{aver} is the average write count defined as

$$W_{\text{aver}} = \frac{\sum_{i=1}^{N} \sum_{j=1}^{M} w_{i,j}}{NM} \tag{9.3}$$

N is the total number of cache sets, and M is the number of cache ways in one set. If $w_{i,j}$ are all the same, both InterV and IntraV are zero. In short, InterV is defined as the coefficient of variation (CoV) of the average write count within cache sets; IntraV is defined as the average of the CoV of the write counts cross a cache set.[1]

Figure 9.1 shows the experimental results of InterV and IntraV in our simulated 4-core system with 32 KB I-L1, 32 KB D-L1, 1 MB L2, and 8 MB L3 caches. The detailed simulation methodology and the setting are described in Sect. 9.8. We compare InterV and IntraV in L2 and L3 caches as we assume that emerging NVM will be first used in low-level caches. From Fig. 9.1, we make two observations:

(1) There are large inter-set write variations. The cache lines in different sets can experience totally different write frequencies because applications can have biased address residency. For instance, *streamcluster* has 189 % InterV in L2, and *swaptions* has 115 % InterV in L3. On average, InterV is 66 % in L2 and 22 % in L3.
(2) There are large intra-set write variations. If just one cache line in a set is frequently visited by cache write hits, it will absorb a large number of cache writes, and

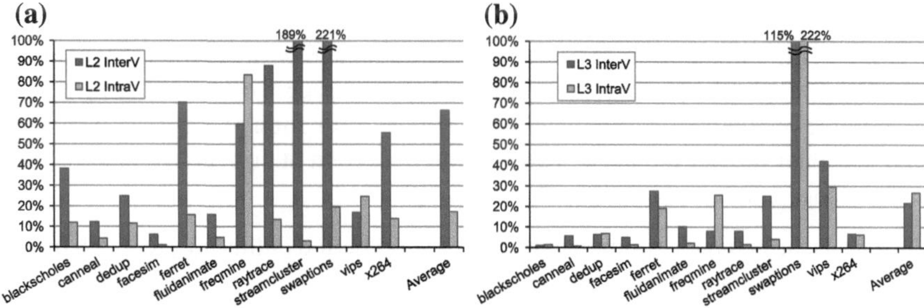

Fig. 9.1 The coefficient of variation for inter-set and intra-set write count of L2 and L3 caches in a simulated 4-core system with 32 KB I-L1, 32 KB D-L1, 1 MB L2, and 8 MB L3 caches

[1] We use average CoV instead of maximum CoV to keep the definitions of intra-set (the CoV of the averages) and inter-set variations (the average of the CoVs) symmetric.

thus the write accesses may be unevenly distributed to the remaining M-1 lines in the set (for an M-way associative cache), e.g., *freqmine* has 84 % IntraV in L2, and *swaptions* has 222 % IntraV in L3. On average, IntraV is 17 % in L2 and 27 % in L3.

We should also notice that write variations in L2 and L3 are greatly different. On average, L3 caches have smaller InterV than L2 caches do. This is because L2 caches are private for each processor but all processors share one L3 cache, and L3 has less write variance since it mixes different requests. However, it is worth noticing that compared to an L3 cache, L2 caches have a larger average write count on each cache line because they are in a higher level of the memory hierarchy and have smaller capacity. Thus, the higher variation of L2 caches makes their limited endurance a more severe problem.

In addition, our results show that IntraV is roughly the same or even larger compared to InterV for some workloads. Combining these two types of write variations together significantly shorten the NVM cache lifetime.

9.4 Cache Lifetime Metrics

Cache lifetime can be defined in two different ways: *raw lifetime* and *error-tolerant lifetime*. We define the raw lifetime by the first failure of a cache line without considering any error recovery effort. On the other hand, we can extend the raw lifetime by using error correction techniques and paying overhead in either memory performance or memory capacity [6, 14, 16, 21], and we call it the error-tolerant lifetime. In this work, we focus on how to improve the raw lifetime at first as it is the base of the error-tolerant lifetime. Later, we discuss the error-tolerant lifetime in Sect. 9.10.3.

The target of maximizing the cache raw lifetime is equivalent to minimizing the worst-case write count to a cache line. However, it is impractical to obtain the worst-case write count throughout the whole product lifetime which might span several years. Instead, in this work, we model the raw lifetime by using three parameters: *average write count*, *inter-set write count variation*, and *intra-set write count variation*. We first sample the statistics of the cache write access behavior during a short period time of simulation, calculate the statistical estimation of these three parameters, and then predict the cache raw lifetime based on statistical estimations. Our methodology can be explained in details as follows:

(1) The cache behavior is simulated during a short period of time t_{sim} (e.g., 10 billion instructions on a 3 GHz CPU).
(2) Each cache line write count is collected to get a average write count W_{aver}. Also, we calculate InterV and IntraV according to Eqs. 9.1 and 9.2.

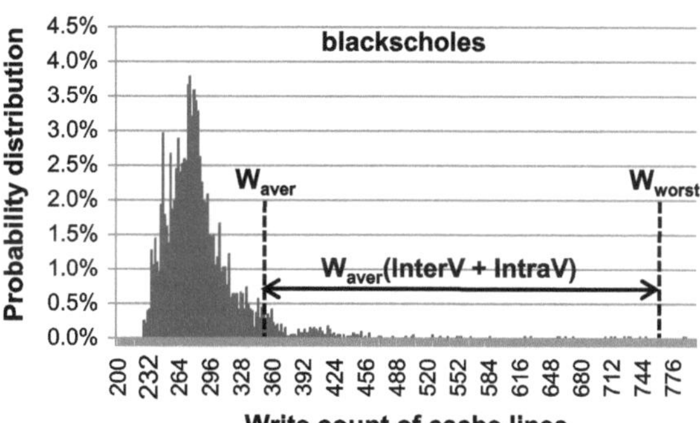

Fig. 9.2 The L2 cache write count probability distribution function (PDF) of blackscholes

(3) Assuming the total write variation of a cache line is the summation of its inter- and intra-set variations,[2] we then have $W_{\text{var}} = W_{\text{aver}} \times (\text{InterV} + \text{IntraV})$.
(4) The worst-case write count is predicted as $W_{\text{aver}} + W_{\text{var}}$ to make sure to cover the vast majority of cases. While this approach is approximate, Fig. 9.2 validates the feasibility of this model.

Assuming the general characteristics of cache write operations for one application do not change with time,[3] the lifetime of the system can be defined as

$$t_{\text{total}} = \frac{W_{\max} \times t_{\text{sim}}}{W_{\text{aver}} + W_{\text{var}}} = \frac{W_{\max} \times t_{\text{sim}}}{W_{\text{aver}}(1 + \text{InterV} + \text{IntraV})} \tag{9.4}$$

Thus, the lifetime improvement (LI) of a cache wear-leveling technique can be expressed as

$$\text{LI} = \frac{W_{\text{aver\_base}}(1 + \text{InterV}_{\text{base}} + \text{IntraV}_{\text{base}})}{W_{\text{aver\_imp}}(1 + \text{InterV}_{\text{imp}} + \text{IntraV}_{\text{imp}})} - 1 \tag{9.5}$$

The objective of cache wear leveling is to increase LI. It needs to reduce inter-set and intra-set variations while not significantly increasing the average write count.

We apply the state-of-the-art least recently used (LRU) cache management policy as the baseline, and Fig. 9.3 shows how bad the baseline is in terms of write variation and cache raw lifetime. Compared to the ideal case where all the write traffic is evenly distributed, some workloads (e.g., *swaptions*) can shorten the cache raw lifetime by 20–30 %. If this is the case of future NVM caches, the system will eventually fail even when most of the NVM cells are still healthy. Therefore, it is extremely critical to design a write variation-aware cache management for the future success of NVM caches, and we need a cache policy that can tackle both inter-set and intra-set write variations.

---

[2] InterV and IntraV are not independent, but the worst-case variation can be modeled as the sum of them.

[3] If system runs different applications over time, the cache write variance can be reduced. However, in this work, we only consider the worst case. It occurs in some practical cases, such as embedded applications in which the data layout could largely remain the same.

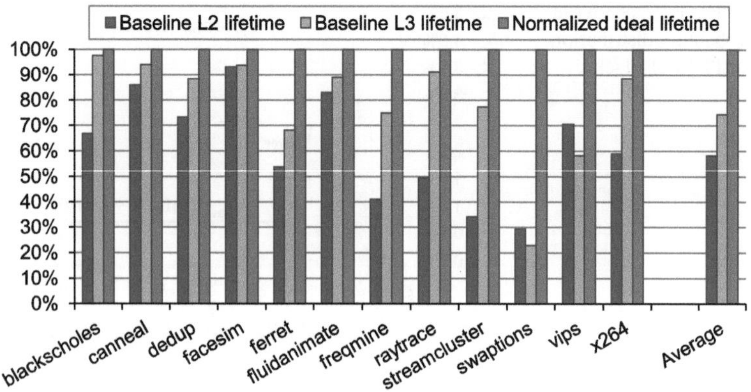

Fig. 9.3 The baseline lifetime of L2 and L3 caches normalized to the ideal lifetime (no write variations in the ideal case)

9.5 Starting from Inter-set Write Variations

9.5.1 Challenges of Inter-set Weal Leveling

Some wear-leveling techniques [12, 13, 15, 22] are focused on increasing the lifetime of NVM main memory. The key principle behind these techniques is to introduce an address remapping layer. This principle remains the same for cache inter-set wear leveling. However, there are some differences between designing wear-leveling policies for NVM main memory and NVM caches.

Using Data Movement: Main memory wear-leveling techniques usually use data movement to implement the address remapping. This is because we cannot afford to lose data in main memory and must move them to a new position after remapping. However, it is not free to move data from one location to another. First, we need temporary data storage to move the data. Second, one cache set movement involves multiple reads and writes. The cache port is blocked during the data movement, and the system performance is degraded. *Start-Gap* [12] is a recently proposed technique for main memory wear leveling. If we directly extend *Start-Gap* to handle the cache wear leveling, it falls into this category.

Using Data Invalidation: Another option to implement set address remapping for NVM caches is data invalidation. We can use cache line invalidations because we can always restore the cache data later from lower-level memories if it is not dirty. This special feature of caches provides us a new opportunity to design a low-overhead cache inter-set wear-leveling technique.

Compared to data movement, data invalidation needs no area overhead. In addition, to quantify the performance difference between these approaches, we design and simulate two systems (see Sect. 9.8 for detailed simulation settings). In the data movement system, the data in one cache set are moved to another after every 100

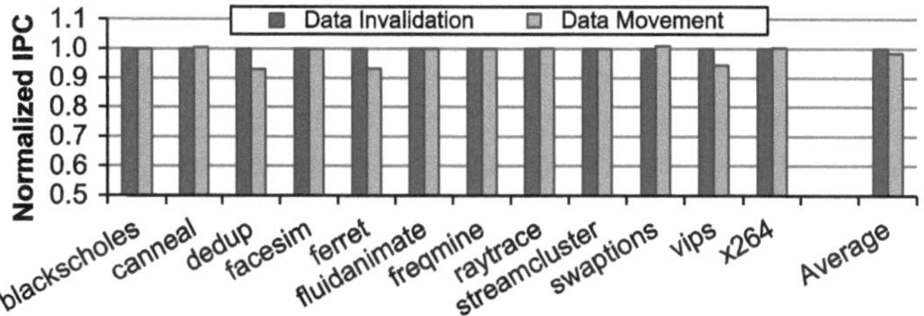

Fig. 9.4 The performance comparison between data invalidation and data movement

writes to the cache, which is extended from the start-gap technique [12]. On the other hand, the data are not moved but invalidated in the second system. Figure 9.4 shows the normalized performance of these two systems. Compared to the data invalidation system, the data movement system has worse performance (i.e., 2 % on average and up to 7 %).

For the data invalidation system, the performance overhead comes from writing back the dirty data in the cache set to main memory and restoring data from lower-level memories which should be hit later. But, we should notice that the latency of writing back and restoring data can often be hidden using write back buffer and MSHR techniques [10]. On the other hand, the time overhead of the data movement system cannot be optimized without adding hardware complexity since the data block movement in one cache needs to be done serially.

Therefore, as the first step of our work, we enhance it by the Swap-Shift (SwS) scheme to reduce the inter-set write variation in NVM caches.

9.5.2 *Swap-Shift*

Considering data invalidation is more favorable in cache inter-set wear leveling, we modify the existing main memory wear-leveling technique and devise a new technique called SwS.

9.5.2.1 SwS Architecture

The basic concept behind SwS is to shift the location of each cache set. However, shifting all the cache sets at once brings a significant performance overhead. To solve this problem, SwS only swaps the data of two neighboring sets at once, and it can eventually shift all the cache sets by one location after $N-1$ swaps, where N is the number of cache sets.

We use a global counter in SwS to store the number of cache writes, and we annotate it as Wr. We also use two registers, SwV (changing from 0 to $N-2$) and shift value (ShV) (changing from 0 to $N-1$), to track the current status of swaps and shifts, respectively. The detailed mechanism is explained as below:

Swap Round (SwapR): Every time Wr reaches a specific threshold (Swap Threshold, ST), a swap between cache set [SwV] and set [SwV+1] is triggered. Note that this swap operation only exchanges the set IDs and invalidates the data stored in these two sets (needs write-back if the data are dirty). After that, SwV is incremented by 1. One SwapR consists of $N - 1$ swaps and indicates that all the cache set IDs are shifted by 1.

Shift Round (ShiftR): ShV is incremented by 1 after each SwapR. At the same time, SwV is reset to 0. One ShiftR consists of N shifts (i.e., SwapR).

Figure 9.5 is an example of how SwS shifts the entire cache by multiple swaps. It shows the SwV and ShV values during a complete ShiftR, which consists of 4 SwapR; it also shows that one SwapR consists of 3 swaps and all cache sets are shifted by 1 after each SwapR. In addition, after one ShiftR, all cache sets are shifted to the original position and all logical set indexes are the same as the physical ones.

The performance penalty of SwS is small because only two sets are swapped at once and the swap interval period can be long enough (e.g., million cycles) by adjusting ST. The performance analysis of SwS is in Sect. 9.10.1.

9.5.2.2 SwS Implementation

The implementation of SwS is shown in Fig. 9.6. We use a global counter to store wr and two registers to store ShV and SwV, respectively. When a logical set number (LS) arrives, the physical set number (PS) can be computed based on three different situations:

(1) If LS $=$ SwV, it means that this logical set is exactly the cache set should be swapped in this ShiftR. Therefore, PS is mapped to the current ShV.
(2) If LS $>$ SwV, it means that this set has not been shifted in this ShiftR. Therefore, PS is mapped to LS $+$ ShV.
(3) If LS $<$ SwV, it means that this set has been already shifted. Therefore, PS is mapped to LS $+$ ShV $+ 1$.

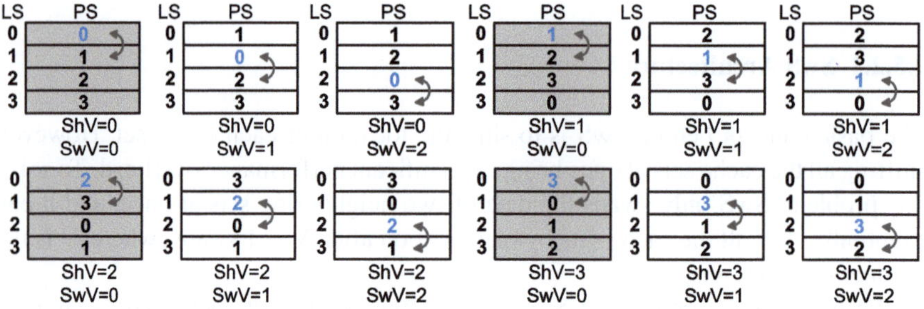

Fig. 9.5 One SwS ShiftR in a cache with 4 sets

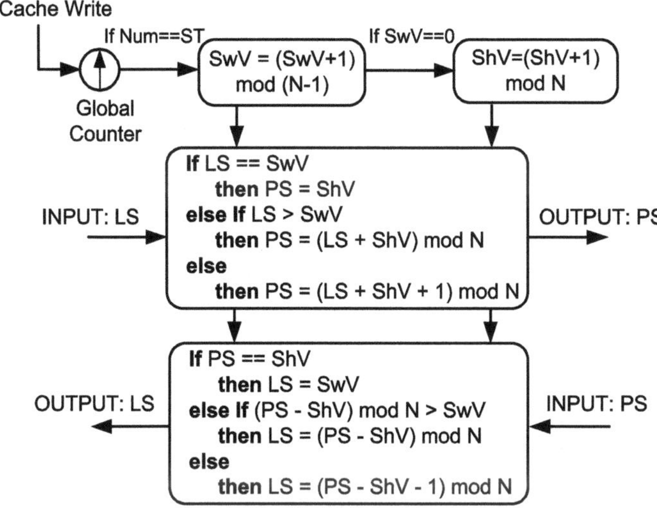

Fig. 9.6 The mapping between logical (LS) and physical set index (PS) in SwS

When a dirty cache line is written back to the next level of cache, the logical set address needs to be regenerated. The mapping from PS to LS is symmetric and is also given in Fig. 9.6. This mapping policy can be verified by the simple example in Fig. 9.5. Because SwV and ShV are changed along with cache writes, the mapping between LS and PS changes all the time. This ensures that the writes to different physical sets are balanced, reducing cache InterV.

Compared to a conventional cache architecture, the set index translation step in SwS only adds a simple arithmetic operation and can be merged into the row decoder. We synthesize the LS-to-PS address translation circuit in a 45 nm technology, and the circuit can handle a LS-to-PS translation within one cycle under a 3 GHz clock frequency.

9.6 Intra-set Variation: A More Severe Issue

SwS can distribute writes to all the cache sets, but it only reduces the inter-set write variation. Our experimental results later in Sect. 9.8.2 show that SwS alone cannot reduce intra-set variations. Therefore, in this section, we start with two straightforward techniques and then follow with a much improved technique, called PoLF, to tackle the cache intra-set variation problem.

9.6.1 Set Line Flush

Intra-set write variations are mainly caused by hot data being written more frequently than others. For example, if one cache line is the frequent target of cache write hits and absorbs a large number of writes, its corresponding set must have a highly unbalanced write distribution.

In a traditional LRU cache policy, every accessed block is marked toward most recently used (MRU) to avoid being chosen as the victim line.

As a result, the LRU policy rarely replaces the hot data since it is frequently accessed by cache write hits. This increases the write count of one block and the intra-set write variation of the corresponding set.

To solve this problem, we first consider a set line flush (LF) scheme. In LF, when there is a cache write hit, the new data are put into the write-back buffer directly instead of writing it to the hit data block, and then the cache line is marked as *INVALID*. This process is called LF. Hence, the block containing the hot data have the opportunity to be replaced by other cold data according to the LRU policy, and the hot data can be reloaded to other cache lines. We invalidate the hot data line instead of moving it to other positions due to the same performance concern explained in Sect. 9.5.1.

LF balances the write count to each cache block, but it flushes every cache write hit no matter whether it contains hot data or not. Obviously, LF causes large performance degradation as the flushed data have to be reloaded if it is hot. To reduce the performance penalty, we need to add intra-set write count statistics and only flush hot cache lines.

9.6.2 Hot Set Line Flush

One of the simplest solutions is to add counters to cache tags, storing the write count of each cache line. We call this scheme hot set line flush (HoLF). Theoretically, if the difference of the largest write count of one line and the average value of the set is beyond a predetermined threshold, we detect a very hot data and should flush it. In this way, the hottest line can be replaced with other data, and the hot data can be reloaded into a relatively cold cache line. The architecture of HoLF is shown in Fig. 9.7.

However, HoLF has significant overhead in both area and performance. The area overhead is large since it requires one counter for every cache line. Considering the typical cache line is 64-byte wide and assuming the write counter is 20-bit, the hardware overhead is more than 3.7 %. In addition, the performance is inevitably impaired because HoLF tracks both maximum and average write count values in every cache set. It is infeasible to initiate multiple arithmetic calculations for every cache write.

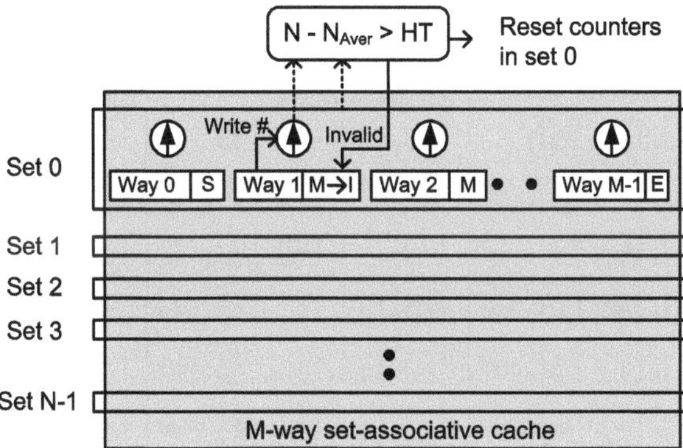

Fig. 9.7 The cache architecture of HoLF. The counters to store the write count are added to every cache line

Therefore, we do not discuss HoLF further in this chapter as it is not a practical solution. Instead, we introduce an improved solution called probabilistic set line flush (PoLF).

9.6.3 Probabilistic Set Line Flush

The motivation of PoLF is to flush hot data probabilistically instead of deterministically.

9.6.3.1 Probabilistic Invalidation

Unlike HoLF, PoLF only maintains one global counter to count the number of write hits to the entire cache, and it flushes a cache line when the counter saturates no matter whether the cache line to be flushed is hot or not. Although there is no guarantee that the hottest data would be flushed as we desire, the probability of PoLF selecting a hot data line is high: the hotter the data are, the more likely it will be selected when the global counter saturates. Theoretically, PoLF is able to flush the hottest data in a cache set most of the time, and the big advantage of PoLF is that it only requires one global counter.

Maintaining LRU: For normal LRU policy, when a cache line is invalidated, the age bits of this line is marked as LRU. However, for PoLF, because hot data are accessed more frequently, it is possible that after invalidating a single hot cache line, the same data will be reinstalled in the very same line on a subsequent miss. Thus, we modify PoLF to maintain age bits for probabilistic invalidations. When PoLF flushes a line, it does not mark it as the LRU line and its age bits are not changed. Later, a subsequent miss will invalidate the actual LRU line and reinstall the hot data in that line. The cache line evicted by PoLF remains invalid until it becomes the LRU line.

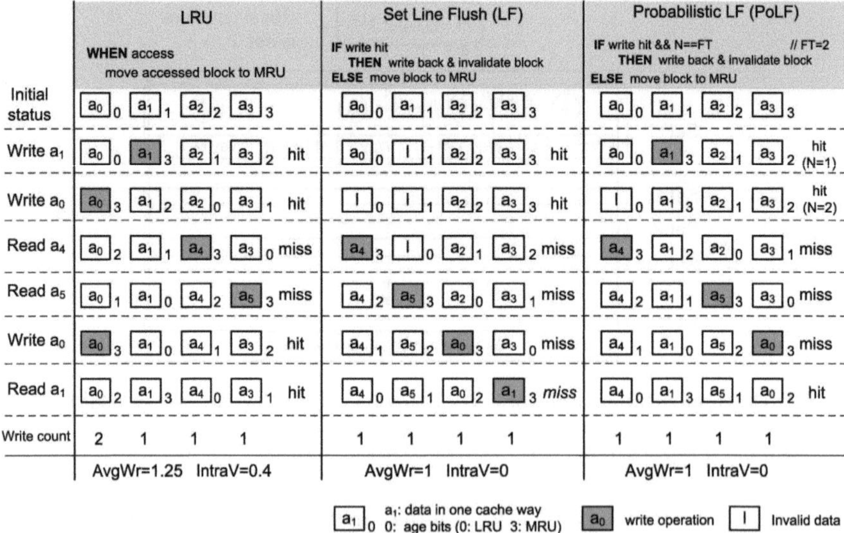

Fig. 9.8 The behavior of one cache set composed of 4 ways under LRU, LF, and PoLF polices for the same access pattern. The total write count of each cache way, the average write count and the intra-set variation are marked, respectively

Comparison with Other Policies: Figure 9.8 shows the behavior of a 4-way cache set under LRU, LF, and PoLF polices for an exemplary access pattern. Our observations from this example are:

(1) For LRU, the hot data a_0 are moved to the MRU position (age bits = 3) after each write hit and are never replaced by other data. Thus, the intra-set variation using LRU is the largest one among all the polices.
(2) For LF, each write hit causes its corresponding cache line to be flushed. The age bits are not changed during write hits. The intra-set variation is reduced compared to the LRU policy because the hot data a_0 are reloaded into another cache line. However, data a_1 are also flushed since every write hit causes one cache line flush, and it brings one additional access miss.
(3) For PoLF, we let every other write hit cause a cache line flush (i.e., line flush threshold FT = 2).[4] Compared to the LRU policy, its intra-set variation is reduced because the hot data a_0 are moved to another cache line. In addition, compared to LF, the number of misses is reduced because a_1 is not flushed.

Thus, PoLF can reduce the intra-set variation as well as ensuring the probability of selecting a hot data line is high.

[4] FT is set as 2 for this illustration. More typical values are much larger and shown in Sect. 9.8.

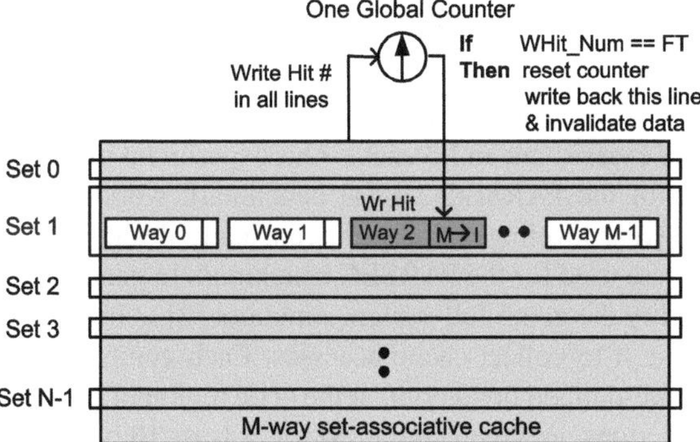

Fig. 9.9 The cache architecture of PoLF. Only one global write hit counter is added to the entire cache

9.6.3.2 PoLF Implementation

We can design the hardware implementation for PoLF as follows. First, we add a write hit counter (one counter for the entire cache). The counter is only incremented at each write hit event. If the counter saturates at one threshold, then the cache will record the write operation that causes the counter saturation and invalidate the corresponding line to the write hit. Figure 9.9 shows the architecture of PoLF. The only hardware overhead of PoLF is a global counter that tracks the total number of write hits to the cache. The tunable parameter of PoLF is the FT.

9.7 i^{2}WAP: Putting Them Together

We combine SwS and PoLF together to form our inter- and intra-set write variation-aware policy, i^{2}WAP. In i^{2}WAP, SwS and PoLF work independently: SwS reduces inter-set variations, and PoLF reduces intra-set variations. The total write variations can be reduced significantly, and the product lifetime can be improved. Moreover, the implementation overhead of i^{2}WAP is quite small: SwS only requires one global counter to store the number of write accesses and two registers to store the current swapping and shifting values; PoLF only needs another global counter to store the number of write hit accesses.

9.8 Experiments

In this section, we first describe our experiment methodology, and then we demonstrate how SwS and PoLF reduce the inter- and intra-set write variations, respectively. Finally, we show how i^{2}WAP improves the NVM cache lifetime.

9.8.1 Baseline Configuration

Our baseline is a 4-core CMP system. Each core consists of private L1 and L2 caches, and all the cores share an L3 cache. Our experiment makes use of a 4-thread OpenMP version of the PARSEC 2.1 [1] benchmark workloads.[5] We run single application since wear-leveling techniques are usually designed for the worst case. The native inputs are used for the PARSEC benchmark to generate realistic program behavior. We modify the gem5 full-system simulator [2] to implement our proposed techniques and use it to collect cache accesses. Each gem5 simulation run is fast forwarded to the pre-defined breakpoint at the code region of interest, warmed up by 100 million instructions, and then simulated for at least 10 billion instructions. The characteristics of workloads are listed in Table 9.1, in which WPKI and TPKI are writes and transactions per kilo-instructions, respectively.

In this work, we use ReRAM L2 and L3 caches as an example. Our techniques and evaluations are also applicable to other NVM technologies. The system parameters are given in Table 9.2.

9.8.2 Write Variations Reduction

9.8.2.1 Effect of SwS on Inter-set Variations

In SwS, the inter-set variation reduction is related to the number of ShiftR during the experimental time. Assuming there are N sets in the cache, one ShiftR includes N SwapR and one SwapR has $N-1$ swaps. After each ShiftR, all the cache sets are

Table 9.1 Workload characteristics in L2 an L3 caches under our baseline configuration

Workload	L2 cache		L3 cache	
	WPKI	TPKI	WPKI	TPKI
Blackscholes	0.07	0.4	0.04	0.3
Canneal	0.04	23	0.01	15
Dedup	1.1	4.8	0.4	0.8
Facesim	3.3	4.7	1.1	1.4
Ferret	1.8	6.3	0.2	0.5
Fluidanimate	0.4	1.4	0.3	0.8
Freqmine	1.3	6.7	0.2	0.4
Raytrace	0.56	0.62	0.03	0.25
Streamcluster	3.7	4.2	0.9	1.1
Swaptions	1.4	2.9	0.02	0.06
Vips	1.1	4.4	0.6	1.0
x264	0.7	16.1	0.2	0.5

[5] A supplementary experiment on multi-program workloads is given in Sect. 9.9.3.

Table 9.2 Baseline configurations

System	4-core, 3 GHz, out-of-order CPU model based on ALPHA 21264
SRAM* I-L1/D-L1 caches	Private, 32/32 KB, 8-way, 64-byte cache line, LRU and write-back, write allocate, 2-cycle
ReRAM L2 cache	Private, 1 MB, 8-way, 64-byte cache line, LRU and write-back, write allocate, 30-cycle
ReRAM L3 cache	Shared, 8 MB, 8-way, 64-byte cache line, LRU and write-back, write allocate, 100-cycle
DRAM main memory	4 GB, 128-entry write buffer, 200-cycle

* We envision L1 is still SRAM due to performance concerns and NVM write endurance limits

shifted through all the possible locations. Thus, the more rounds the cache is shifted, the more evenly the write accesses are distributed to each cache set.

We annotate the round number of ShiftR as RRN, and it can be computed as follows:

$$\text{RRN} = \frac{W_{\text{total}}}{\text{ST} \times N \times (N-1)} = \frac{\text{WPI} \times I_n}{\text{ST} \times N \times (N-1)} \tag{9.6}$$

in which ST is the swap threshold, W_{total} is the product of WPI (write access per instruction) and I_n (the number of simulation instructions). For the same application, if the execution time is longer, which means I_n is larger, we can use a larger ST value to get the same RRN.

To illustrate the relationship between the inter-set variation reduction and RRN, we run simulations using different configurations with different execution lengths. Figure 9.10 shows the result. We can see when RRN is increased, the inter-set variation is reduced significantly. When RRN is larger than 100, the inter-set variation can be reduced to smaller than 5 % of its original value.

For a 1 MB cache running an application with WPKI equal to 1, if we want to reduce its inter-set variation by 95 % within 1 month, then the ST can be set larger than 100,000 according to Eq. 9.6 by setting RRN as 100. However, simulating a system within 1-month wall clock time is never realistic. To evaluate the effectiveness of SwS, we use a smaller ST (e.g., ST = 10) in a relatively shorter period of execution time (e.g., 100 billion instructions) to get a similar RRN. Figure 9.10b shows the inter-set variation of an L2 cache after adopting SwS when RRN equals to 100. The average inter-set variation is significantly reduced from 66 to 1.2 %.

In practice, ST can be scaled along with the entire product lifespan since our wear-leveling goal is to balance the cache line write count in the scale of several months if not years. Thus, the swap operation in SwS is infrequent enough to hide its impact on system performance.

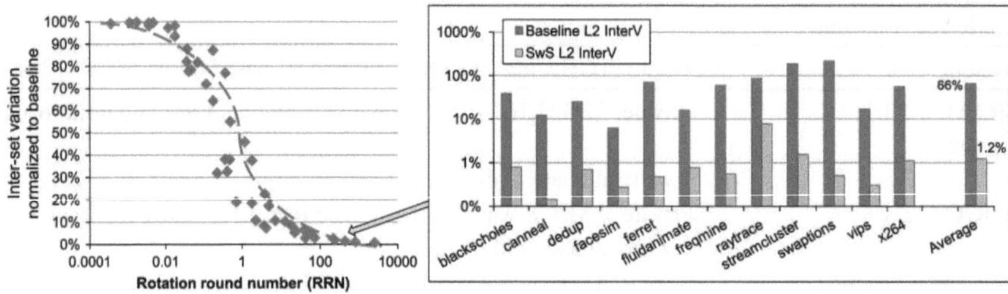

Fig. 9.10 Inter-set variations normalized to baseline when RRN increases in SwS scheme. The zoom-in sub-figure shows the detailed L2 inter-set variation of different workloads after adopting the SwS scheme when RRN equals to 100

9.8.2.2 Effect of PoLF on Intra-Set Variations

Figure 9.11 shows how PoLF affects intra-set variations and average write counts for L2 and L3 caches. It can be seen that PoLF reduces the intra-set variation significantly and the strength of PoLF is changed with different FT values.

When FT equals to 1, the PoLF scheme flushes every write hit and it is equivalent to the LF scheme. Figure 9.11 shows that LF can further reduce intra-set variations compared to PoLF. However, the average write count of LF is increased significantly. Thus, considering the impact on both intra-set variations and average write counts, we choose PoLF with FT that equals to 10. The results show that by adopting PoLF, the average intra-set variation of L2 cache can be reduced from 17 to 4 %, and the average intra-set variation of L3 cache can be reduced from 27 to 6 % with a FT value of 10. The average write count is increased by less than 2 % compared to the baseline.

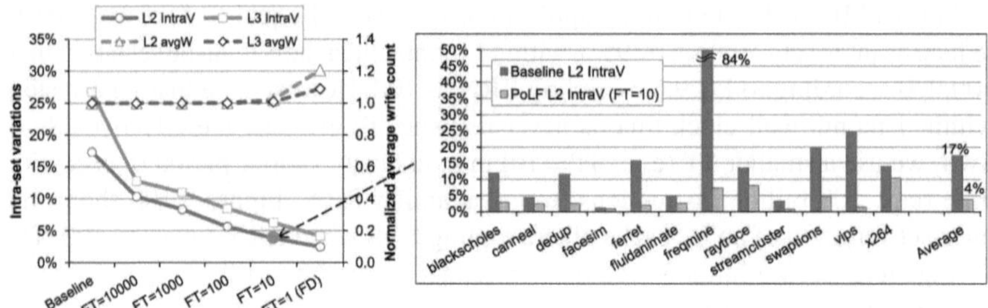

Fig. 9.11 The average intra-set variation and the average write count normalized to baseline for L2 and L3 caches after adopting a PoLF scheme. The zoom-in sub-figure shows the detailed L2 intra-set variation for different workloads after adopting an PoLF scheme with line flush threshold (FT) of 10

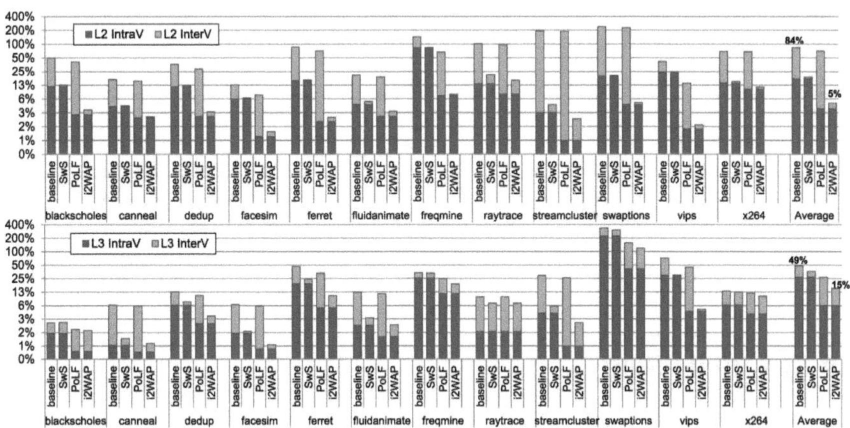

Fig. 9.12 The total variation for L2 and L3 caches under the baseline configuration, SwS scheme (RRN = 100), PoLF scheme (FT = 10) and i^{2}WAP policy. Each value is broken down to the inter-set variation and the intra-set variation. Note that a log scale is used to cover a large range of variations

9.8.2.3 Effect of i^{2}WAP on Total Variations

Figure 9.12 shows the total variations of L2 and L3 caches under different policies. Compared to the baseline, SwS reduces inter-set variations across all workloads, but it does not affect intra-set variations. On the other hand, PoLF reduces intra-set variations with small impact on inter-set variations. By combining SwS and PoLF, i^{2}WAP reduces the total variations significantly. Figure 9.12 shows that on average the total variation is reduced from 84 to 5 % for L2 caches and from 49 to 15 % for L3 caches.

9.8.3 Lifetime Improvement of i^{2}WAP

After reducing inter-set and intra-set variations, the lifetime of NVM caches can be improved. Figure 9.13 shows the LI of L2 and L3 caches after adopting SwS only, PoLF only, and the combined i^{2}WAP policy, respectively. The LI varies based on the workload. Basically, the larger the original variation value is, the bigger the improvement a workload has. The overall LI is 75 % (up to 224 %) for L2 caches and 23 % (up to 100 %) for L3 caches.

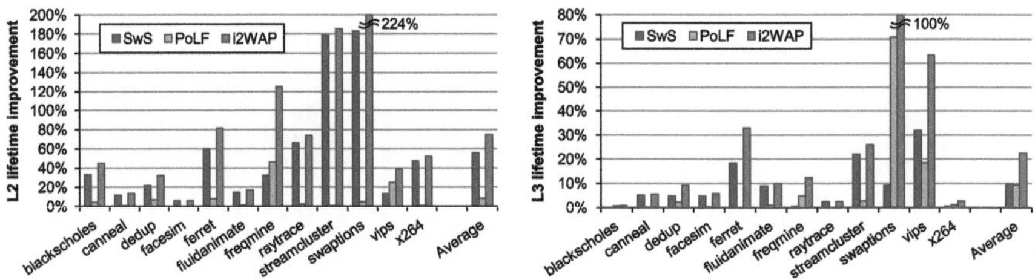

Fig. 9.13 The lifetime improvement after adopting i^{2}WAP using Eq. 9.5 (*Left* L2, *Right* L3)

9.9 Sensitivity Study

9.9.1 Sensitivity to Cache Associativity

As shown in Table 9.2, we use 8-way associative L2 and L3 caches in the baseline system. To study on the effectiveness of i^{2}WAP on different cache configurations, we do a sensitivity study on different associativity numbers ranging from 4 to 32. All the other system parameters remain the same.

Figure 9.14 shows the total variations of L2 and L3 caches under different policies when the associativity is changed. For both L2 and L3 caches in the baseline system, with the increase in the cache associativity, InterV is decreased and IntraV is increased. The reason is that when the cache capacity is fixed, the set number decreases as the associativity increases. Thus, more writes are merged into one cache set and the write variation between different sets becomes smaller. Furthermore, IntraV is amplified since the number of cache block in one set is increased and the write imbalance becomes worse.

Regardless of how the associativity changes, adopting i^{2}WAP reduces the total variations significantly by combining SwS and PoLF. Figure 9.14 shows that on average the total variation of the 4-way system is reduced from 109 to 17 % for L2

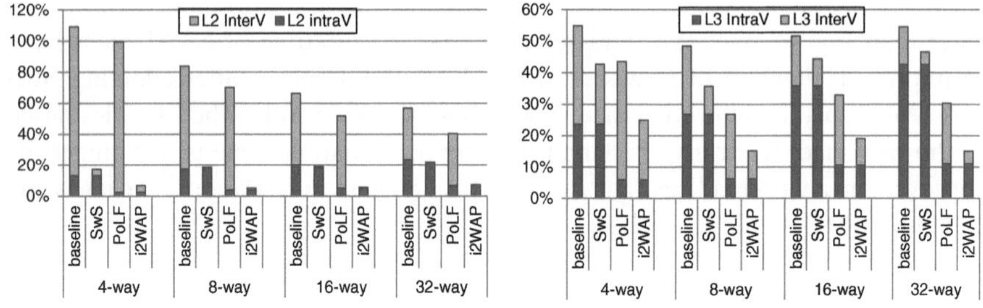

Fig. 9.14 The total variation for L2 and L3 caches with 4-way, 8-way, 16-way, and 32-way under the baseline configuration, SwS scheme (RRN = 100), PoLF scheme (FT = 10), and i^{2}WAP policy. Each value is broken down to the inter-set variation and the intra-set variation

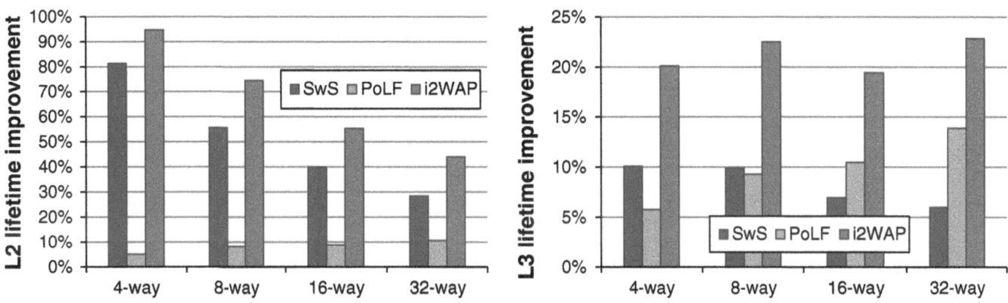

Fig. 9.15 The lifetime improvement after adopting i^{2}WAP with different cache associativity using Eq. 9.5 (*Left* L2, *Right* L3)

caches and from 55 to 25 % for L3 caches; for the 16-way system, the total variation is reduced from 66 to 6 % for L2 caches and from 52 to 19 % for L3 caches; for the 32-way system, the total variation is reduced from 57 to 7 % for L2 caches and from 55 to 15 % for L3 caches.

The NVM cache lifetime is improved after reducing InterV and IntraV. Figure 9.15 shows the LI of L2 and L3 caches after adopting SwS only, PoLF only, and the combined i^{2}WAP policy, respectively. On average, the LI is 95 and 20 % for L2 and L3, respectively, in a 4-way system; it is 55 and 20 % for L2 and L3, respectively, in a 8-way system; it is 44 and 23 % for L2 and L3, respectively, in a 16-way system. In general, the larger the original variation value is, the bigger the improvement i^{2}WAP can bring.

9.9.2 Sensitivity to Cache Capacity

Another sensitivity study is targeted to the cache capacity. In Sect. 9.8, we use 1 MB L2 and 8 MB L3 caches as shown in Table 9.2. We expect that i^{2}WAP also works effectively on caches with different capacity. We have experiments on different L2 capacity ranging from 512 KB to 4 MB and different L3 capacity ranging from 4 to 32 MB. Figure 9.16 shows the result. On average the total variation is reduced by 90–95 % for L2 caches and 58–73 % for L3 caches, respectively.

Figure 9.17 is the according result of the LI. On average, the lifetime improvement is 66–153 % and 22–26 %, respectively. These results validate that i^{2}WAP works effectively regardless of the cache capacity. For L2 caches, we can see that as capacities increase, the value of LI also increases. The reason is that the write imbalance is worse in larger capacity caches and the baseline variation value is also increased. Thus, larger caches provide more space for i^{2}WAP to decrease the variation and improve the lifetime. For L3 caches, the trend of the variation growth is much smaller than the ones in L2 caches, and the intra-set variations occupy a larger proportion in the total baseline variations. Thus, the LI of L3 caches when the capacity is changed shows smaller differences than L2 caches.

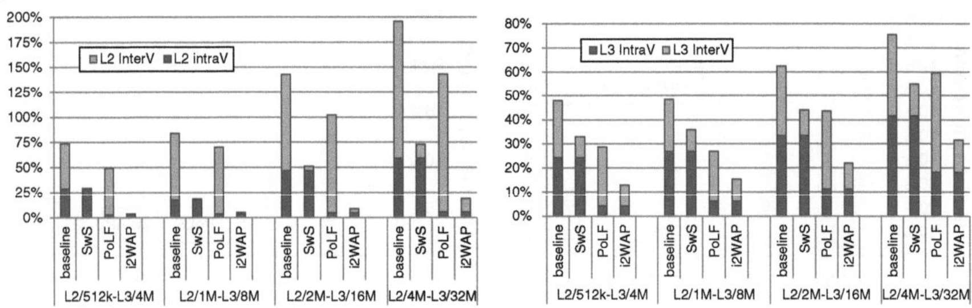

Fig. 9.16 The total variation for L2 and L3 caches with different capacities under the baseline configuration, SwS scheme (RRN = 100), PoLF scheme (FT = 10), and i^{2}WAP policy. Each value is broken down to the inter-set variation and the intra-set variation

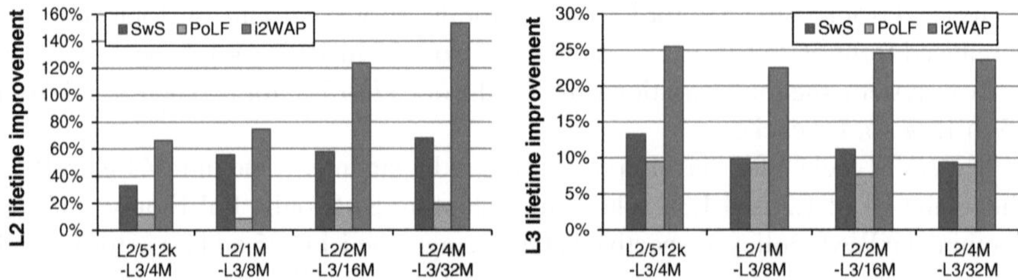

Fig. 9.17 The lifetime improvement after adopting i^{2}WAP with different cache capacities using Eq. 9.5. (*Left*: L2, *Right*: L3)

9.9.3 Sensitivity to Multi-program Applications

All the previous simulations are based on multi-thread workloads. To study the effectiveness of i^{2}WAP on multi-program applications, we simulate workload mixtures from SPEC CPU2006 benchmark suite [19]. The other experimental configurations are the same as described in Sect. 9.8.

Figure 9.16 shows the variations of L2 and L3 caches for multi-program applications using mixed SPEC CPU2006 workloads in a 4-core system. Table 9.3 lists the combination of the workload mixtures. Intuitively, multiple cores share one L3 cache and run different programs, and the access traffic to the L3 cache should be well mixed and thus balanced. However, Figure 9.16 shows that both InterV and IntraV in the shared L3 cache are still very large in some cases. On average, InterV and IntraV of L3 caches are 28 % (up to 100 %) and 26 % (up to 64 %), respectively. Similar to the results of multi-thread experiments, the variations in L2 caches is larger than the ones in L3 caches since L2 only serves one program and has more inbalanced write. On average, the total variation is reduced from 132 to 24 % for L2 caches and from 54 to 15 % for L3 caches, respectively (Fig. 9.18).

Figure 9.19 shows the LI for the multi-program workloads. For L2 and L3 caches, the overall LI is 88 % (up to 387 %) and 33 % (up to 136 %), respectively. Therefore, i^{2}WAP is effective in reducing the write variations and improving the cache lifetime for multi-program workloads.

Table 9.3 The workload list in the mixed groups

Mixed group	Workloads
Mix 1	astar+bwaves+bzip2+gcc
Mix 2	bzip2+astar+gobmk+h264ref
Mix 3	gromacs+bzip2+gcc+gobmk
Mix 4	gcc+gromacs+hmmer+namd
Mix 5	gobmk+h264ref+hmmer+gromacs
Mix 6	h264ref+hmmer+milc+namd
Mix 7	milc+namd+omnetpp+wrf
Mix 8	namd+h264ref+gcc+astar
Mix 9	omnetpp+wrf+astar+bwaves
Mix 10	wrf+milc+bwaves+gromacs

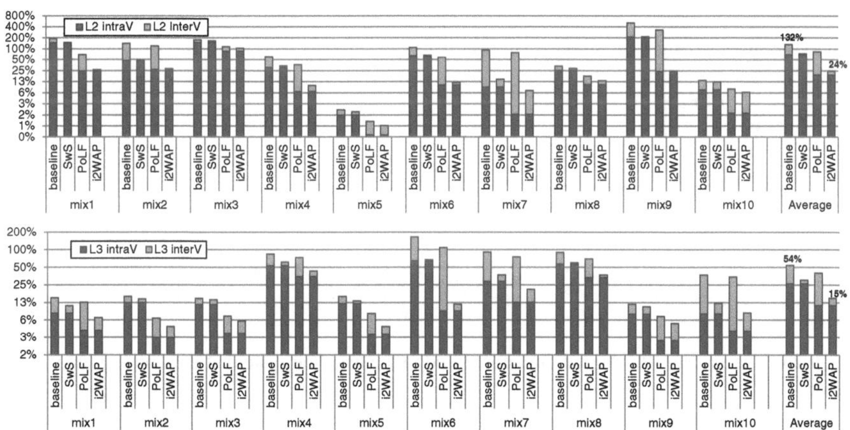

Fig. 9.18 The total variation for L2 and L3 caches for multi-program applications using mixed SPEC CPU2006 workloads under the baseline configuration, SwS scheme (RRN = 100), PoLF scheme (FT = 10), and i^{2}WAP policy. Each value is broken down to the inter-set variation and the intra-set variation

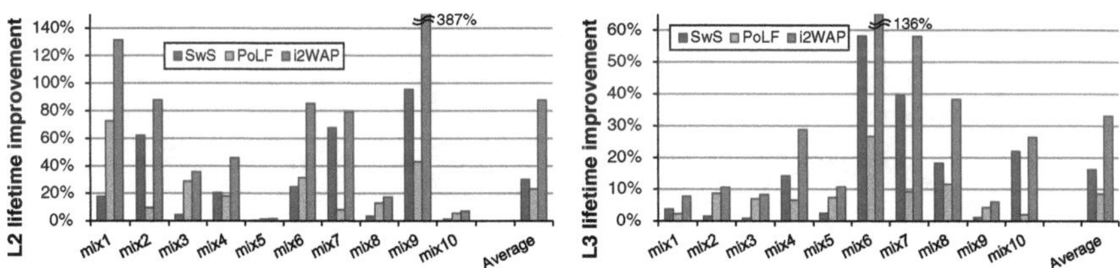

Fig. 9.19 The lifetime improvement after adopting i^{2}WAP for multi-program applications using Eq. 9.5. (*Left* L2, *Right* L3)

9.10 Analysis of Other Issues

9.10.1 Performance Overhead of i²WAP

Since i²WAP causes extra cache invalidations, it is necessary to compare its performance to a baseline system without wear leveling.[6] Figure 9.20 shows the performance overhead of a system in which L2 and L3 caches using i²WAP with ST = 100,000 and FT = 10 compared to a baseline system in which an LRU policy is adopted.

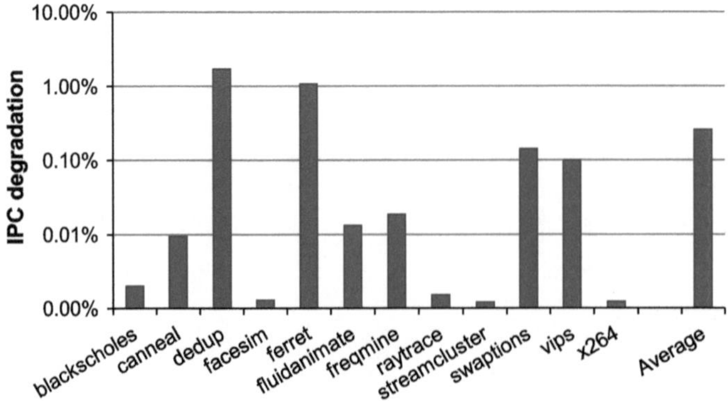

Fig. 9.20 The system IPC degradation compared to the baseline system after adopting the i²WAP policy

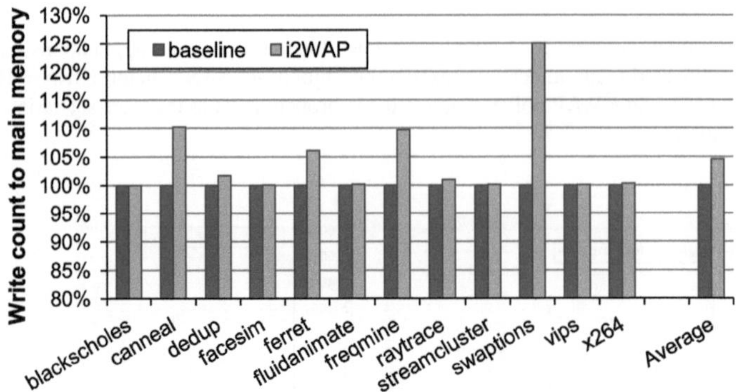

Fig. 9.21 The write count to the main memory normalized to the baseline system after adopting i²WAP

6 The simulator has a protocol to ensure cache coherency when invalidations occur; thus, the performance overhead of this part is included.

As shown in Fig. 9.20, on average, the IPC of the system using i^{2}WAP is reduced only by 0.25 % compared to the baseline. The performance penalty of i^{2}WAP is very small because of two reasons:

- In SwS, the interval of swap operations is long (e.g., 10 million instructions), and only two cache sets are remapped for each operation.
- In PoLF, write hit accesses are infrequent enough to ensure the frequency of LF operations is low (e.g., once per 10^5 instructions), and only one cache line is flushed each time. In addition, designers can trade-off between the number of flush operations and the variation value by adjusting FT.

9.10.2 Impact on Main Memory

We also evaluate the impact of extra write backs on main memory. Figure 9.21 shows the write count to the main memory compared to the baseline system after adopting the i^{2}WAP policy. The result shows that its impact on the write count is very small, only increasing about 4.5 % on average. For most workloads, the write count is increased by less than 1 %. In addition, because most writes can be filtered by caches and the write count of main memory is much smaller than that of caches, the endurance requirement for nonvolatile main memory is much looser. In the worst case, although the write count to the main memory of *swaptions* is increased by 25 %, its absolute value of write backs is very small (about 0.001 writes to the main memory per kilo-instructions). Thus, it does not significantly degrade the lifetime of the main memory even though its write access frequency is relatively higher.

9.10.3 Error-tolerant Lifetime

While our analyses are all focused on the raw cache lifetime, this lifetime can be easily extended by tolerating partial cell failures. There are two factors causing the different failure time of cells. The first one is the variation of write counts, which is addressed mainly in this work. The second one is the inherent variation of the cell's lifetime due to process variations, which needs another type of techniques to solve (discussed in Sect. 9.1). For both factors, the system lifetime can be extended by tolerating a small number of cell failures.

It is much simpler to extend i^{2}WAP and tolerate the failed cache lines comparing to tolerating main memory failures [6, 14–16, 21]. We can force the failed cache lines to be tagged *INVALID*, so that no further data would be written to the failed cache lines. In this case, the number of ways in the corresponding cache set is only reduced by 1, e.g., from 8-way associative to 7-way associative.

The error-tolerant lifetime is at least the same as the raw lifetime and may be much longer. However, the performance is degraded because the cache associativity

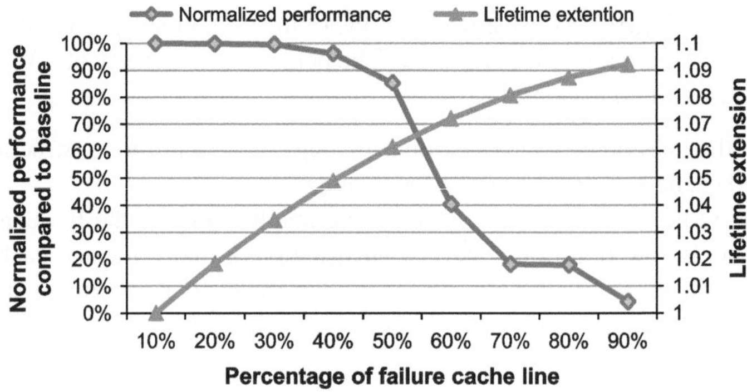

Fig. 9.22 The performance degradation and lifetime extension during gradual cache line failure on a nonvolatile cache hierarchy

is reduced. Figure 9.22 shows an analysis of an ReRAM-based cache hierarchy with 32 KB L1 caches, 1 MB L2 caches, and 8 MB L3 caches. It shows that if the system can tolerate the failure of 50 % of the cache lines at all levels, the lifetime can be extended by 6 % and the performance penalty is 15 %.[7]

9.10.4 Security Threat Analysis

Thus far, the workloads we have considered are only typical. However, memory technologies with limited write endurance always pose a unique security threat. An adversary might design an attack that stresses a few lines in the cache to reach the endurance limit and then cause the system fail.

The security threat models for main memory and caches are different. The cache address space is a subset of the total memory space, and the result of cache replacement policies is difficult to predict accurately. This makes random attacks, such as birthday paradox attack [17] difficult to implement in caches. Thus, a more effective attack for nonvolatile caches is repeated address attack (RAA) [12].

For a simple RAA, the attacker can write a data line in cache repeatedly and then cause cache line failure. It is easy to implement in a cache without an endurance-aware policy. For L1 data caches, when a cache line is repeated written, it would not be replaced under an LRU policy since it is continuously referenced and the address of the attacked line is never changed during the repeated writing. The mechanism is similar in a lower-level cache, and the only difference is that the attacker would use the higher-level cache's write-back to repeatedly write the lower-level cache. For example, circularly writing $NUM_{ways} + 1$ of data in one set of L1 data caches would force it to write back to the L2 cache repeatedly. Assuming it takes 10–50 cycles to

[7] The value of performance degradation and the lifetime extension depend on the cache hierarchy and capacity, but the trends for different configurations are similar.

write back to one L2 cache line, then the time required to make a L2 cache line fail is given by

$$\text{Time to Fail} = \frac{\text{Cycle per write} \times \text{Cell endurance}}{\text{Cycles per second}} = 6\text{–}30\,\text{min} \tag{9.7}$$

Thus, the traditional cache policy opens a serious security problem for nonvolatile caches with limited write endurance.

However, our proposed i^{2}WAP policy can mitigate this problem. By adopting i^{2}WAP, the injected attacks would be distributed in every different cache line because of the following two reasons:

(1) The mapping between physical sets and logical sets is shifted by the SwS policy and according to the write count, which is hard to predict. In addition, it becomes more difficult for the attacker to guess the shift period if there are other processes competing the same cache resource.
(2) Cache line invalidation is quickly triggered by repeatedly write hits. PoLF guarantee a high probability to invalidate the attacked cache line, and data are then loaded in another random location, and the attacker cannot predict the new location.

Thus, the attacks would be distributed in every different cache line. For a 1 MB L2 cache with i^{2}WAP, the time to make a cache line fail would be

$$\begin{aligned}\text{Time to Fail} &= \frac{\text{Number of cache line} \times \text{Cycle per write} \times \text{Cell endurance}}{\text{Cycles per second}} \\ &= 2\text{–}10 \text{ months}\end{aligned} \tag{9.8}$$

In lower-level caches with a larger number of cache lines and longer write cycles, the time to attack the system till failure is even longer. Thereby, such a long duration is sufficient to detect the abnormal attack accesses and make the system safe.

9.11 Conclusion

Modern computers require larger memory system, but the scalability of traditional SRAM and DRAM is constrained by leakage and cell density. Emerging NVM is a promising alternative to build large main memory and on-chip caches. However, NVM technologies usually have limited write endurance, and the write variation would cause heavily written memory blocks to fail much earlier than their expected lifetime. Thus, wear-leveling techniques are studied to eliminate memory write variations.

In this chapter, we use i^{2}WAP as an example to show how to design an endurance-aware cache management policy. i^{2}WAP uses SwS to reduce the inter-set variation and PoLF to reduce the intra-set variation. i^{2}WAP can balance the write traffic to each cache line, greatly reducing both intra-set and inter-set variations, and thus improving the lifetime of NVM on-chip L2 and L3 caches. The implementation overhead of i^2WAP is small, only needing two global counters and two global registers, but it can improve the lifetime of NVM caches by 75 % on average and up to 224 % over the conventional cache management policies.

References

1. Bienia, C., Kumar, S., Singh, J. P., & Li, K. (2008). The PARSEC benchmark suite: Characterization and architectural implications. In PACT. pp. 72–81.
2. Binkert, N., et al. (2011). gem5: A multiple-ISA full system simulator with detailed memory model. Computer Architecture News.
3. Chen, B., Lu, Y., Gao, B., et al. (2011). Physical mechanisms of endurance degradation in TMO-RRAM. 12.3.1-12.3.4. http://dx.doi.org/10.1109/IEDM.2011.6131539.
4. Huai, Y. (2008). Spin-transfer torque MRAM (STT-MRAM): Challenges and prospects. *Association of Asia Pacific Physical Societies, Bulletin*, *18*(6), 33.
5. International Technology Roadmap for Semiconductors. (2012). Process integration, devices, and structures 2012 update. http://www.itrs.net/.
6. Ipek, E., Condit, J., Nightingale, E. B., Burger, D., & Moscibroda, T. (2010). Dynamically replicated memory: Building reliable systems from nanoscale resistive memories. *ASPLOS*, 3–14.
7. Joo, Y., Niu, D., Dong, X., Sun, G., Chang, N., & Xie, Y. (2010). Energy- and endurance-aware design of phase change memory caches. In Automation and Test in Europe Conference and Exhibition, Design, pp. 136–141.
8. Kim, K., & Ahn, S. J. (2005). Reliability investigations for manufacturable high density PRAM. In IRPS, pp. 157–162.
9. Kim,Y.-B., Lee, S. R., Lee, D., Lee, C.B., et al. (2011). Bi-layered RRAM with unlimited endurance and extremely uniform switching. In VLSI. pp. 52–53.
10. Kroft, D. (1998). Lockup-free instruction fetch/prefetch cache organization. In *25 Years of the International Symposia on Computer architecture (selected papers)*, pp. 195–201.
11. Lin,W. S., Chen, F. T., Chen, C. H. L., & Tsai, M.-J.. (2010). Evidence and solution of over-RESET problem for HfO_x based resistive memory with sub-ns switching speed and high endurance. In *Proceedings of the International Electron Devices Meeting*. 19.7.1–19.7.4.
12. Qureshi, M. K., Karidis, J. P., et al. (2009a). Enhancing lifetime and security of PCM-based main memory with start-gap wear leveling. In *Proceedings of the International Symposium on Microarchitecture*, pp. 14–23.
13. Qureshi, M. K., Srinivasan, V., & Rivers, J. A. (2009b). Scalable high performance main memory system using phase-change memory technology (pp. 24–33). Architecture: In *Proceedings of the International Symposium on Computer*.
14. Schechter, S., Loh, G. H., Straus, K., & Burger, D. (2010). Use ECP, not ECC, for hard failures in resistive memories (pp. 141–152). In *Architecture: Proceedings of the International Symposium on Computer*.
15. Seong, N. H., Woo, D. H., & Lee, H.-H. S. (2010a). Security refresh: Prevent malicious wear-out and increase durability for phase-change memory with dynamically randomized address mapping (pp. 383–394). In *Architecture: Proceedings of the International Symposium on Computer*.
16. Seong, N. H., Woo, D. H., Srinivasan, V., Rivers, J. A. & Lee, H.-H. S. (2010b). SAFER: Stuck-at-fault error recovery for memories (pp. 115–124). In *Proceedings of the International Symposium on Microarchitecture*.

17. Seznec, A. (2010). A phase change memory as a secure main memory. *Computer Architecture Letters*, *9*(1), 5–8.
18. Slaughter, J. M., Rizzo, N. D., Mancoff, F. B., Whig, R., Smith, K., Aggarwal, S., et al. (2010). Toggle and spin-torque MRAM: Status and outlook. *Journal of the Magnetic Society of Japan*, *5*, 171.
19. Spec, CPU. (2006). SPEC CPU2006. http://www.spec.org/cpu2006/.
20. Wang, J., Dong, X., Xie, Y., & Jouppi, N. P. (2013). i^2WAP: Improving non-volatile cache lifetime by reducing inter- and intra-set write variations. In *Proceedings of the International Symposium on High-Performance Computer Architecture*.
21. Yoon, D. H., Muralimanohar, N., Chang, J., Ranganathan, P., et al. (2011). FREE-p: Protecting non-volatile memory against both hard and soft errors (pp. 466–477). In *Proceedings of the International Symposium on High-Performance Computer Architecture*.
22. Zhou, P., Zhao, B., Yang, J., & Zhang, Y. (2009). A durable and energy efficient main memory using phase change memory technology (pp. 14–23). In *Architecture: Proceedings of the International Symposium on Computer*.

Chapter 10
A Circuit-Architecture Co-optimization Framework for Exploring Nonvolatile Memory Hierarchies

Xiangyu Dong, Norman P. Jouppi and Yuan Xie

Extension of Conference Paper: This submission is extended from "A Circuit-Architecture Co-optimization Framework for Evaluating Emerging Memory Hierarchies" published on ISPASS'13. The additional material provided in the submission includes a detailed explanation of the circuit-level and the architecture-level models, a new case study of using PCRAM, and a sensitivity study on processor core counts.
This work is supported in part by SRC grants, NSF 1218867, 1213052, 0903432 and by DoE under Award Number DE-SC0005026.

10.1 Introduction

The state-of-the-art memory hierarchy design with SRAM on-chip caches and DRAM off-chip main memory is now being challenged from two aspects. First, both SRAM and DRAM technologies are leaky. The SRAM leakage power and the DRAM refresh power will start to dominate if memory capacities keep growing. Some data already show that 25–40 % of total power is attributed to the memory system [1] and some embedded processor caches can compromise over 40 % of the total

---

X. Dong
Qualcomm Technology, Inc., Qualcomm, USA
e-mail: xiangyud@qualcomm.com

N. P. Jouppi
Google Inc., CA, USA
e-mail: jouppi@acm.org

Y. Xie (✉)
Pennsylvania State University and AMD Research, Pennsylvania, USA
e-mail: yuanxie@cse.psu.edu

Y. Xie (ed.), *Emerging Memory Technologies*,
DOI: 10.1007/978-1-4419-9551-3_10,

chip power budget [2]. Second, SRAM and DRAM are facing many difficulties in scaling down. For example, it is hard to scale down DRAM below a 20-nm process node due to the difficulty in keeping an adequate amount of cell capacitance [3]. The recent shift of some L3 on-chip caches from SRAM to eDRAM [4] and the research momentum in replacing DRAM main memory with various emerging nonvolatile memories [5–13] reflect the responses to such challenges in designing an energy-efficient and cost-effective memory hierarchy.

Recently, many alternative memory technologies, such as phase-change RAM (PCRAM)[1] [14–16], spin-torque transfer RAM (STTRAM)[2] [17, 18], and resistive RAM (ReRAM)[3] [19–21], have been demonstrated. These emerging nonvolatile memory technologies have attractive properties of high density, fast access, good scalability, and nonvolatility, and they have drawn the attention of the computer industry and challenged the role of SRAM and DRAM in the mainstream memory hierarchy for the first time in more than 30 years.

Since each of the emerging memory technologies has their pros and cons and the peripheral circuit design can vary the memory module property greatly, the future memory hierarchy will have a much larger design space. Therefore, it is necessary to have an estimation framework that can quickly find the optimal memory technology choice and the corresponding circuit design style in terms of performance, energy, or area.[4] But, there are two challenges before doing that.

First, unlike SRAM whose cells and macro-designs are highly standardized, emerging memory technologies only have prototypes whose performance and energy properties can vary greatly. Such circuit variation can already be observed from the related literature [14–21], where some of the memory prototypes show extremely fast access speed while others show extremely dense structure. In order to model this variety and the circuit-level trade-offs, we build a circuit-level performance, energy, and area model.

Second, in order to build an optimization loop covering circuit- and architecture-level design options, we require models that reflect how architectural metrics (e.g., IPC and power consumption) change as we tune the underlying memory hierarchy design knobs (i.e., cache capacity, cache associativity, and cache read or write latency). Conventionally, such model is built through simulations; however, it is impractical to run time-consuming simulations for each possible design input. To surmount this difficulty, we apply statistical analysis and effectively use limited simulation runs to approximate the entire architectural design space.

After modeling the circuit- and architecture-level trade-offs, we combine them into a circuit-architecture joint design space exploration framework and use this framework to optimize different memory hierarchy levels by adopting emerging memory technologies. In this work, we show that combined with SRAM L1 or L2 caches, the versatility of emerging memory technologies can excel in the remaining memory hierarchy levels from L2 or L3 caches to main memories, and such a hybrid hierarchy has significant benefits in energy and area reduction with insignificant performance degradation overhead. As an example, our analysis shows that using

[4] Silicon area is the main factor that determines chip cost.

ReRAM in L3 caches can achieve overall improvements on EDP (by 28 %) and EDAP (by 39 %) on an 8-core chip multiprocessor (CMP) system. Finally, we propose a simulated annealing approach that can quickly find the near-optimal solution in designing a energy-efficient or cost-efficient memory hierarchy.

10.2 Related Work

Since our work involves both circuit- and architecture-level models, we first describe prior work on circuit-level memory design space exploration and predictive performance models.

10.2.1 Circuit Model for Memory Modules

Many circuit-level models have been provided to enable SRAM or DRAM design explorations. For example, CACTI [22, 23] is to estimate the performance, energy, and area of SRAM and DRAM caches. However, as CACTI is originally designed to model an SRAM-based cache, and some of its fundamental assumptions do not match the actual emerging nonvolatile memory circuit implementations. Besides CACTI, Amrutur and Horowitz [24] introduced an analytical model for estimating the SRAM array speed and power scaling. Another circuit-level model [25] uses a logic synthesis tool to build a circuit library and relies on curve fitting to represent the circuit design trade-offs.

10.2.2 Predictive Performance Model

Statistical models [25–30] can be used to infer the impact of architectural input configurations on overall performance metrics. Although it is time-consuming to collect sufficient sample data from conventional simulation, this is a one-time effort, and all the later outputs can be generated with the statistical model. Different fitting models have been used in the inference process that fits a predictive model through regression. Joseph et al. [26] used linear regression, Lee and Brooks [27] used cubic splines, whereas Azizi et al. [25] applied polynomial functions to create architecture-level models. Artificial neural network (ANN) [28–30] is another popular approach to build a predictive architecture model, which can efficiently explore exponential-size architectural design spaces with many interacting parameters.

10.3 Circuit Model: An RC Approach

After more than 30 years of application, both SRAM and DRAM memory modules have defined their own typical design styles. For example, 6T or 8T SRAM cells are widely adopted in the on-chip cache designs, and 1T DRAM cells are also typical. Moreover, on-chip SRAM designs are mainly supported by standard libraries or even memory compiler tools, and commodity DRAM is highly standardized as well. Due to the technology maturity, both SRAM and DRAM memory modules now have less variety.

However, such design consistency cannot be found in the emerging memory module design. Recently, many STTRAM, PCRAM, and ReRAM prototype chips have been designed and demonstrated [14–21], but few of them show consistency in reporting the performance, energy, and area data. This is actually common for any new technology. Due to the still evolving states of these emerging technologies, there is no single design, standard design option can balance the trade-off among chip performance, energy consumption, and chip area. Therefore, researchers have made various decisions on design and manufacturing, and thus, it causes a large variation among designs.

Such variation brings challenges as well as opportunities for using emerging memory technologies in future memory hierarchies. The opportunity is that we can still freely bias the optimal design options toward different optimization targets on the different memory hierarchy levels, especially when large prototype chip variations tell us these technologies can cover a wide design spectrum from highly latency-optimized microprocessor caches to highly density-optimized secondary storage. But, the challenge is to build a performance, energy, and area model for these emerging memory technologies even before they become mature. Therefore, we first build a circuit-level model for nonvolatile memory technologies.

10.3.1 Modeling Philosophy

We apply the modeling philosophy used in CACTI [31, 32] to establish a library of emerging memory technologies spanning from ultra-fast to ultra-dense memory designs. Similar to CACTI, we capture the device-level RC property of memory cells and use traditional RC analysis to estimate their performance and energy consumption. We follow standard design rules to predict the silicon area occupied by each circuit components. We obtain the process-related data of transistors and metal layers from the ITRS report [3] and the MASTAR tool [33]. The data cover the process nodes from 22 to 180 nm and support three transistor types, which are *high performance*, *low operating power*, and *low standby power*.

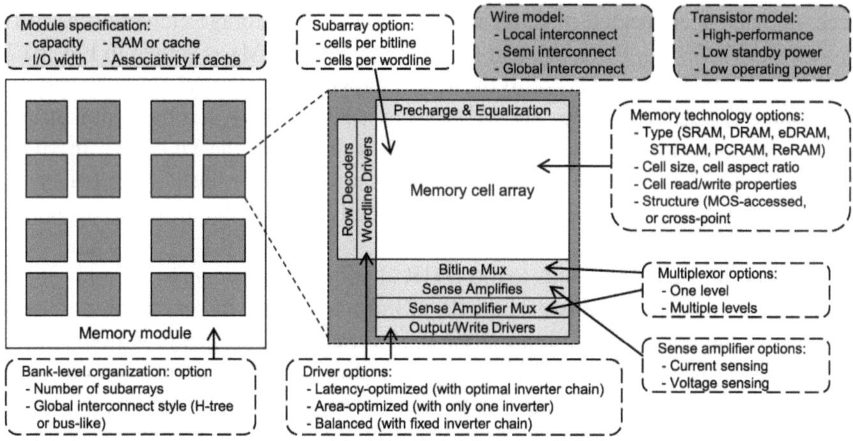

Fig. 10.1 The circuit-level model for memory module timing, power, and area estimations

10.3.2 Circuit Components and Tuning Knobs

Figure 10.1 shows the basic components abstracted in this circuit-level model. Each memory module is modeled as a set of banks, every bank can contain multiple subarrays, and a memory operation is fulfilled by simultaneous accesses to multiple subarrays in a bank. Depending on the design requirement, a bank can be partitioned into subarrays with different granularity. The rule of thumbs is that smaller subarrays are faster and larger subarrays are area-efficient.

A subarray is the elementary structure, in which there are a set of peripheral circuit including row decoders, column multiplexers, output drivers, and so on. There are a large amount of design knobs that can be tuned in the subarray design, especially regarding the choice of peripheral circuit. For example, the output driver design can have the option of following logical effort [34] and use optimal levels and sizes of inverters for high performance, or it can be simply designed as a single inverter for area efficiency. For sense amplifiers, the voltage-sensing scheme is straightforward but slower, while the current-sensing one incurs two-level sensing but much faster and more suitable for sensing the resistance difference of emerging memory cells. Other tuning knobs such as the multiplexer design are also shown in Fig. 10.1.

10.3.3 Memory Array Structure

There are two types of memory arrays modeled in this work: MOS-accessed and cross-point.

MOS-accessed cells correspond to the typical 1T1R (1-transistor-1-resistor) structure used by many nonvolatile memory prototype chips [17, 18], in which an NMOS access device is connected in series with the nonvolatile storage element (i.e., MTJ in STT-RAM, GST in PCRAM, and metal oxide in ReRAM) as shown in Fig. 10.2. Such an NMOS device turns on/off the access path to the storage element by varying the voltage applied to its gate. The MOS-accessed cell usually has the best isolation between neighboring cells due to the high OFF resistance of the MOSFET. In MOS-accessed cells, the size of access transistor is bounded by the current needed by the write operation. This NMOS needs to be sufficiently large so that it can drive enough write current.

Cross-point cell corresponds to the 1D1R (1-diode-1-resistor) [14, 15, 35, 36] or the 0T1R (0-transistor-1-resistor) [19, 20, 37] structures used by several high-density nonvolatile memory chips. Figure 10.3 shows a cross-point array without diodes (i.e., 0T1R structure). For a 1D1R structure, a diode is inserted between the word line and the storage element. Such cells either rely on the one-way connectivity of diode (i.e., 1D1R) or the material's nonlinearity (i.e., 0T1R) to control the memory access path.

Compared to MOS-accessed cells, cross-point cells have much smaller cell sizes. The area efficiency benefit of the cross-point structure is evident in the comparison between Figs. 10.2 and 10.3. The removal of MOS access devices leads to a memory cell size of only $4F^2$, where F is the process feature size. Unfortunately, the cross-point structure worsens the isolation among memory cells and thus brings challenges to peripheral circuitry designs. Several design issues such as half-select write, two-step sequential write, and external sensing [38] are included in our model.

10.3.4 Model Accuracy

We validate our circuit model against STTRAM [18], PCRAM [14], and ReRAM [21] prototypes. In general, the performance (i.e., read latency, write latency) estimation error is within 20 %, and the area estimation error is below 10 %.

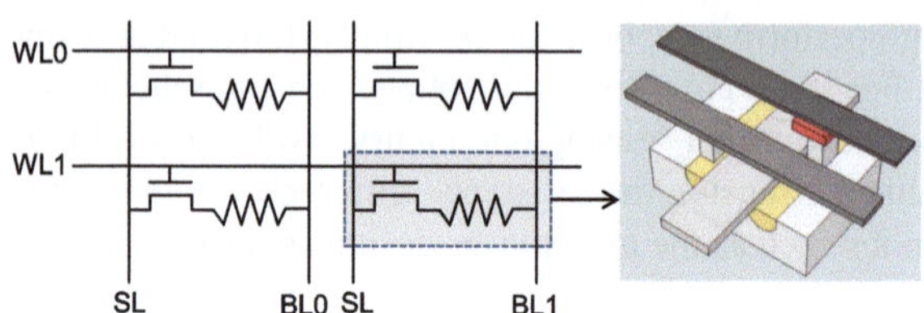

Fig. 10.2 The schematic view of MOS-accessed ReRAM arrays (*WL* wordline, *BL* bitline, *SL* sourceline)

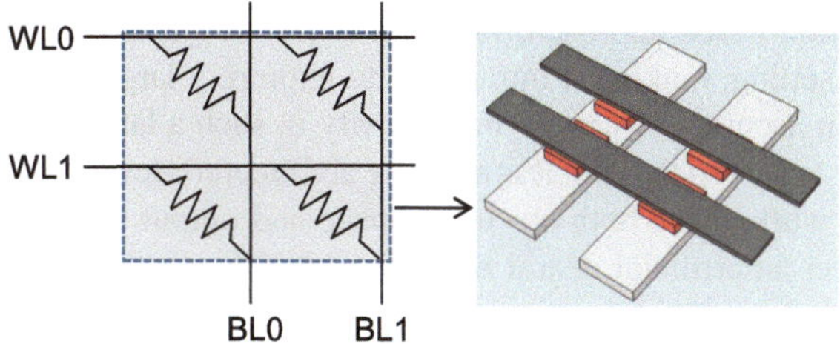

Fig. 10.3 The schematic view of cross-point ReRAM arrays without access devices (*WL* wordline, *BL* it bitline)

10.3.5 Circuit-Level Model Summary

In summary, our circuit-level model takes the memory design parameters such as technology node, memory capacity, associativity, block size, and cell type as the inputs, and it gives circuit-level outputs such as read/write latency, read/write dynamic energy per access, leakage power, and silicon area. Our circuit-level model provides 8 optimizations targets, which are read latency, write latency, read energy, write energy, read EDP, write EDP, silicon area, and leakage power, and each of these optimized designs is evaluated in the later circuit-architecture joint design space exploration. The optimization is achieved by tuning the design knobs including, but not limited to, subarray size, global interconnect style, driver design, sense amplifier design, multiplexer design, and memory cell structure.

10.4 Architecture Model: An ANN Approach

At the architectural level, we need performance models that predict the architectural performance of the overall system such as IPC and access counts at all cache levels as we change the underlying memory hierarchy. The input parameters at the architectural level are the parameters such as cache capacity, cache associativity, read latency, and write latency.

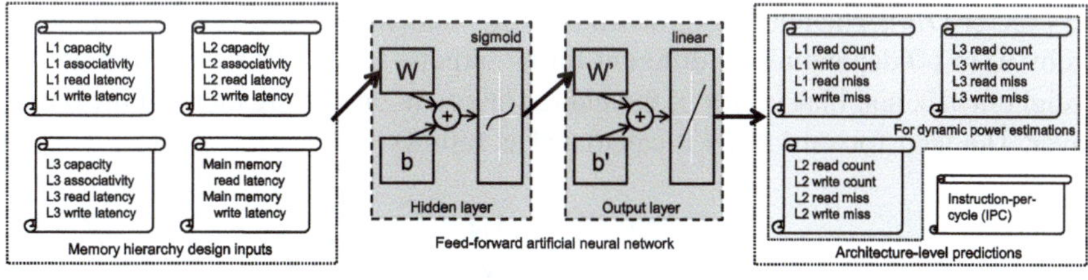

Fig. 10.4 The basic organization of a two-layer feed-forward artificial neural network (ANN)

In a simulation-based approach, long run times are necessary to simulate each possible input setting, making it intractable to explore a large design space. However, simulation accuracy is not the first priority in such a large-scale design space exploration. Instead, a speedy but less accurate architecture-level model is a preferred choice. In this work, since both our input space and output space are high dimensional, we select an artificial neural network (ANN) to fit the sampled simulation results into a predictive performance model.

10.4.1 Artificial Neural Network

Figure 10.4 shows the simplified diagram of a two-layer artificial neural network with one sigmoid hidden layer (which uses sigmoid functions as the calculation kernel) and one linear output layer (which uses linear functions as the calculation kernel). The input and output design parameters are also shown in Fig. 10.4. The essential architectural outputs for energy-performance-area evaluation are the read/write access counts and the read/write miss counts of every level of caches, and read/write access counts of the main memory, and the number of instructions that each microprocessor core has processed. To feed the architectural model, the inputs of the architectural design space are the capacity, associativity, read/write latency of all the cache modules and the main memory, which can be generated from the aforementioned circuit-level model. The statistical architectural model makes an output estimate from given input sets, and it can be treated as a black box that generates predicted outputs as a function of the inputs,

$$\mathrm{L1_{readCount}} = f_1(\mathrm{L1_{capacity}}, \mathrm{L1_{assoc}}, \mathrm{L1_{readLatency}}, \ldots, \mathrm{L3_{capacity}}, \ldots, \mathrm{Mem_{writeLatency}}) \tag{10.1}$$

$$\ldots = \ldots$$

$$\mathrm{L3_{writeMiss}} = f_{n-1}(\mathrm{L1_{capacity}}, \mathrm{L1_{assoc}}, \mathrm{L1_{readLatency}}, \ldots, \mathrm{L3_{capacity}}, \ldots, \mathrm{Mem_{writeLatency}}) \tag{10.2}$$

$$\mathrm{IPC} = f_n(\mathrm{L1_{capacity}}, \mathrm{L1_{assoc}}, \mathrm{L1_{readLatency}}, \ldots, \mathrm{L3_{capacity}}, \ldots, \mathrm{Mem_{writeLatency}}) \tag{10.3}$$

In our model, the input dimension is 14 (vector I_{14}), and the output dimension is 13 (vector O_{13}). The number of neurons in the hidden layer (X) is S, which ranges from 30 to 60 depending on different fitting targets. In Fig. 10.4, W and b are the weight matrix and bias vector of the hidden layer; W′ and b′ are those of the output layer. The feed-forward ANN is calculated as follows:

$$\mathrm{X}_S = \sigma\left(\mathrm{W}_{S\times 14}\mathrm{I}_{14} + \mathrm{b}_S\right) \tag{10.4}$$

$$\mathrm{O}_{13} = \psi\left(\mathrm{W}'_{13\times S}\mathrm{X}_S + \mathrm{b}'_{13}\right) \tag{10.5}$$

where $\sigma(\cdot)$ and $\psi(\cdot)$ are sigmoid and linear functions.

10.4.2 Training and Validation

ANN is able to fit multidimensional mapping problems given consistent data and enough neurons in the hidden layer. The accuracy of the statistical architectural model depends on the number of training samples achieved from actual full-system simulations. In this work, 3,000 cycle-accurate full-system simulation results are collected for each workload. Among each set of 3,000 samples, 2,400 data samples are used for training, 300 are used for testing, and the other 300 are used for validation during the training procedure to prevent over-training [39]. To reduce variability, multiple rounds of cross-validation during which data are rotated among the training, testing, and validation sets are performed using different partitions, and the validation results are averaged over the rounds. Every ANN is configured to have 30–60 hidden neurons and trained using the Levenberg-Marquardt algorithm [40]. The Levenberg-Marquardt algorithm trains the ANN by adjusting the weight matrices and bias vectors based on the data iteratively until the ANN accurately predicts the outputs from the input parameters.

10.4.3 Sample Collection

In this work, we collect samples to evaluate an 8-core chip multiprocessor (CMP).[5] Each core is configured to be a scaled 32nm in-order SPARC-V9-like processor core with a 3.2 GHz frequency. A private L1 instruction cache (I-L1), an L1 data cache (D-L1), and a unified L2 cache (L2) are associated with each core. Eight cores together share an on-die L3 cache. The detailed input design space is listed in Table 10.1. In this work, I-L1 and D-L1 are assumed to have the same design specification.

We use NPB [41] and PARSEC [42] as the experimental workloads. The workload size of the NPB benchmark is CLASS-C (DC has no CLASS-C setting, and CLASS-B is used instead), and the native inputs are used for the PARSEC benchmark to generate realistic program behavior. In total, 23 benchmark applications are evaluated, and we build 23 separate ANN models for the 8-core CMP architecture-level model.[6] Later, all the experimental results are based on the average value of these 23 workloads. We randomly pick design configurations per benchmark and use the Simics full-system simulator [43] to collect sample data. Each Simics simulation is fast forwarded to the pre-defined breakpoint at the code region of interest, warmed up by 1 billion instructions, and then simulated in the detailed timing mode for 10 billion cycles.

[5] We also use the same methodology to collect the data for a 4-core CMP performance model, and the result is shown in Sect. 10.6.4.

[6] Another 23 ANN models are built for the 4-core CMP model.

10.4.4 Model Accuracy

Figure 10.5 illustrates the IPC prediction errors of the architecture-level performance model after training. The x axis shows the relative error between the predicted and the actual values, and the y axis presents the cumulative distribution function. In Fig. 10.5 (middle), we can find that 8 benchmarks in PARSEC have a probability of 80 % to achieve an IPC prediction error of only 0.1 %, and the probability of achieving IPC prediction errors of less than 0.2 % is very close to 100 %.

To measure the model accuracy, we use the metric $error = |predicted - actual|/actual$. The average prediction error is 4.29 %. Figures 10.6, 10.7, 10.8 show three examples of the ANN fitting results: a very accurate fit (0.15 % error), a typical fit (3.06 % error), and the worst fit in this work. Even in the worst case, the prediction error is under 18.71 %.

The prediction results of other output parameters (e.g., L1 read count, L1 write count, L2 read count, L2 write count) are similar to the IPC prediction.

10.5 Joint Design Space Exploration Framework

In this section, we describe how the circuit- and the architecture-level models are combined into a joint design space exploration framework.

Table 10.1 Input design space parameters

Parameter	Range
Processor frequency	3.2 GHz
Processor core	8-core, in-order
I-L1 (D-L1) capacity	8KB to 64 KB
I-L1 (D-L1) associativity	4-way to 8-way
I-L1 (D-L1) read latency	2-cc to 40-cc
I-L1 (D-L1) write latency	2-cc to 700-cc
L2 capacity	64 KB to 512 KB
L2 associativity	8-way or 16-way
L2 read latency	5-cc to 80-cc
L2 write latency	5-cc to 800-cc
L3 capacity	512 KB to 16 MB
L3 associativity	8-way to 32-way
L3 read latency	20-cc to 100-cc
L3 write latency	20-cc to 900-cc
Memory read latency	30-cc to 300-cc
Memory write latency	30-cc to 1,000-cc

10.5.1 Framework Overview

Figure 10.9 shows an overview of this joint circuit-architecture exploration framework. As mentioned, 3,000 randomly generated architecture-level inputs per benchmark workload are used to produce 3,000 corresponding samples in the architectural design space. The samples are then fed into the ANN trainer to establish the architecture-level performance model for each benchmark workload. The trained ANN is used as the architecture-level performance model. The circuit-level inputs are first passed through the memory module performance, energy, and area model, and then fed into the ANN-based architecture-level performance model to generate the predicted architecture-level results such as IPC and power consumption together with the silicon area estimates. When the predicted result does not meet the design requirement, feedback information containing the distance between the design optimization target and the current achieved result is sent to a simulated annealing [44] optimization engine, and a new design trial is generated for the optimization loop. This optimization procedure steps forward iteratively until the design requirement (e.g., best EDP or best EDAP) is achieved or a near-optimal solution is reached. We use a simulated annealing engine to conduct this optimization step, and this is described in Sect. 10.7 in detail.

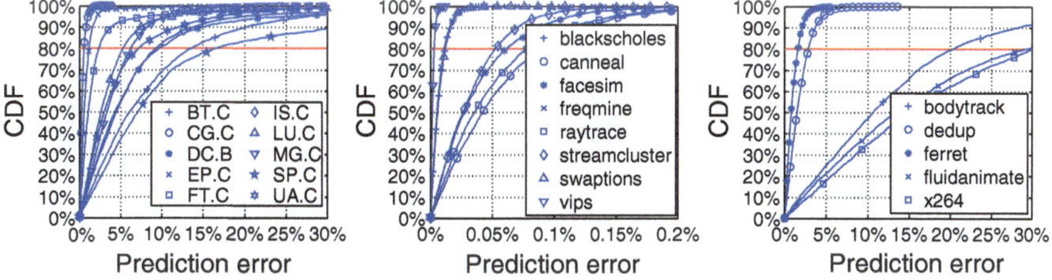

Fig. 10.5 CDF plots of error on IPC prediction of NPB and PARSEC benchmark applications: The x-axis shows the prediction error; the y-axis shows the percentage of data points that achieve the prediction error less than each x value

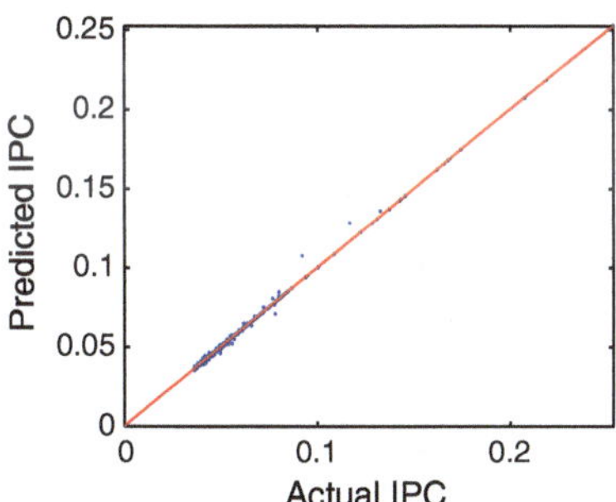

Fig. 10.6 An accurate ANN fitting example: MG from NPB

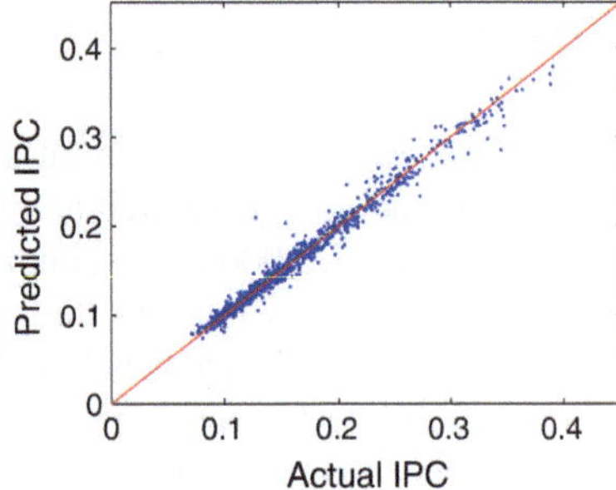

Fig. 10.7 A typical ANN fitting example: dedup from PARSEC

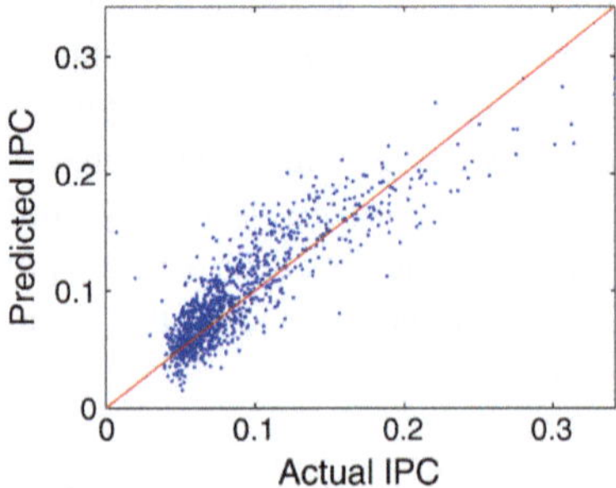

Fig. 10.8 The worst ANN fitting example: x264 from PARSEC

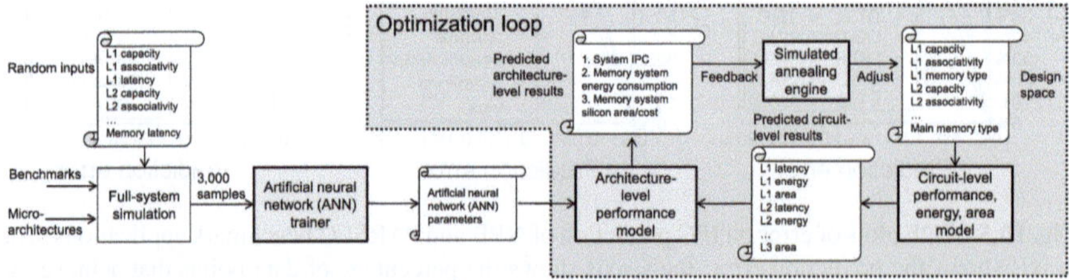

Fig. 10.9 Overview of the optimization framework. Architecture-level models are generated using sampling and an ANN trainer. Circuit-level models are used to estimate the latency, energy, and area of each memory module in the hierarchy. A simulated annealing engine is applied to find the near-optimal solution without exhaustive search

10.5.2 Circuit-Architecture Combination

After obtaining the access activities of each cache level, the memory subsystem power consumption can be calculated. Because the dynamic energy consumption of main memory is proportional to the last-level cache miss rate, we include it as a part of the memory subsystem power consumption for a fair comparison. The power consumption of logic components including the processor cores, on-chip memory controller, and inter-core crossbar are estimated by McPAT [45]. We use a 32 nm technology in this work.

From McPAT, the logic components have 7.41 W leakage power ($\mathrm{P}_{\mathrm{logic,leakage}}$) and 10.98 W peak dynamic power. The run-time dynamic power consumption ($\mathrm{P}_{\mathrm{logic,dynamic}}$) is scaled down from the peak dynamic power according to the actual IPC value. The total power consumption of the processor chip is calculated as follows:

$$\begin{aligned} \mathrm{E}_{\mathrm{memory,dynamic}} &= \sum_{i=1}^{3} \left[N_{\mathrm{readHit}_i} E_{\mathrm{hit}_i} + N_{\mathrm{readMiss}_i} E_{\mathrm{miss}_i} + \left(N_{\mathrm{writeHit}_i} + N_{\mathrm{writeMiss}_i}\right) E_{\mathrm{write}_i}\right] \\ &\quad + N_{\mathrm{readMiss}_3} E_{\mathrm{read}_4} + N_{\mathrm{writeMiss}_3} E_{\mathrm{write}_4} && (10.6) \\ \mathrm{P}_{\mathrm{memory,leakage}} &= 2N_{\mathrm{core}} P_1 + N_{\mathrm{core}} P_2 + P_3 && (10.7) \\ \mathrm{P}_{\mathrm{processor,total}} &= \mathrm{E}_{\mathrm{memory,dynamic}}/T + \mathrm{P}_{\mathrm{logic,dynamic}} + \mathrm{P}_{\mathrm{logic,leakage}} && (10.8) \end{aligned}$$

In Eq. 10.6, N_{readHit_i}, N_{readMiss_i}, N_{writeHit_i}, and $N_{\mathrm{writeMiss}_i}$ are the read count, read miss count, write count, and write miss count of the Level-i cache, which are generated from the ANN-based architecture-level model. E_{hit_i}, E_{miss_i}, and E_{write_i} are the dynamic energy consumption of a hit, miss, and write operation in the Level-i cache, and they are obtained from the RC-based circuit-level model. E_{read_4} and E_{read_4} are the dynamic energy consumption of main memory read and write operations, since we label the main memory as the fourth level of the memory hierarchy. In Eq. 10.7, N_{core} is the number of cores, and P_i represents the leakage power consumption of each cache level. The coefficient 2 is because of the identical data and instruction L1 caches (D-L1 and I-L1) in this work. Equation 10.8 gives the total power consumption where T is the simulation time ($T = 10\,\mathrm{B}/3.2\,\mathrm{GHz} = 3.125$ s according to our experimental setup).

10.6 Design Exploration: An ReRAM Case Study

In this section, we demonstrate how to perform a circuit-architecture joint memory hierarchy design space exploration by adopting emerging ReRAM technology.

10.6.1 ReRAM Technology

ReRAM is an emerging nonvolatile memory technology that involves electro- and thermochemical effects in the resistance change of a metal-oxide-metal system[7]. An ReRAM cell consists of a metal oxide layer sandwiched between two metal electrodes as shown in Fig. 10.10. The electronic behavior of metal/oxide interfaces depends on the oxygen vacancy concentration of the metal oxide layer. Typically,

[7] There are other models explaining the ReRAM working mechanism.

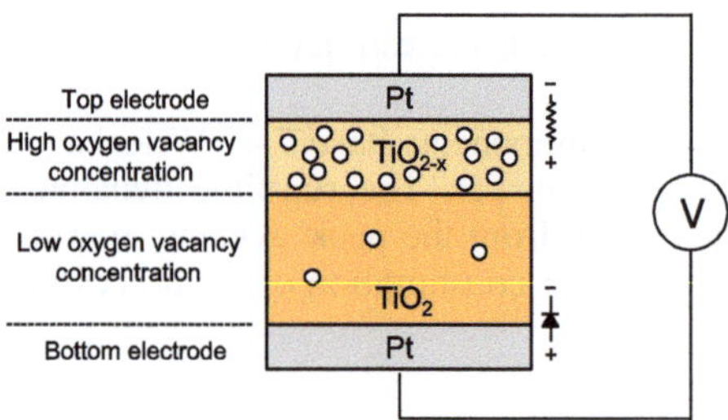

Fig. 10.10 An ReRAM cell example which uses Pt and Ti

Table 10.2 ReRAM technology assumptions

<table>
<tr><th></th><th colspan="2">ReRAM [38]</th></tr>
<tr><th></th><th>MOS-accessed</th><th>Cross-point</th></tr>
<tr><td>Cell size</td><td>20 F^2</td><td>$4F^2$</td></tr>
<tr><td>Write pulse duration</td><td>One pulse 50 ns per pulse</td><td>Two pulses 50 ns per pulse</td></tr>
<tr><td>State-0 resistance</td><td colspan="2">10 kΩ</td></tr>
<tr><td>State-1 resistance</td><td colspan="2">500 kΩ</td></tr>
<tr><td>Half-select resistance</td><td>–</td><td>100 kΩ</td></tr>
<tr><td>Write endurance</td><td colspan="2">10^{12}</td></tr>
</table>

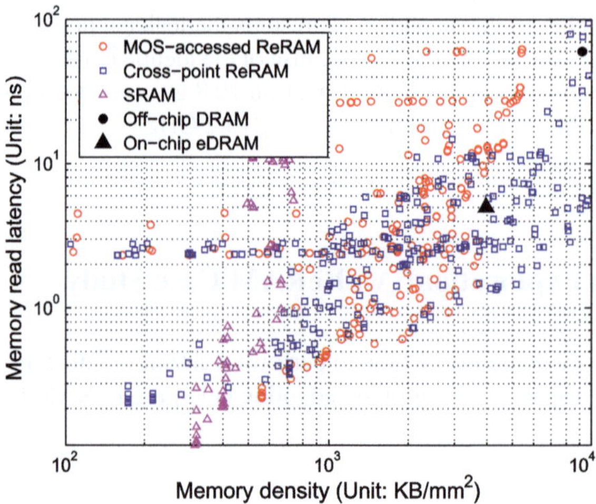

Fig. 10.11 The design spectrum of 32 nm ReRAM: read latency versus density

the metal/oxide interface shows Ohmic behavior in the case of very high doping and rectifying in the case of low doping [46]. In Fig. 10.10, the TiO_x region is semi-insulating indicating lower oxygen vacancy concentration, while the TiO_{2-x} is conductive indicating higher concentration.

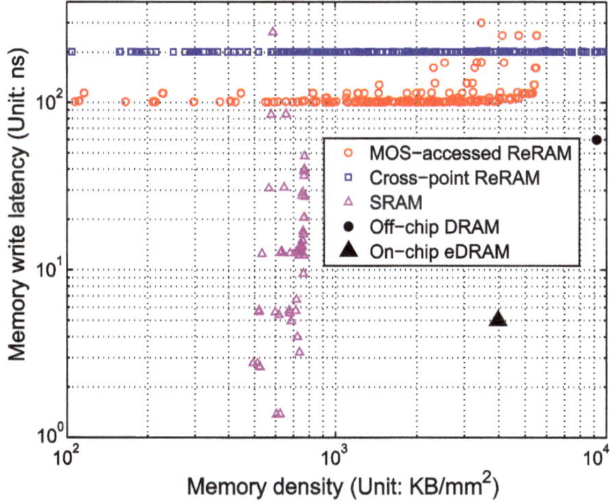

Fig. 10.12 The design spectrum of 32 nm ReRAM: write latency versus density

As an example, we use the ReRAM device parameters shown in Table 10.2 and explore the circuit-level design space at first. Figures 10.11 and 10.12 demonstrate the design spectrum of the emerging ReRAM memory technology. For comparison, the design spectrum of SRAM and DRAM is also shown. Note that MOS-accessed ReRAM and cross-point ReRAM are more than 10X denser than SRAM, and cross-point ReRAM can be as dense as DRAM. In terms of speed, ReRAM has comparable read speed to that of SRAM, but significantly slower write speed. The write latency of MOS-accessed ReRAM is dominated by the switching pulse duration, which is 50 ns in our experiments, and the latency of cross-point ReRAM is twice this due to two-step writes.

10.6.2 Wear-Leveling Assumption

Similar to NAND flash, ReRAM technology has limited write endurance (i.e., the number of times that an ReRAM cell can be overwritten). Many recent techniques [6, 7, 10] have been developed for extending the endurance of PCRAM-based memories, which can be directly borrowed for ReRAM wear leveling. In this case study, we simply assume that these techniques can give us perfect wear leveling. Given this assumption and the current write endurance level of 10^{10}–10^{12} [21, 47, 48], we can explore the possibility of using ReRAM technology in L2 and L3 caches. In later experiments, we add an additional 2 ns access latency due to the consideration of wear-leveling hardware overhead.

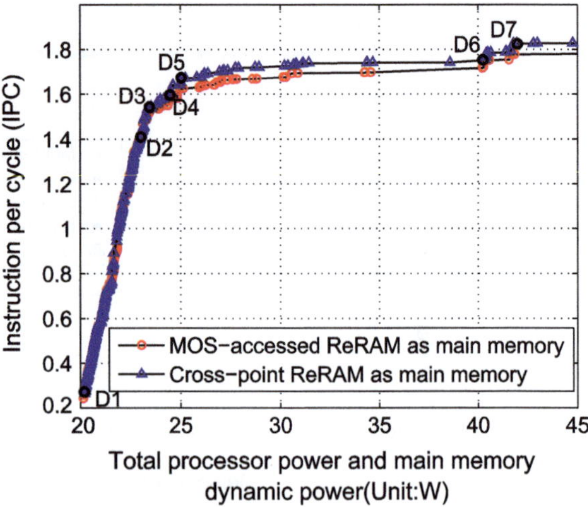

Fig. 10.13 *Pareto curves*: energy and performance trade-off of the memory hierarchy. Main memory dynamic power is included for a fair comparison

10.6.3 Memory Hierarchy Design Exploration

We next explore architecture-level memory hierarchy design space via ANN to show the Pareto-optimal curves of the trade-off range after adopting ReRAM technology. In this step, we separate the cache design space (L1, L2, and L3) and the memory design space. We assume the main memory is built by either cross-point ReRAMs that are optimized for density or MOS-accessed ReRAMs that are optimized for latency. Table 10.3 lists the timing and area parameters of MOS-accessed and cross-point ReRAM main memory solutions.

Focusing on the design space exploration of ReRAM-based memory hierarchies, Fig. 10.13 shows the Pareto-optimal curves of the power-performance trade-off. The x-axis is the total power consumption of the processor chip, and the y-axis is the IPC performance. It can be observed from Fig. 10.13 that a great amount of power consumption can be reduced by only taking a small amount of performance degradation. For instance, as shown in Fig. 10.13, design option D4 (using SRAM L1 and L2 caches but ReRAM L3 cache) reaches 1.610 IPC by consuming 24.50 W total power. Compared to design option D7 (using SRAM-only cache hierarchy) that reaches 1.826 IPC but consumes 42.00 W power, the achieved power reduction is 42 % but the performance degradation is only 12 %. The design option also meets the

Table 10.3 MOS-accessed and cross-point ReRAM main memory parameters (1 Gb, 8-bit, 16-bank)

	MOS-accessed	Cross-point
Die area	129 mm^2	48 mm^2
Read latency	6.2 ns	10.0 ns
Write latency	54.9 ns	107.1 ns
Burst read latency	4.3 ns	4.3 ns
Burst write latency	4.3 ns	4.3 ns

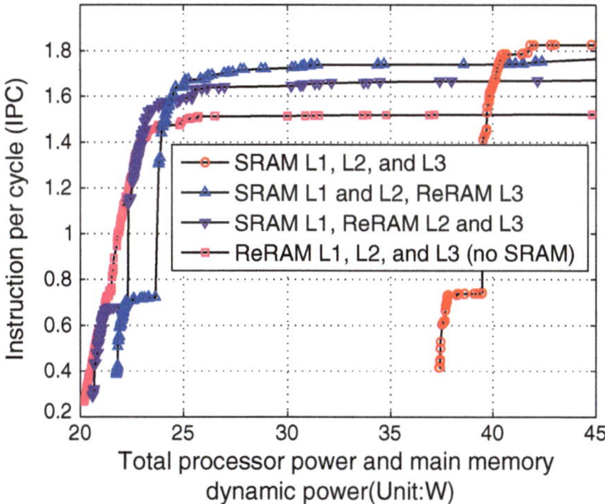

Fig. 10.14 *Pareto curves* (cross-point ReRAM as main memory): energy and performance trade-off under different constraints on ReRAM deployment

constraint set by 10^{10} write endurance as discussed in Sect. 10.6.2. If the ReRAM write endurance is assumed to be 10^{12}, more aggressive options (e.g., using L2 ReRAM caches) can further reduce the power consumption. For example, design option D3 (using ReRAM L2 and L3 caches) reaches 1.545 IPC by consuming only 23.52 W total power. To show how different cache hierarchy designs have been explored, we list the design parameters of 7 design options (D1 to D7) in Table 10.4.

We find that the Pareto-optimal curves are composed of several segments, such as D1-to-D2, D4-to-D6. The joints between every two segments represent the place where SRAM/ReRAM replacement occurs. Such replacements can be found in Fig. 10.14. In general, IPC improvements are achieved by adding more SRAM resources, and greater reductions in power consumption come from replacing SRAM with ReRAM. Figure 10.14 shows an ReRAM-only cache hierarchy is on the global Pareto-front but it achieves less than 0.30 IPC, and that segment has a large slope. Thus, it suggests SRAM should still be deployed at least in L1 caches. But, starting form L2, ReRAM cache deployment can achieve considerable additional power reduction by only sacrificing a small amount of performance. This is especially true for a hybrid on-chip cache hierarchy with SRAM L1/L2 caches and ReRAM L3 caches. Figure 10.14 shows that in this region, the total power consumption can be lowered to 24.50 W but the IPC is only degraded from 1.826 to 1.610 (i.e., design option D4 in Fig. 10.13).

Another benefit obtained from using ReRAM caches is silicon area reduction. Figure 10.15 shows the Pareto-optimal curves of cache area and performance trade-offs, which have similar shapes to the ones in the power-performance trade-off as shown in Fig. 10.15. The processor core area (including memory controller and crossbar) is 45.6 mm^2 as estimated by McPAT [45]. Achieving the highest performance using pure-SRAM caches costs at least another 12 mm^2 of silicon area while

Table 10.4 On-die cache hierarchy design parameters of 7 design options

	D1	D2	D3	D4	D5	D6	D7
L1 capacity	32 KB	8 KB	8 KB	8 KB	32 KB	8 KB	8 KB
L1 associativity	8	8	4	4	4	4	4
L1 memory type	M-ReRAM	SRAM	SRAM	SRAM	SRAM	SRAM	SRAM
L1 optimized for	L	WP	RL	WP	RL	RP	RL
L1 sensing scheme	EX	IN	IN	IN	IN	IN	IN
L2 capacity	512 KB	512 KB	512 KB	64 KB	64 KB	64 KB	64 KB
L2 associativity	8	16	8	16	8	8	8
L2 memory type	M-ReRAM	M-ReRAM	M-ReRAM	SRAM	SRAM	SRAM	SRAM
L2 optimized for	L	L	L	L	WE	WE	RE
L2 sensing scheme	IN	IN	IN	IN	IN	IN	IN
L3 capacity	16 MB	16 MB	8 MB	8 MB	64 MB	8 MB	8 MB
L3 associativity	16	16	8	8	8	8	8
L3 memory type	M-ReRAM	M-ReRAM	M-ReRAM	X-ReRAM	M-ReRAM	SRAM	SRAM
L3 optimized for	L	L	L	L	RE	L	WP
L3 sensing scheme	IN	IN	EX	IN	EX	IN	IN
IPC	0.265	1.410	1.545	1.610	1.677	1.751	1.826
Power consumption (W)	20.20	22.98	23.52	24.50	25.08	40.31	42.00
Silicon area (mm^2)	49.45	49.64	48.25	48.16	62.92	57.41	58.23

*Memory type abbreviation: M-ReRAM = MOS-accessed ReRAM, X-ReRAM = Cross-point ReRAM; Optimization abbreviation: *RL* read latency, *WL* write latency, *RE* read energy, *WE* write energy, *RP* read EDP, *WP* write EDP, *L* leakage, *A* area; sensing scheme abbreviation: *IN* internal, *EX* external

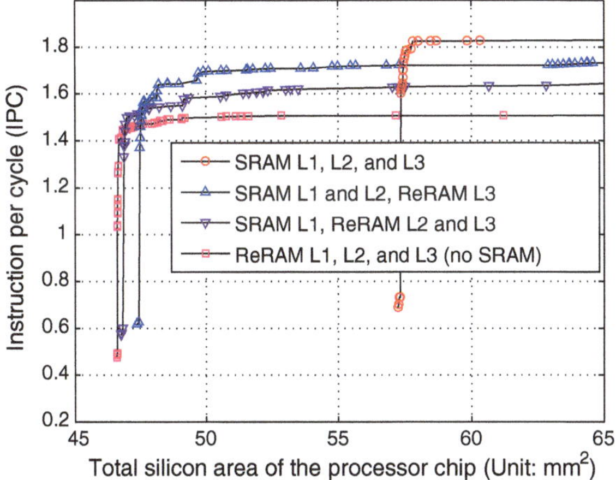

Fig. 10.15 *Pareto curves* (cross-point ReRAM as main memory): cache area and performance trade-off under different ReRAM deployments

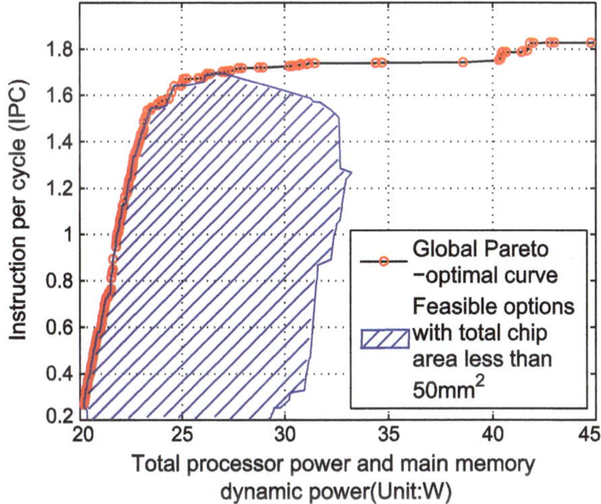

Fig. 10.16 The global Pareto-optimal curve (cross-point ReRAM as main memory) and feasible design options with total chip area less than 50 mm^2

replacing the SRAM L3 cache with ReRAM can save more than 7 mm^2 in chip area by degrading performance from an IPC of 1.82–1.70. We show the feasible region of designs with less than 50 mm^2 total cache area in Fig. 10.16. This result is extremely useful in the low-cost computing segment where the performance requirement is just-in-time but the chip cost is the first priority. Figure 10.16 also indicates using ReRAM caches can reduce power consumption and silicon area at the same time, which further improves EDAP.

10.6.4 Other Case Studies

As a general-purpose tool, our circuit-architecture joint space exploration framework is not limited to the case study of ReRAM-based cache hierarchy on an 8-core CMP system. Using the same circuit-level model, we can build another circuit library for PCRAM-based cache modules (PCRAM parameters are listed in Table 10.5). Furthermore, using the same architecture-level model, a 4-core CMP performance model can also be established.

Figures 10.17, 10.18 demonstrate how the framework can be easily adapted and help the design space exploration for an 8-core PCRAM-based and a 4-core ReRAM-based cache hierarchy in terms of the power-performance trade-off and the area-performance trade-off, respectively. The tool quickly provides the estimation that:

- In a 8-core CMP system, adopting *PCRAM* technology in L3 caches can achieve an EDP improvement of 16 % and an EDAP improvement of 25 %.
- In a *4-core* CMP system, adopting ReRAM technology in L3 caches can achieve an EDP improvement of 25 % and an EDAP improvement of 35 %.
- In a 8-core CMP system, adopting ReRAM technology in L3 caches can achieve an EDP improvement of 28 % and an EDAP improvement of 39 %.

From the 8-core PCRAM/ReRAM comparison, we can quickly find that ReRAM's fast switching property makes ReRAM outperform PCRAM in terms of IPC.

10.7 Design Optimization

Running full design space exploration using an exhaustive search is time-consuming and may not be necessary in most cases. Therefore, to use this joint circuit-architecture model as a practical memory hierarchy design assistant, an efficient optimization method is required. In this work, we use a simplified simulated annealing [44] algorithm to find a near globally optimal solution. The simulated annealing heuristic is described in Algorithm 1.

In this optimization methodology, we first randomly choose an initial design option, $s0$, and calculate its annealing energy function from the joint circuit-architecture model. The annealing energy function can be EDP, EDAP, or any other energy-performance-area combination. The optimization loop continuously tries neighbor-

Table 10.5 PCRAM technology assumptions [16]

Cell size	$36F^2$
Reset pulse duration	100 ns
Set pulse duration	300 ns
State-0 resistance	5 kΩ
State-1 resistance	500 kΩ

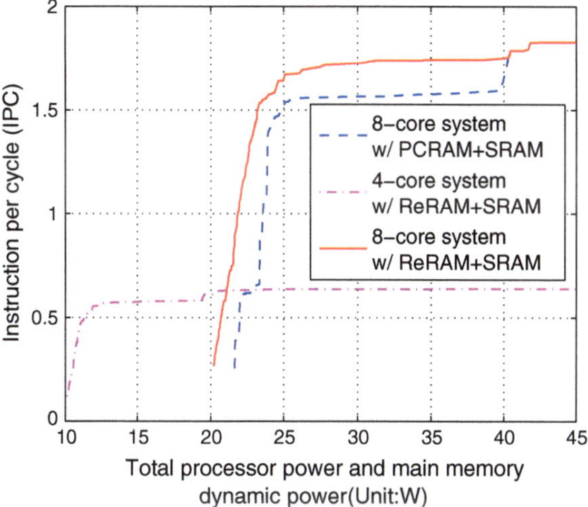

Fig. 10.17 The power-performance *Pareto curves*: (*1*) an 8-core system with PCRAM+SRAM; (*2*) a 4-core system with ReRAM+SRAM; (*3*) an 8-core system with ReRAM+SRAM

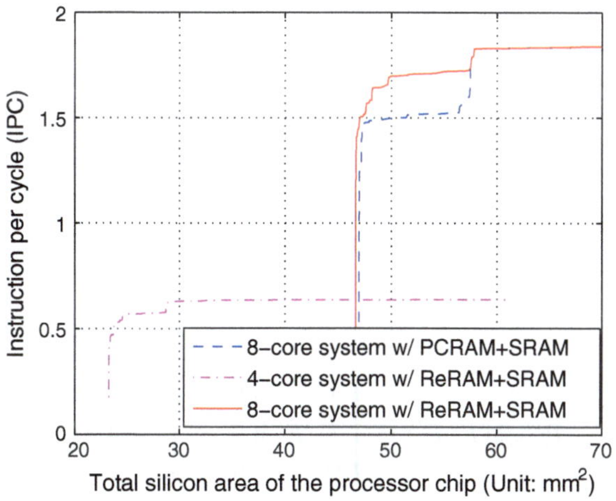

Fig. 10.18 The area-performance *Pareto curves*: (*1*) an 8-core system with PCRAM+SRAM; (*2*) a 4-core system with ReRAM+SRAM; (*3*) an 8-core system with ReRAM+SRAM

ing options[8] of the current one. If the new design option is better than the previous one, it is adopted unconditionally; if not, it is adopted with probability depending on an *acceptance* function. The *acceptance* probability, P_{accept}, is defined as,

---

[8] In this work, a neighboring option is generated by changing two parameters from the parameter set of L1 capacity, L1 associativity, L1 memory type, L2 capacity, L2 associativity, L2 memory type, L3 capacity, L3 associativity, and L3 memory type.

Algorithm 1 Design space optimization algorithm

```
state=s0, energy=E(state)
repeat
  new_state = neighbor(state), new_energy = E(new_state)
  if new_energy < energy then
    state = new_state, energy = new_energy {Accept unconditionally}
  else if T(energy, new_energy) > random() then
    state = new_state, energy = new_energy {Accept with probability}
  end if
until energy stops improving in the last K rounds
return state
```

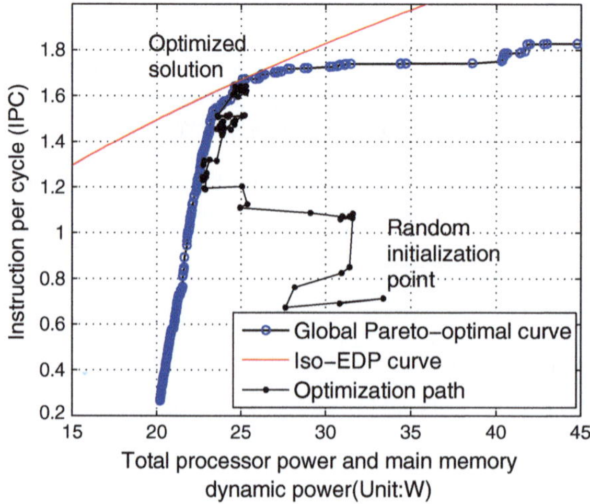

Fig. 10.19 The path of EDP optimization

$$P_{\text{accept}}(E, E') = \begin{cases} 1 & \text{if} E' < E \\ E/E' & \text{if } E \leq E' < 1.3E \\ 0 & \text{otherwise} \end{cases} \tag{10.9}$$

where E is the old energy, and E' is the new energy.

Figure 10.19 shows how the simulated annealing algorithm eventually evolves an initial random design option to a near globally optimal one in terms of EDP. In addition, Fig. 10.20 shows the EDAP optimization path.

Compared to exhaustive search of the same design space that takes more than 8 hours on an 8-core Xeon X5570 microprocessor, the proposed optimization methodology usually finds the near-optimal values in less than 30 seconds. This optimization scheme provides an almost instant design decision given any specified performance, energy, or area requirements. Furthermore, it becomes feasible to integrate this model into higher level tools that consider not only the memory system design trade-offs but also design trade-offs within microprocessor cores [25].

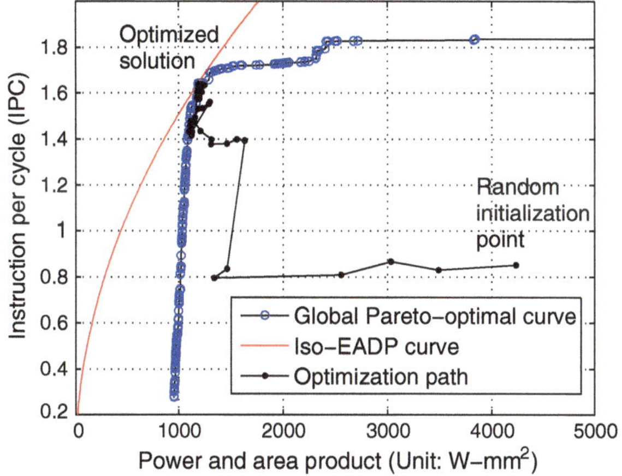

Fig. 10.20 The path of EDAP optimization

10.8 Discussion

Power consumption has been an issue for many years. Our exploration and optimization results demonstrate that both the EDP and the EDAP optimal points are close to the y-axis on the IPC versus power plot. In this design space region, ReRAM resources are adopted in the memory hierarchy design (for example, using ReRAM L3 caches, or more aggressively using ReRAM L2 and L3 caches). Even if performance constraints are applied, using ReRAM starting with the L3 cache always brings energy efficiency. Moreover, locating these energy-optimal points on the IPC versus area plot also shows significant silicon area savings achieved from ReRAM without incurring much performance degradation. Compared to the best values of pure-SRAM designs, the introduction of ReRAM in L3 caches improves EDP and EDAP by 28 and 39 % on a scaled 32 nm 8-core SPARC-V9-like processor chip, respectively. The memory technology shift from SRAM to ReRAM achieves these improvements for the following reasons:

Table 10.6 Overview of the proposed universal memory hierarchy

Level	L1 cache	L2 cache
Memory type	SRAM	SRAM or MOS-accessed ReRAM
Endurance requirement	10^{13} [Section 10.6.2]	10^{11} [Section 10.6.2]
Level	L3 cache	Main memory
Memory type	MOS-accessed or cross-point ReRAM	MOS-accessed or cross-point ReRAM
Endurance requirement	10^{10} [Section 10.6.2]	10^{8} [7]

- The compact ReRAM module size greatly reduces the silicon area used for on-chip memories (EDAP improvement) or allows more on-chip memory resource deployment to improve the performance (EDP and EDAP improvement);
- The relatively smaller ReRAM size implies shorter wordlines and bitlines in the ReRAM cell array and thus reduces the dynamic energy consumption per memory access (EDP and EDAP improvement);
- The nonvolatility property of ReRAM eliminates the leakage energy consumption of memory cells (EDP and EDAP improvement).

Therefore, we envision a heterogeneous memory hierarchy as summarized in Table 10.6. In such a hierarchy, SRAM is used in L1 and L2 caches, MOS-accessed ReRAM may be used in L3 or even in L2 caches if ReRAM technology keeps improving (e.g., improvement on write speed and write endurance), and low-cost cross-point ReRAM may be used in L3 caches and main memory.

10.9 Conclusion

In the next era of computing, we need more energy-efficient and cost-effective computing. However, conventional SRAM and eDRAM technologies used in memory hierarchy designs have problems in reducing power consumption and silicon area with scaling. On the other hand, many emerging nonvolatile memory technologies such as STTRAM, PCRAM, and ReRAM have been researched and the corresponding prototypes have been demonstrated. These new memory technologies bring desired features such as high density, fast access, good scalability, and nonvolatility, and they are potentially useful in many levels of future energy-efficient and cost-effective memory hierarchies. However, such emerging memory technologies are still new, and there are too many uncertainties in evaluating their actual impact on the future memory hierarchy design. Therefore, it is necessary to have a framework that can model the circuit-level trade-offs among performance, energy, and area and can leverage such design variety into the architecture-level memory hierarchy.

In this work, we first build a circuit-level performance, energy, and area estimation model for emerging memory technologies, then use this model to explore a wide-range of memory module implementations, and generate a memory module library with various optimized designs. After that, we integrate this circuit-level model into an ANN-based architecture-level model and create a general performance-energy-area optimization framework for the memory hierarchy design in a joint circuit-architecture design space. Our validation results show that the proposed framework is sufficiently accurate for the purpose of design space exploration, and by using this framework, we are able to rapidly explore a very large space of memory hierarchy designs and find good solutions in terms of energy-performance-area trade-offs. Moreover, we use this framework to evaluate new memory technologies such as ReRAM. Our experimental results reveal the memory design preference for ReRAM in an 8-core CMP setting when the design targets EDP or EDAP goals. Our results show using ReRAM starting from L3 caches can achieve a 28 % EDP improvement and a 39 % EDAP improvement, which means the trade-offs in designing ReRAM memory hierarchy can greatly boost the energy efficiency or cost efficiency with only a slight impact on the IPC.

In general, this work is the initial effort to study the feasibility of building an energy-efficient or cost-efficient memory hierarchies by adopting emerging memory technologies. We believe this work is only the first step toward a new generation of energy-efficient and cost-efficient heterogeneous computer memory hierarchies.

References

1. Udipi, A. N., et al. (2010). Rethinking DRAM design and organization for energy-constrained multi-cores. *In Proceedings of the International Symposium on Computer Architecture* (pp. 175–186).
2. Meng, Y., et al. (2005). On the limits of leakage power reduction in caches. *Architecture: In Proceedings of the International Symposium on High-Performance Computer* (pp. 154–165).
3. International technology roadmap for semiconductors. Process integration, devices, and structures 2010 update. (2010). http://www.itrs.net/.
4. Kalla, R., et al. (2010). POWER7: IBM's next-generation server processor. *IEEE Micro, 30*(2), 7–15.
5. Lee, B. C., et al. (2009). Architecting phase change memory as a scalable DRAM alternative. *Architecture: In Proceedings of the International Symposium on Computer* (pp. 2–13).
6. Zhou, P., et al. (2009). A durable and energy efficient main memory using phase change memory technology. *Architecture: In Proceedings of the International Symposium on Computer* (14–23).
7. Qureshi, M. K., et al. (2009b). Scalable high performance main memory system using phase-change memory technology. *Architecture: In Proceedings of the International Symposium on Computer* (pp. 24–33).
8. Qureshi, M. K. (2009a). Enhancing lifetime and security of PCM-based main memory with start-gap wear leveling. *In Proceedings of the International Symposium on Microarchitecture* (pp. 14–23).
9. Seong, N. S., et al. (2010). Security refresh: Prevent malicious wear-out and increase durability for phase-change memory with dynamically randomized address mapping. *In Proceedings of the International Symposium on Computer Architecture* (pp. 383–394).
10. Schechter, S. (2010). Use ECP, not ECC, for hard failures in resistive memories. *Architecture: In Proceedings of the International Symposium on Computer* (pp. 141–152).
11. Dong, X., et al. (2008). Circuit and microarchitecture evaluation of 3D stacking magnetic RAM (MRAM) as a universal memory replacement. *In Proceedings of the Design Automation Conference* (pp. 554–559).
12. Sun, G., et al. (2009). A novel 3D stacked MRAM cache architecture for CMPs. *Architecture: In Proceedings of the International Symposium on High-Performance Computer* (pp. 239–249).
13. Smullen, C. W., et al. (2011). Relaxing non-volatility for fast and energy-efficient STT-RAM caches. *In Proceedings of the International Symposium on High Performance, Computer Architecture* (pp. 50–61).
14. Lee, K.-J., et al. (2008). A 90 nm 1.8 V 512 Mb diode-switch PRAM with 266MB/s read throughput. *IEEE Journal of Solid-State Circuits, 43*(1), 150–162.
15. Yoshitaka, S., et al. (2009). Cross-point phase change memory with $4F^2$ cell size driven by low-contact-resistivity poly-Si diode. *In Proceedings of the Symposium on VLSI Technology* (pp. 24–25).

16. De Sandre, G., et al. (2010). A 90nm 4Mb embedded phase-change memory with 1.2V 12ns read access time and 1MB/s write throughput. *In Proceedings of the International Solid-State Circuits Conference* (pp. 268–269).
17. Kawahara, T., et al. (2007). 2 Mb spin-transfer torque RAM (SPRAM) with bit-by-bit bidirectional current write and parallelizing-direction current read. *In Proceedings of the International Solid-State Circuits Conference* (pp. 480–617).
18. Tsuchida, K., et al. (2010). A 64 Mb MRAM with clamped-reference and adequate-reference schemes. *In Proceedings of the International Solid-State Circuits Conference* (pp. 268–269).
19. Chen, Y.-C., et al. (2003). An access-transistor-free (0T/1R) non-volatile resistance random access memory (RRAM) using a novel threshold switching, self-rectifying chalcogenide device. *In Proceedings of the International Electron Devices Meeting* (pp. 750–753).
20. Kim, K.-H., et al. (2010). Nanoscale resistive memory with intrinsic diode characteristics and long endurance. *Applied Physics Letters*, *96*(5), 053106.1-053106.3.
21. Sheu, S.-S., et al. (2011). A 4 Mb embedded SLC resistive-RAM macro with 7.2 ns read-write random-access time and 160 ns MLC-access capability. *In Proceedings of the IEEE International Solid-State Circuits Conference* (pp. 200–201).
22. Wilton, S. J. E., & Jouppi, N. P. (1996). CACTI: An enhanced cache access and cycle time model. *IEEE Journal of Solid-State Circuits*, *31*, 677–688.
23. Thoziyoor, S., et al. (2008b). CACTI 5.1 technical report. Technical report HPL-2008-20. HP labs.
24. Amrutur, B. S., & Horowitz, M. A. (2000). Speed and power scaling of SRAM's. *IEEE Journal of Solid-State Circuits*, *35*(2), 175–185.
25. Azizi, O., et al. (2010). Energy-performance tradeoffs in processor architecture and circuit design: A marginal cost analysis. *In Proceedings of the International Symposium on Computer Architecture* (pp. 26–36).
26. Joseph, P. J., et al. (2006a). Construction and use of linear regression models for processor performance analysis. *In Proceedings of the International Symposium on High-Performance Computer Architecture* (pp. 99–108).
27. Lee, B. C., & Brooks, D. M. (2006). Accurate and efficient regression modeling for microarchitectural performance and power prediction. *In Proceedings of the International Conference on Architectural Support for Programming Languages and Operating Systems* (pp. 185–194).
28. Ipek, E., et al. (2008). Efficient architectural design space exploration via predictive modeling. *ACM Transactions on Architecture and Code Optimization*, *4*(4), 1:1–1:34.
29. Joseph, P. J., et al. (2006b). A predictive performance model for superscalar processors. *In Proceedings of the International Symposium on Microarchitecture* (pp. 161–170).
30. Dubach, C., et al. (2007). Microarchitectural design space exploration using an architecture-centric approach. *In Proceedings of the International Symposium on Microarchitecture* (pp. 262–271).
31. Muralimanohar, N. (2008). Architecting efficient interconnects for large caches with CACTI 6.0. *IEEE Micro*, *28*(1), 69–79.
32. Thoziyoor, S., et al. (2008a). A comprehensive memory modeling tool and its application to the design and analysis of future memory hierarchies. *In Proceedings of the International Symposium on Computer Architecture* (pp. 51–62).
33. International technology roadmap for semiconductors. The model for assessment of cmoS technologies and roadmaps (MASTAR). (2010). http://www.itrs.net/models.html.
34. Sutherland, I. E., et al. (1999). Logical effort: Designing fast CMOS circuits. Morgan Kaufmann.
35. Zhang, Y., et al. (2007). An integrated phase change memory cell with ge nanowire diode for cross-point memory. *In Proceedings of the IEEE Symposium on VLSI, Technology* (pp. 98–99).
36. Lee, M.-J., et al. (2007). 2-stack 1D–1R cross-point structure with oxide diodes as switch elements for high density resistance RAM applications. *In Proceedings of the IEEE International Electron Devices Meeting* (pp. 771–774).
37. Kau, D. C., et al. (2009). A stackable cross point phase change memory. *In Proceedings of the IEEE International Electron Devices Meeting*, 27.1.1-27.1.4.

38. Xu, C., et al. (2011). Design implications of memristor-based RRAM cross-point structures. *In Proceedings of the Design, Automation and Test in, Europe* (pp. 1–6).
39. Sarle, W. S. (1995). Stopped training and other remedies for overfitting. *In Proceedings of the Symposium on the Interface of Computing Science and, Statistics* (pp. 55–69).
40. Marquardt, D. W. (1963). An algorithm for least-squares estimation of nonlinear parameters. *Journal of the Society for Industrial and Applied Mathematics*, *11*(2), 431–441.
41. NASA advanced supercomputing (NAS) division. The NAS parallel benchmarks (NPB) 3.3. http://www.nas.nasa.gov/Resources/Software/npb.html.
42. Bienia, C., et al. (2008). The PARSEC benchmark suite: characterization and architectural implications. *In Proceedings of the International Conference on Parallel architectures and Compilation Techniques* (pp. 72–81).
43. Magnusson, P. S., et al. (2002). Simics: A full system simulation platform. *Computer*, *35*(2), 50–58.
44. Kirkpatrick, S., et al. (1983). Optimization by simulated annealing. *Science Magazine*, *220*(4598), 671–680.
45. Li, S., et al. (2009). McPAT: An integrated power, area, and timing modeling framework for multicore and manycore architectures. *In Proceedings of the International Symposium on Microarchitecture* (pp. 469–480).
46. Yang, J. J., et al. (2008). Memristive switching mechanism for metal/oxide/metal nanodevices. *Nature Nanotechnology*, *3*(7), 429–433.
47. Kim, Y.-B., et al. (2011). Bi-Layered RRAM with unlimited endurance and extremely uniform switching. *In Proceedings of the Symposium on VLSI Technology* (pp. 52–53).
48. Eshraghian, K., et al. (2010). Memristor MOS content addressable memory (MCAM): Hybrid architecture for future high performance search engines. *IEEE Transactions on Very Large Scale Integration (VLSI) Systems*, *99*, 1–11.

Chapter 11
Ferroelectric Nonvolatile Processor Design, Optimization, and Application

Yongpan Liu, Huazhong Yang, Yiqun Wang, Cong Wang, Xiao Sheng, Shuangchen Li, Daming Zhang and Yinan Sun

Abstract Nonvolatile processor (NVP) is one of the most promising techniques to realize energy-efficient computing systems with zero standby power, instant-on features, high resilience to power failures, and fine-grained power management. As flip-flops as well as static random access memories (SRAM) should be replaced by nonvolatile memory in an NVP, it puts rigid requirements on the nonvolatile memories, such as nearly unlimited operation cycles, ultra-short access time and easy integration to CMOS technology. Ferroelectric memory can meet those metrics with good energy efficiency, which is appropriate to realize an NVP. However, there are several major design problems, such as the unknown design flow of a ferroelectric NVP, the nontrivial area overheads as well as the absent of the real application systems. To overcome those challenges, we present the first fabricated NVP with zero standby power, 7 μs sleep time and 3 μs wake-up time, consisting of a flip-flop controller (FFC), a distributed memory architecture and a voltage detection system. Compared with an existing industry processor, it can achieve over 30–100× speedup on the wake-up/sleep time and 70× energy savings on the data backup and recall operations. Meanwhile, the ferroelectric NVP exhibits comparative performance and power consumption in normal operations. To attack its area challenges, we design a parallel compare-and-compress architecture (PaCC) and an appropriate vector selecting method to reduce the number of nonvolatile registers by 70–80 % with less than 1 % overflow possibility, which leads to up to 30 % processor area savings. Another segment-based parallel compression (SPaC) architecture is proposed to trade off the chip area and the backup speed. It divides the system vector into several segments and compresses them in parallel. Compared with the PaCC solution, it can improve the backup speed by 83 % with 16 % area savings over the full replacement architecture. Finally, we demonstrate two kinds of battery-less sensor nodes based on the NVP for the first time.They aimed at the moving object detection and the body sensor appli-

Y. Liu (✉) · H. Yang · Y. Wang · C. Wang · X. Sheng · S. Li · D. Zhang · Y. Sun
Department of Electronic Engineering, Tsinghua University, Beijing, Peoples Republic of China
e-mail: ypliu@tsinghua.edu.cn

Y. Xie (ed.), *Emerging Memory Technologies*,
DOI: 10.1007/978-1-4419-9551-3_11,

cations. As both systems are powered by energy-harvesting systems, they eliminate the battery lifetime constraints and work reliably under frequency power failures.

11.1 Introduction

In embedded applications, low-power processors can be shut down to eliminate leakage power. However, their states in volatile storage elements, such as flip-flops, registers, and static random access memories (SRAM), will be lost, because the charge on the capacitors is quickly drained without power supply. To keep those states, they need to backup the states into the secondary nonvolatile storages, such as flash memory or hard disks. But transferring data between a volatile processor and secondary nonvolatile storages leads to long delays, large switching energy consumption, and vulnerability to power failures. Therefore, efficient alternatives are needed to remove such limitations.

Nonvolatile processor (NVP) is one of the most promising alternatives, which replaces flip-flops, registers, and SRAMs in processors with nonvolatile ones. As the nonvolatile memory can store the system states locally, data transfer overheads are significantly reduced. Researchers have evaluated various nonvolatile memories in processors [1–4] to show the following advantages: (1) zero standby power: the processor can retain its state without power, while the traditional ones suffer from the increasing leakage power to keep data; (2) instant-on and instant-off features: the processor can resume its work within several cycles from the stalled point, while the traditional one needs millions of initializing cycles; (3) high resilience to power failures: the processor can work reliably under the environments with frequent power interrupts, such as energy-harvesting and wireless-powered applications [2]; (4) fine-grained power management supported [1]: the processor can be shut down whenever possible due to the ultra-low energy and fast recovering characteristics. Therefore, research on NVPs is interesting.

To implement an NVP, appropriate memory technologies are needed. Flash is a mature high-density nonvolatile memory in commercial microcontrollers [5, 6]. However, it has drawbacks of low endurance, slow writing speed, block erasing pattern, and high mask cost as distributed nonvolatile registers. Phase-change random access memory (PCRAM) has the potential to be used as a cache with SRAMs [7]. However, its asymmetric reading/writing characteristics and limited lifetime hinder its application as flip-flops and registers. Among existing nonvolatile memories [8], ferroelectric RAM (FeRAM) and magnetic RAM (MRAM) emerge as the most promising candidates for NVPs because of their nearly unlimited operation cycles, ultra-short access time and easy integration to CMOS technology. As FeRAM shows better energy efficiency and mature fabrication process, this chapter will focus on the ferroelectric NVP.

There are several major design challenges from the ferroelectric NVP. First, though previous work had reported the implementation of nonvolatile flip-flops (NVFF) [9–11], the complete design flow to implement a ferroelectric NVP is still absent. Besides NVFFs and RAMs, it is necessary to build flip-flop controllers (FFCs), distributed memory architecture, peripheral power manage circuits, and specific design tools. However, none of previous work solved those problems. Second, as most of popular NVFFs adopt hybrid architecture, researchers observed nontrivial area overheads. In [2, 10, 12], they reported over 40 % increasing of chip area due to the replacement of nonvolatile registers. It will have a great impact on the cost and yield. Therefore, investigating techniques to realize area-efficient NVPs is necessary. Finally, the absent of a real NVP hinders the exploration of its application. There are no reports on how to use NVPs and combine their characteristics with appropriate applications [13–15], such as smart sensors and energy-harvesting devices. To overcome those challenges from NVP design, optimization, and applications, our contributions are listed as below:

1. We present the first fabricated NVP with zero standby power, 7 μs sleep time, and 3 μs wake-up time, consisting of FFC, distributed memory architecture and voltage detection system.
2. We demonstrate that the NVP can achieve over 30–100× speedups on the wake-up/sleep time and 70× energy savings on the data backup and recall operations compared with an existing industry processor. Meanwhile, the NVP exhibits comparative performance and power consumption in normal operations.
3. We propose a parallel compare-and-compress architecture (PaCC) to reduce the area of nonvolatile registers. With an appropriate vector selecting method, the compression codec can reduce the number of nonvolatile registers by 70–80 % with less than 1 % overflow possibility, which leads to up to 30 % processor area savings.
4. We present a SPaC architecture to trade off the chip area and the backup speed. With a hybrid off-line/online partition method, it divides the system vector into several segments and compresses them in parallel. Compared with the PaCC solution, it can improve the backup speed by 83 % with 16 % area savings over the full replacement architecture.
5. We demonstrate two kinds of battery-less sensor nodes based on the NVP. They aimed at moving vehicle detection and body sensor applications. As both systems are powered by energy-harvesting devices, they eliminate the battery lifetime constraints and work reliably under frequency power failures.

The chapter is organized as follows: Sect. 11.2 discusses the mechanism of ferroelectric memory and the models. We introduce the design of NVP in Sect. 11.3 and present the related area optimizing techniques in Sect. 11.4. We discuss two applications of NVP in Sect. 11.5 and present the related work in Sect. 11.6. Section 11.7 concludes the chapter.

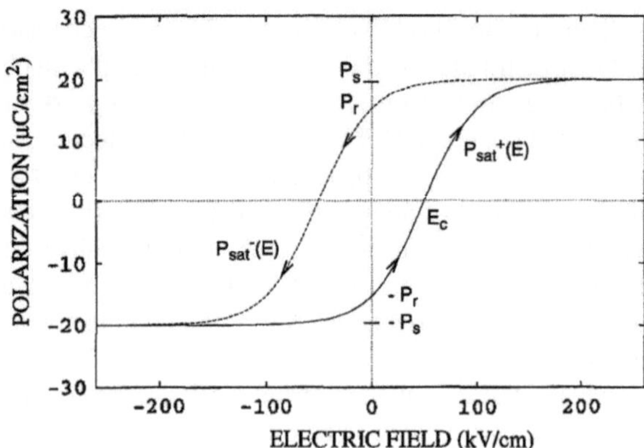

Fig. 11.1 Polarization–voltage hysteretic loop [16]

11.2 Background

This section describes the mechanism and the model of ferroelectric capacitor (FeCap) in the memory cells. It is an indispensable background for the ferroelectric NVP design.

11.2.1 Ferroelectric Material Mechanism

A ferroelectric material refers to a material with two stable polarization states. It switches from one state to another if an external strong electric field is applied. Furthermore, the relationship between the electric field E and the electric displacement D of a ferroelectric material is hysteretic. Figure 11.1 shows a typical polarization–voltage (P–V) hysteretic loop. After removing the external electric field, the remanent polarization $+/-P_r$ represents logic "0" or "1". Those polarization information can be stored reliably without power supply.

With the stable polarization states of ferroelectric materials, we can store information in FeCaps by applying an external field. Figure 11.2 shows the popular designs of a ferroelectric memory cell, including the 1T-1C and the 2T-2C architecture. The 1T-1C design uses one FeCap to store the information, while there is a complementary FeCap pair in the 2T-2C architecture [17]. When we need to read a specific FeCap in a 1T-1C cell, a poled reference cell is used. By measuring the voltage/current between the capacitor and the reference cell, we can determine the stored value. The 1T-1C cell is area efficient but less reliable. The 2T-2C cell uses two FeCaps to store complementary polarization states. It can store and recall the information more reliably but with much larger area overheads.

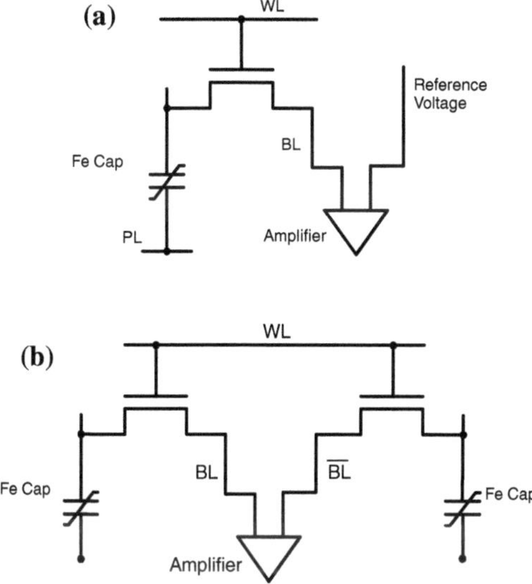

Fig. 11.2 Memory design with FeCaps [17]. **a** 1T1C, **b** 2T2C

11.2.2 Ferroelectric Capacitor Simulation Model

To better design and optimize the ferroelectric memory, we need a circuit model to efficiently describe the electrical characteristics of a FeCap. There are several models for FeCaps, such as zero-switching-time macro-modeling [16, 18]. However, they are computationally expensive. Furthermore, previous models generally omitted to model the capacitor during the power-off state, which is desirable for the unified simulation framework.

For circuit design, we need a behavior model for the SPICE simulator and characterize the nonvolatile property when power failures occur. Therefore, we built the FeCap model in Fig. 11.3. It consists of two nonlinear voltage-control capacitors, a voltage-control voltage source, and some voltage-control switches. A Schmitt trigger generates the control signals for switches S1–S4. Two voltage-control resistances mimic the nonvolatile property. Their values are zero when the power is on and tend to be infinite when the power is off to keep the charge in the capacitors.

This model can be easily implemented in SPICE, and the simulation results of *P*–*V* hysteresis loop are shown in Fig. 11.4. By tuning the circuit parameters of the model, the curve coincides with the measured data. The behavior model of ferroelectric capacitors can be used in the SPICE simulation for the NVFF design.

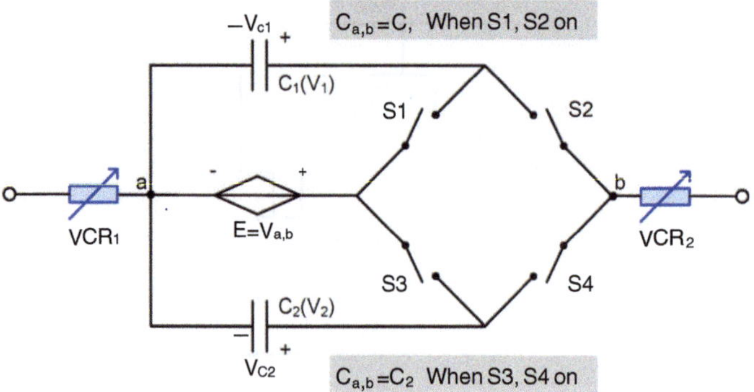

Fig. 11.3 Ferroelectric capacitance SPICE model

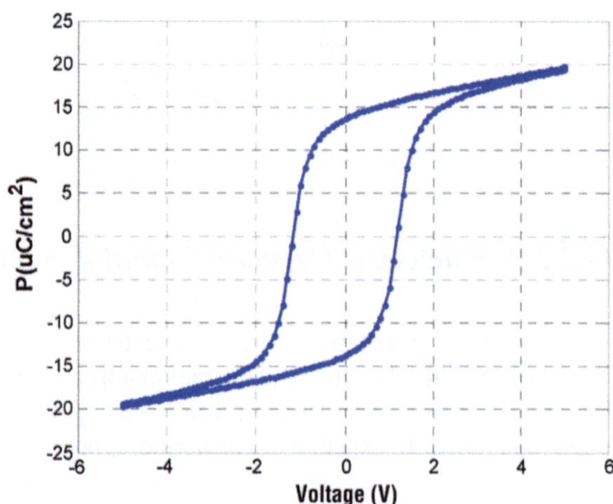

Fig. 11.4 Simulation result of P–V hysteresis loop

11.3 Nonvolatile Processor Design

In this section, we will introduce an actual NVP design based on ferroelectric flip-flops. First, we show the overall architecture of our design. Second, we describe two special nonvolatile circuits differing from a volatile processor: the ferroelectric-capacitor-based flip-flops and the FFC. Third, we discuss how NVP works under power failures. Finally, we give out some measured results from the fabricated NVP.

11.3.1 Overall Architecture

Figure 11.5 shows the block diagram of the fabricated NVP. It contains an MCS51 instruction set compatible core, a mode-selecting module, a JTAG debugging module,

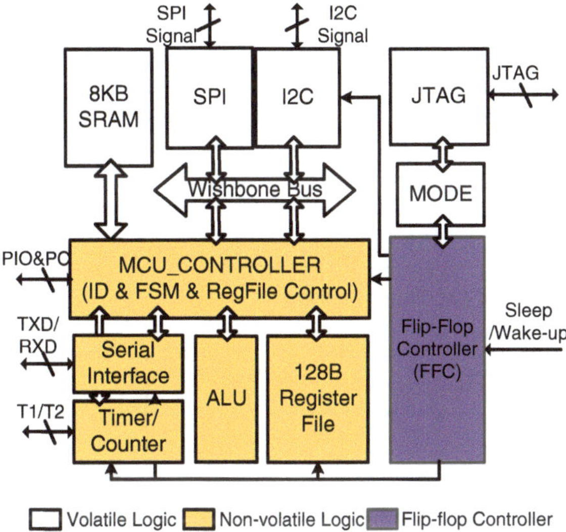

Fig. 11.5 Overall architecture of NVP

and two peripheral interfaces. The core consists of an MCU controller, an 8-bit arithmetic logic unit, a serial interface, a 128-byte register file, and an 8 K-byte SRAM. We have replaced the 128-byte register file and all flip-flops in the core with FeCap-based flip-flops (FeFFs). As the SRAM is used as a secondary data memory, we do not replace it in this implementation. However, it can be replaced with FeRAM easily [19]. The mode-selecting module controls the operating mode of the processor, including the volatile mode, the nonvolatile mode, and the debug mode. The JTAG module supports the user to debug the processor via scan chains. The SPI and I2C interfaces are connected to the core via a Wishbone bus. They are realized in the volatile logic domain. The FFC provides controlling signals to all FeFFs and volatile flip-flops in parallel after receiving the sleep or wake-up signals. As the FeFF and FFC are two main parts differing from a volatile processor, their detail structures are described as follows.

11.3.2 FeFF Design and Optimization

Figure 11.6a [12] shows the FeFF diagram. It adopts a hybrid CMOS and ferroelectric process, consisting of a standard master-slave D flip-flop (DFF) and a backup FeCap pair. They are isolated by two CMOS switches M1/M2 controlled by the "RW" signal. In the normal operating mode, the switches M1/M2 are open so the FeFF works as a standard DFF. Therefore, no performance losts will be introduced to the NVP during the normal operations. When a sleep/wake-up signal is detected, the

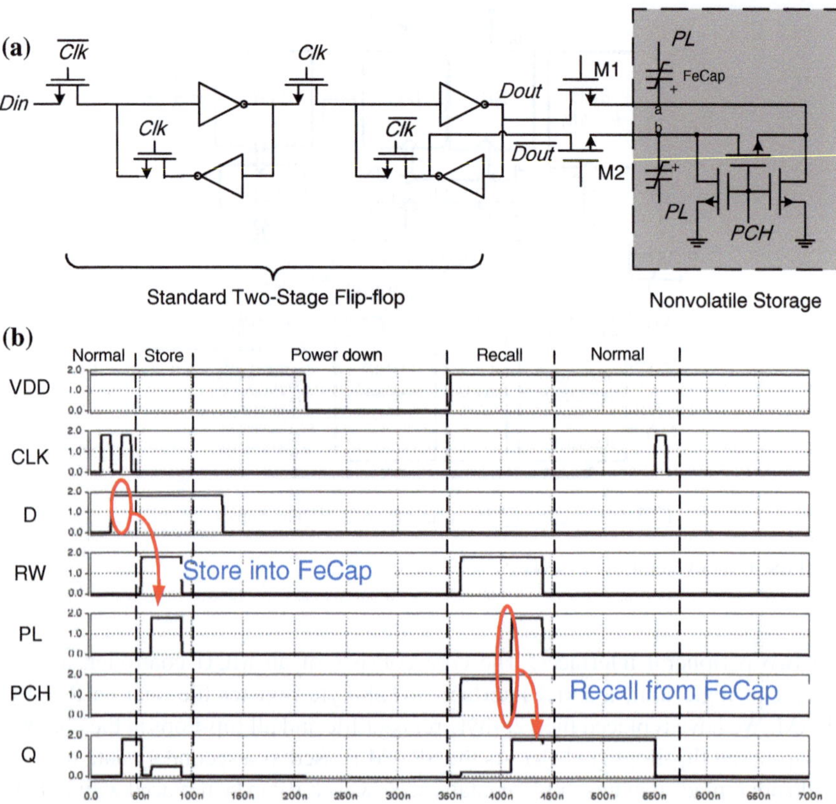

Fig. 11.6 Modeling and simulation of FeFF. **a** Proposed ferroelectric flip-flop structure. **b** Timing char of store and recall

controller will generate signals to make FeFFs store and restore their states and close the switches M1/M2. In the store operation, the complementary outputs of the DFF are connected to the positive plates of the FeCaps. "PL" is pulled up to polarize the FeCap pair into different complementary values to remember the present state. In the restore operation, "PCH" is pulled up to shorten nodes "a" and "b". The back-to-back inverter pair operates in a semi-stable state. After that, "PL" is pulled up and "PCH" is pulled down simultaneously. The FeCap pair drives nodes "a" and "b" with different currents until nodes "a" and "b" stay at stable complementary outputs "0/1" or "1/0". The differential architecture improves the reliability and performance with area overheads. We simulate this circuit in HSPICE, and Fig. 11.6b shows the waveform of the circuit during store and restore operations.

11.3.3 Flip-Flop Controller

The FFC generates controlling signals to FeFFs during sleep and wake-up actions. Figure 11.7a shows the block diagram of a FFC. It consists of a timing block and a finite-state machine (FSM). The timing block is self-timed by the inverter chain and the three timers provide overflow signals (Tov1-Tov3) to the FSM. The FSM generates the controlling signals ("RW," "PL," "PCH") based on Tov1–Tov3 to meet the timing requirements in Fig. 11.7c. The "Sleep/Wake-up" signal triggers the FSM to execute the sleep sequence or the wake-up sequence. The output "CG" is the clock-gating signal to both FeFFs and DFFs. Figure 11.7b shows the interconnections between the FFC and the flip-flops. "CG" gates the clock of both the FeFFs and the volatile flip-flops during the store and recall actions in Fig. 11.7c. It prevents writing uncertainty when the FeFFs executing store operations and halts the system before the FeFFs complete the data recall.

11.3.4 Power Management Circuit and Optimization

This section will discuss how to generate the sleep/wake-up signal for the NVP to backup its system state when power failures happen. An configurable voltage detection system (CVDS) is proposed to generate those signals.

The CVDS architecture is shown in Fig. 11.8a, which contains two configurable units. The first one is a switched capacitor array attached to the power line. The configurable capacitor, denoted as C_{PL}, provides the data backup energy to the NVP by

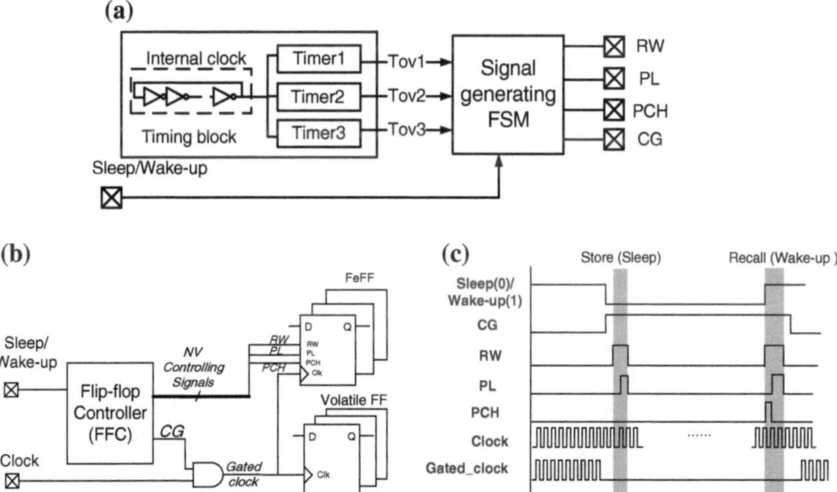

Fig. 11.7 Block diagram of flip-flop controller. **a** Block diagram of FFC. **b** Interconnection between FFC and FFs. **c** Timing chart in store and recall actions

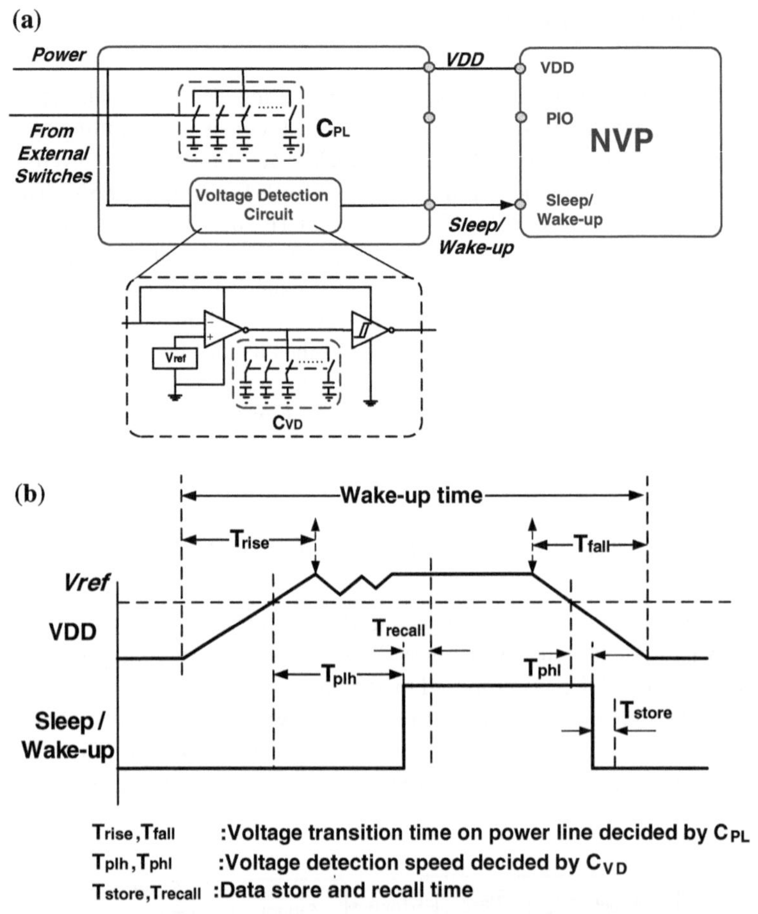

Fig. 11.8 Architecture and controlling timing chart of configurable voltage detection system. **a** Architecture of configurable voltage detection system. **b** Timing chart of system level sleep/wake-up process

keeping the voltage above the operating threshold after power down. The other one, denoted as C_{VD}, is another switched capacitor array used in the voltage detection circuit. The voltage detection circuit detects the power failure and recovery. It generates the sleep/wake-up signal (0/1) within a specific time (T_{plh}) after the voltage of the power line is above the threshold. We set the value of T_{plh} considering both the system reliability and the backup speed. The controlling words for those switched capacitor arrays are given by external input switches. Figure 11.8b shows the signal waveform during the sleep/wake-up actions and the timing diagram influenced by C_{PL} and C_{VD}.

We show the measured results of the wake-up time with different capacitance of C_{PL} and C_{VD} under different power conditions in Fig. 11.9. The wake-up time

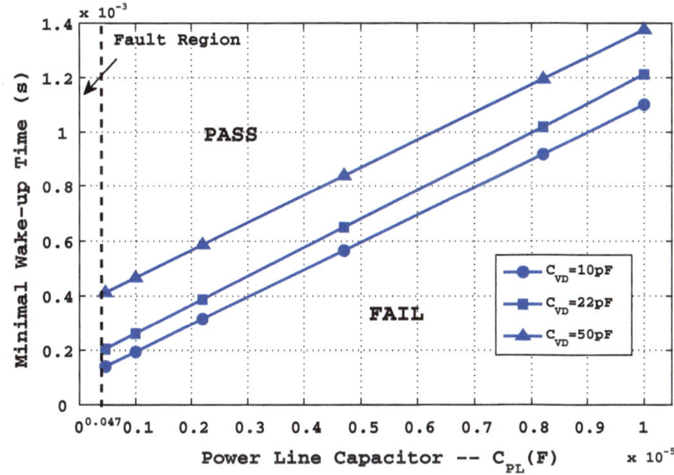

Fig. 11.9 The minimal wake-up time for system accuracy under different capacitance values

should be larger than the sum of power-stabilizing time, voltage detection time, and on-chip intrinsic wake-up time. C_{PL} has an impact on the power-stabilizing time via the RC-constant, where the minimal requested wake-up time decreases with smaller C_{PL}. However, when C_{PL} is smaller than 470 nF, it cannot provide sufficient energy for the NVP to backup the states so that the system goes to the false region, where the NVP can not backup its states correctly. C_{VD} affects the voltage detection time via its RC-constant, where the wake-up time decreases with smaller C_{VD}. C_{VD} cannot be smaller than 10 pF for the power and clock signals to be stabilized. As we can see, the minimal system wake-up time (>100 μs) is much larger than the on-chip intrinsic sleep and wake-up time (<10 μs), because the most dominating factor is the time for the power and clock signals to be stabilized, instead of the on-chip intrinsic circuit delay.

11.3.5 Chip Testing

The fabricated NVP, named as **THU1010N**, uses the ROHM 0.13 μm CMOS-ferroelectric hybrid process. Figure 11.10 shows its photomicrograph and general statistics. It has two unique nonvolatile characteristics: 1,607-bit FeFFs and zero standby power. The maximum operating frequency and the active power consumption are measured under a suite of embedded benchmarks for sensor networks, including FFT, FIR, and ZigBee protocols. Specially, we show the properties of sleep and wake-up actions, as well as some comparison results in Table 11.1.

Table 11.1 compares the NVP with a popular industrial processor "MSP430" [5] and its variants with FeRAM "MSP430FR series" [19]. We show the sleep/wake-up

Die Area	3.4mm×3.4mm (Pad Dominated)
Pin Number	100
Max Clock Feq.	25MHz
Core Voltage	1.2V
Capacity of Memory	1607-**bit FeFF** 8KB SRAM
Nonvolatile Level	Register level
Active Power	106uW/MHz
Standby Power	0

Fig. 11.10 Micrograph and general design statistics of THU1010N

Table 11.1 Comparison results of sleep and wake-up properties between THU-1010N and emerging commercial microprocessor

Microprocessor type		THU1010N	TI-MSP430-5 series with Flash [5]	TI-MSP430 with FRAM [19]
Sleep/Wake-up time	Sleep time (μs)	7	6×10^3	212
	Wake-up time (μs)	3	3×10^3	310
Sleep/Wake-up energy	Sleep energy (pJ/bit)	14.4	2.76×10^5	N/A
	Wake-up energy (pJ/bit)	5.04	373	N/A

time and the energy consumption during the store and recall operations. The sleep and wake-up time of "MSP430" come from the switching time between LPM4.5 mode and active mode [5, 19]. The results show that the NVP has tremendous advantages in the sleep/wake-up time and energy. It achieves over 100–$1,000\times$ speedup in the sleep time, 30–$100\times$ speedup in the wake-up time, and $19,000\times$ saving in the sleep energy. In the chip level, we need a few microseconds to switch from the power-down mode to the active mode. Therefore, we conclude that the distributed nonvolatile architecture of the NVP provides much better performance than the existing centralized nonvolatile storage. It is quite promising for the energy-harvesting and power management applications to use the NVP technique.

Moreover, we discover that the core voltage has an effect on the sleep/wake-up speed. Figure 11.11 shows the sleep/wake-up time under different supply voltages. Lower voltages lead to a slower speed. It is because the delays of both FeFFs and the FFC circuit become larger under a lower voltage. However, the sleep/wake-up time can still be smaller than 20 and 10 μs under a 0.8-V supply voltage. It implies that the NVP can keep its instant-on feature under low voltages. Compared with the several milliseconds backup time in the centralized structure, the microseconds switching time will provide more power saving opportunities for the fine-grained on-chip power management.

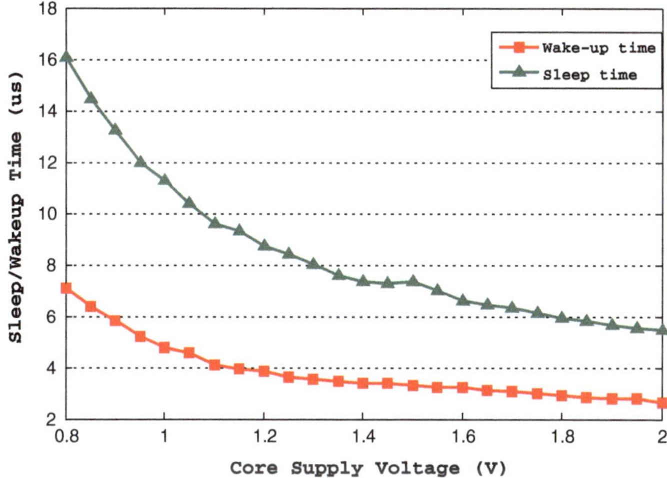

Fig. 11.11 Sleep and wake-up time in an NVP under different supply voltages

11.4 Compression Codec for Nonvolatile Processor

Though the NVP can sleep and wake-up very quickly, the nonvolatile storage elements induce large area overheads. This section will describe some area-efficient techniques to tackle this challenge. First, we propose an area-efficient architecture for the NVP. Furthermore, we describe two specific implementations referred to as PaCC and SPaC.

11.4.1 Area Challenges and Solutions

The nonvolatile memory cells, such as FeCaps (see Fig. 11.6), magnetic tunnel junctions (MTJ), and floating-gate transistors, usually occupy a quite large area. To remove the performance losts in the normal operations, a hybrid flip-flop is usually adopted, containing a standard flip-flop and nonvolatile cells. It makes the area of NVFFs even bigger. Table 11.2 shows the area overheads of a single NVFF and an NVP under different nonvolatile techniques. It shows that all of them introduce nontrivial area overheads. Note that the area overheads do not consider the variation and reliable issues, which make the chip area even larger. For example, the FeCap has a sandwich structure with a ferroelectric film layer between two metal layers. To reduce error rates during the read and write operations, the ferroelectric film should be large enough. Measurements show that nearly five times area overhead is observed in a commercial FeFF. Recently, Beach et al. [20] pointed out that the MTJ-based memory demonstrates statistical read and write behaviors and requires extra error-correcting units.

Table 11.2 Area challenges under different nonvolatile techniques

Approach	Area of NVFF in DFFs	Area overhead of entire chip (%)
Floating gate [2]	1.4x	19
Magnetic RAM [10]	2x	40
FeRAM [21]	4–5x	90

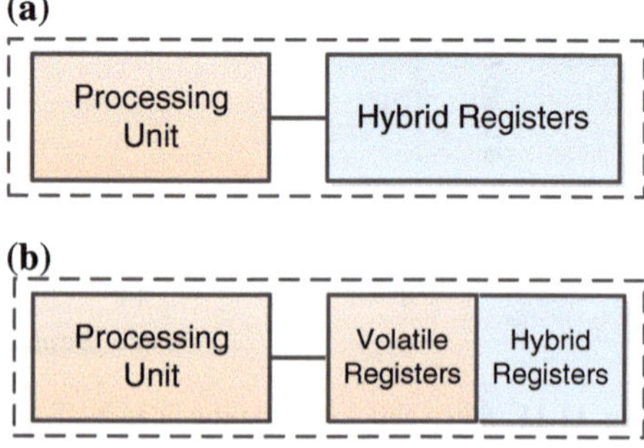

Fig. 11.12 Hybrid architectures for NVP. **a** Hybrid architecture of the fabricated NVP. **b** Partial hybrid architecture for NVP

To quantify the area overheads in an NVP, we assume that an NVFF is α times larger than the original flip-flop and all the NVFFs occupy $\beta(0 < \beta < 1)$ of the total chip area. The area overhead S_{ov} equals to $\beta \times (\alpha - 1)$ when replacing standard flip-flops with NVFFs. In a fabricated NVP [21], β approximates to 20 %, α is near to 5, and $S_{ov} \approx 80\,\%$. Therefore, efficient design techniques are needed to alleviate those area impacts.

Figure 11.12b presents a partial hybrid nonvolatile architecture to deal with the area challenge. In this architecture, we only replace parts of flip-flops with NVFFs so as to reduce the chip area. However, there may be insufficient NVFFs to backup the system states. Data compression and corresponding techniques are proposed to ensure the correct backup and recovery operations. The data compression can be realized in either a software or a hardware approach. In the software approach, the application stores critical data into NVFFs, while keeping other data in the volatile flip-flops. Only critical data are stored when power off. The advantage of the software approach is its flexibility. However, it puts the burden on the application developers to determine which data should be stored. Another solution is to design a specific hardware codec to perform data compression. It can compress data fast and is transparent to the application designers. Thus, we present two concrete designs of compression codec, named PaCC and SPaC, as follows.

11.4.2 An Parallel Compare-and-Compress Codec for Nonvolatile Processors

This section presents an area-efficient architecture referred to as parallel compare-and-compress codec (PaCC) to reduce the number of the nonvolatile registers, and thus the area of the NVP. It consists of an improved RLE algorithm and the corresponding hardware implementation. Furthermore, a state table selection solution is proposed to improve the PaCC's compression ratio.

11.4.3 PaCC Overview

First, we analyze the redundancy in processor states stored in registers, which reveals the potential of register reduction by compression. Let $v_i = 0/1$ represents the value of the ith flip-flop in a processor at a given time. We use $\mathbf{V} = (\nu_1, \nu_2, ..., \nu_n)$ to denote the processor state at that point. Simulations show that over 80 % of the v_i's are unchanged during a program execution. If we construct a reference vector $\mathbf{V}_{\text{ref}}$ such that each $v_i \in \mathbf{V}_{\text{ref}}$ equals to the most common value for the corresponding flip-flop, a differential vector $\mathbf{V}_{\text{diff}} = \mathbf{V} \oplus \mathbf{V}_{\text{ref}}$ can be obtained for a state vector $\mathbf{V}$. As $\mathbf{V}_{\text{diff}}$ may consist of many consecutive zeros, we can compress $\mathbf{V}_{\text{diff}}$ to a much shorter vector $\mathbf{V}^{\text{c}}_{\text{diff}}$ and use fewer NVFFs to store it. We define the compression ratio as

$$CR = \frac{|\mathbf{V}^{\text{c}}_{*}|}{|\mathbf{V}_{*}|}, \tag{11.1}$$

where $\mathbf{V}_*$ is any original vector and $\mathbf{V}^{\text{c}}_*$ is the compressed version.

The comparison of a conventional NVP and the PaCC architecture is shown in Fig. 11.13. Figure 11.13a shows a straightforward way to implement an NVP which simply augments a volatile processor with FeFFs and a FFC. Such a processor stores the system state $\mathbf{V}$ without compression, and the FeFFs increase area significantly. The PaCC architecture (see Fig. 11.13b) consists of volatile registers which stores the current system state $\mathbf{V}$, a state table, a compression codec (divided into a PaCC encoder and a PaCC decoder), and a small set of FeFFs. The state table is used to store $\mathbf{V}_{\text{ref}}$, and the compression codec is used to make conversions between $\mathbf{V}_{\text{diff}}$ and $\mathbf{V}^{\text{c}}_{\text{diff}}$, and the comparison is done by the bit-wise XORs. Though the volatile registers, the state table, and the compression codec bring in additional area, the significant reduction in the number of FeFFs leads to a much smaller overall chip area. The operation of a PaCC can be partitioned into the encoding procedure and the decoding one. The encoding procedure accomplishes the following:

- halt the clock to maintain the state vector $\mathbf{V}$ when a power interruption is detected,
- select a reference vector $\mathbf{V}_{\text{ref}}$ from the state table, which can generate the most consecutive zeros after comparison,
- calculate $\mathbf{V}_{\text{diff}}$ by XORing $\mathbf{V}$ and $\mathbf{V}_{\text{ref}}$,

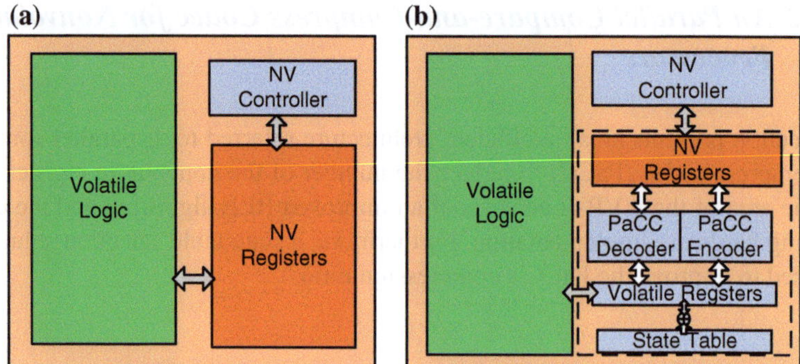

Fig. 11.13 Conventional architecture versus PaCC architecture. **a** Conventional Architecture of NVP. **b** Parallel Compare-and-Compress (PaCC) Architecture of NVP

- compress $\mathbf{V}_{\text{diff}}$ using the PaCC encoder to get $\mathbf{V}^{c}_{\text{diff}}$, and
- store $\mathbf{V}^{c}_{\text{diff}}$ into the nonvolatile storages.

The decoding procedure works in the opposite direction. Obviously, the PaCC codec design and the $\mathbf{V}_{\text{ref}}$ selecting are two critical issues in the architecture. We will discuss them as follows.

11.4.4 PaCC Codec

To achieve a fast compression process along with a good CR, we use a threshold-based parallel run-length encoding (PRLE) algorithm to compress the $\mathbf{V}_{\text{diff}}$. Instead of processing the input data stream bit-by-bit, this algorithm introduces a parallel observation mechanism into the traditional RLE to bypass continuous k-bit 0(1) in parallel. In order to further improve the compression ratio, we adopt a threshold value L_{th} to deal with the short 0(1) chains as in [22]. We only encode the 0(1) chains longer than the L_{th} bit because short chains cannot be compressed by PRLE.

Algorithm 10.1 presents the details of PRLE algorithm. The inputs of the algorithm are the differential vector $\mathbf{V}_{\text{diff}}$, the observation window width (OWW) k, and the threshold Lth. The output is the compression result $\mathbf{V}^{c}_{\text{diff}}$. Variable $\mathbf{S}_{\text{uni}}$ denotes a uniform sequence of all 0's or 1's and $\mathbf{S}_{\text{non}}$ denotes a nonuniform sequence such as {0100101 ...} which does not contain any 0/1 chains longer than the L_{th} bit.

After the initialization (Line 1), the main body of the algorithm is a "while" loop (Line 2) which checks the end of the input vector. We append the following 0/1 chain to $\mathbf{S}_{\text{uni}}$ as the temporary uniform sequence. In this step, we execute the parallel observation to check whether the following k bits or the remaining bits are all 0/1's. If so, we append all of them to $\mathbf{S}_{\text{uni}}$ to bypass them in one step (Line 4–6). Otherwise we only add the current bit to $\mathbf{S}_{\text{uni}}$ (Line 7–9). Subsequently, if a 0/1 transition is detected (Line 10), we decide how to process $\mathbf{S}_{\text{uni}}$. We check whether the length of $\mathbf{S}_{\text{uni}}$ exceeds L_{th}. If not, which means that the current $\mathbf{S}_{\text{uni}}$ actually belongs to a nonuniform sequence, we append $\mathbf{S}_{\text{uni}}$ to $\mathbf{S}_{\text{non}}$ and clear $\mathbf{S}_{\text{uni}}$ (Line 11–13). If the length of $\mathbf{S}_{\text{uni}}$ is larger than L_{th}, we first call function $Process\_ShortSEQ$ to encode $\mathbf{S}_{\text{non}}$ and $Process\_LongSEQ$ to encode $\mathbf{S}_{\text{uni}}$, then re-initialize $\mathbf{S}_{\text{non}}$ and $\mathbf{S}_{\text{uni}}$ (Line 16–20). If reaching the end of $\mathbf{V}_{\text{diff}}$ ($s == n+1$), we process $\mathbf{S}_{\text{uni}}$ as above (Line 11–13, 16–20). Specially, we call function $Process\_ShortSEQ$ to encode the remaining $\mathbf{S}_{\text{non}}$ (Line 14–15) when $|\mathbf{S}_{\text{uni}}| <= L_{\text{th}}$ occurs. The results of the functions $Process\_ShortSEQ$ and $Process\_ShortSEQ$ follow the format shown in Fig. 11.14. We call them copied segments and encoded segments, respectively.

Algorithm 1: Threshold-based parallel RLE algorithm

Input: $\mathbf{V}_{diff} = \{a_1, a_2, ..., a_s, ..., a_n\}, L_{th}, k$
Output: $\mathbf{V}_{diff^c}$
Variables: $\mathbf{S}_{uni}, \mathbf{S}_{non}, s$

```
1  Initialization: V^c_diff = φ, S_uni = φ, S_non = φ, s = 1;
2  while s ≤ n do
3      w = min{s + k − 1, n};
4      if {a_s, ..., a_w} == {00...0}or{11...1} then
5          {a_s, ..., a_w} → S_uni;
6          s = w + 1;
7      else
8          a_s → S_uni;
9          s = s + 1;
10     if (a_s ≠ a_s+1 && s ≤ n)||(s == n + 1) then
11         if |S_uni| <= L_th then
12             S_uni → S_non;
13             S_uni = φ;
14             if s == n + 1 then
15                 Process_ShortSEQ(S_non) → V^c_diff;
16         else
17             Process_ShortSEQ(S_non) → V^c_diff;
18             Process_LongSEQ(S_uni) → V^c_diff;
19             S_non = φ;
20             S_uni = φ;
```

In this section, we present the hardware design of a PaCC Codec. Figure 11.15 shows the block diagram of our PaCC encoder which is the hardware realization of PaCC algorithm. It consists of an input-end shifting network which shifts n-bit $\mathbf{V}_{\text{diff}}$ from the volatile registers to the RLE encoding module. The n-bit output is the shifted value for updating the volatile registers. Similarly, the output-end shifting network shifts the m-bit compression results to the NV registers. Besides the shifting networks, the "all 0/1 detector" block helps to execute k-bit parallel observation and generate a bypass signal to the RLE encoder and length controller. The length controller provides the shifting length to the input-end shifting network according to

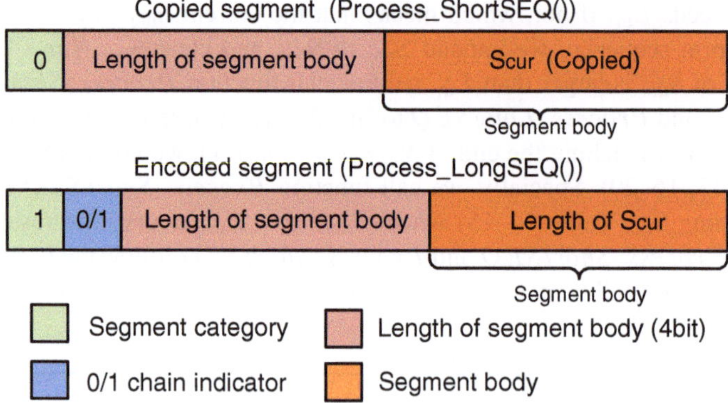

Fig. 11.14 Encoded and copied segment structure. The copied segment is the format used by chains with mixed 0's and 1's of lengths shorter than L_{th}. The encoded segment is used by the chains with lengths longer than L_{th}

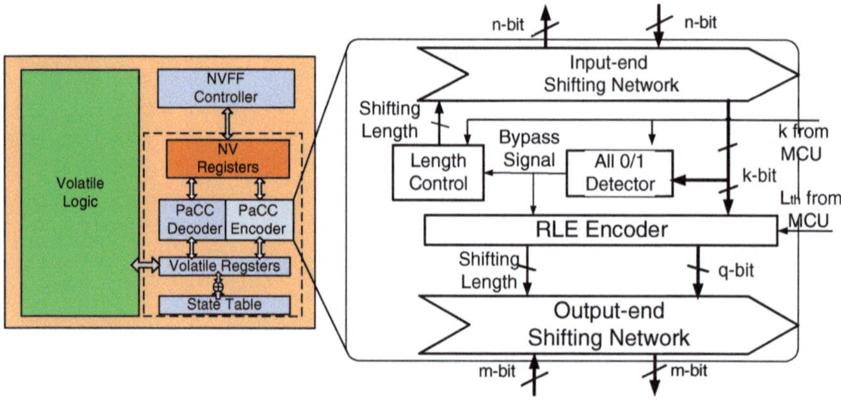

Fig. 11.15 The block diagram of PaCC encoder and its location in the PaCC architecture

the bypass signal as well as OWW k. The RLE encoder compresses the k-bit input serially when the bypass signal is disabled, otherwise bypasses the k-bit input. The format of the compression result is based on the given L_{th} (see Fig. 11.14). The OWW k and threshold L_{th} are given by the microcontroller (MCU) based on actual applications. Once the RLE encoder accomplishes the compression, it sends the q-bit compressed segment (see Fig. 11.14) to the output-end shifting network.

The PaCC decoder is similar to the encoder. Since encoding and decoding are opposite operations, the decoder can reuse the two shifting networks. The input-end and output-end part are exchanged and the data flow in the opposite direction. The only difference in the PaCC decoder is that it contains a RLE decoding module instead of the RLE encoding module. Thus, we omit the detail discussion on the decoder part.

11.4.5 State Table Optimization

The state table stores and provides $\mathbf{V}_{\text{ref}}$ for generating $\mathbf{V}_{\text{diff}}$. Since different applications may have very different $\mathbf{V}_{\text{ref}}$'s, only one $\mathbf{V}_{\text{ref}}$ is insufficient. The state table thus contains multiple $\mathbf{V}_{\text{ref}}$'s as well as a selection mechanism.

The overall design of the state table is shown in Fig. 11.16 which consists of a reference vector array and a selection unit. Since encoding and decoding use the same $\mathbf{V}_{\text{ref}}$, the choice of $\mathbf{V}_{\text{ref}}$ should be retained when power is off. To meet this requirement, we propose two methods. One method adopts external inputs such as switches to hold the choice of $\mathbf{V}_{\text{ref}}$. This method simplifies hardware design and control, but the selection is not so flexible. The other method uses additional nonvolatile storage (e.g., 2 3-bit NVFF) to memorize the choice of $\mathbf{V}_{\text{ref}}$ generated dynamically from the MCU according to the actual application. This method requires more complicated software control, but achieves more flexible selection. In real cases, we can use a hybrid selection mechanism combining these two methods such that if MCU only runs one application, we can use the external switch method to reduce the control complexity while in other cases, we use the dynamic NVFF based selection.

Then we will decide how to choose an appropriate reference vector for a specific application. We can formulate the problem as following: assuming β system-state vectors $\{\mathbf{V}_1, \mathbf{V}_2, ..., \mathbf{V}_\beta\}$ should be stored, we need to determine an optimal reference vector $\mathbf{V}_{\text{opt}}$ to minimize the length $L_{\mathbf{V}_{\text{comp}}}$ of the compressed vector under the worst case:

$$\mathbf{V}_{\text{ref, opt}} = \arg\min_{\mathbf{V}_{\text{ref}}} \left(\max_{i=1}^{\delta} L(C(\mathbf{V}_i \oplus \mathbf{V}_{\text{ref}})) \right) \tag{11.2}$$

where $\mathbf{V}_i \oplus \mathbf{V}$ represents XOR two vectors bit-by-bit; function P calculates the length of the vector after compressing $\mathbf{V}$ by PRLE algorithm.

The direct searching space of $\mathbf{V}_{\text{ref}}$ is very large and function P does not have an analytical pattern, so we develop a naive heuristic to find a solution. Intuitively, the CR of RLE algorithm is related to the number of 0's in $\mathbf{V}_{\text{diff}}$, so we can get a

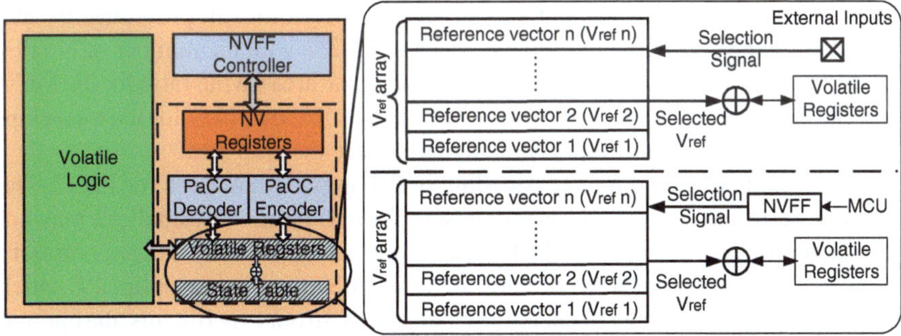

Fig. 11.16 State table structure and the two proposed reference vector selection methods

suboptimal reference vector $\mathbf{V}_{\text{sub,opt}}$ resulting in the most 0's in $\mathbf{V}_{\text{diff}}$ in the global vision. Therefore, we set the ith bit of $\mathbf{V}_{\text{sub,opt}}$ as follows:

$$\mathbf{V}_{\text{sub,opt}}(i) = M(\{j \in 1, 2, ..., \beta | \mathbf{V}_j(i)\}) \tag{11.3}$$

$M(\mathbf{S})$ equals to the majority element in the set $\mathbf{S}$. In our experiments, this method can achieve quite good compression ratio in most cases; however, it may lead to poor results for some special vectors because it ignores the continuity of 0's in $\mathbf{V}_{\text{diff}}$. Some better heuristic algorithms can be explored to address the optimization problem in our future work.

11.4.5.1 Evaluation Results

In this part, we will show the evaluation results of PaCC in chip area and compression speed. We use Cadence NC-Verilog to sample the system-state vectors and evaluate the clock cycles statistics. The area statistics is obtained from Synopsys Design Compiler under Rohm's 0.13 μm ferroelectric-CMOS hybrid process. To simulate the processor behavior in real embedded applications, we use the benchmark programs of Fibonacci, sorting and square root from Dalton Project [23]; Rijndael and FFT from MiBench [24], and ZigBee MAC Protocol from Z-Stack [25].

We evaluate the area efficiency of PaCC for the programs. We randomly select 50 state vectors for each program and calculate the optimized reference vector $\mathbf{V}_{\text{ref,opt}}$ based on heuristics in Eq. 11.2. We get the desired number of NVFFs L_{nv} and the area reduction numbers in Table 11.3. Each row represents the results from one program. The columns give out the compression ratio of PRLE, the number of NVFFs, the area reduction ratio of MCU (both MCU only and the whole chip), and the overflow possibility. All data are obtained under the optimal threshold L_{th} for each program. The optimal threshold is the one which results in the smallest number of NVFFs among all the threshold values in [4,50]. In the programs considered, the optimal L_{th} is always 9 or 10. This is due to the fix encoding format in Fig. 11.14.

As shown in Table 11.3, different programs may lead to different numbers of NVFFs (see column 3). Thus, the area savings vary for different programs. By utilizing PaCC, the compression ratio can reach to 19.2 % with the number of NVFFs reduced from 1607 to 308. Based on this reduction, the area saving ratio for the MCU only can be 23.4–30.2 % and the worst case ratio is still above 15 % for the total chip. We conclude that the algorithm is effective to reduce the chip area.

The run-time of encoding and decoding is also important metrics for PaCC. The encoding performance depends on the chosen OWW k. Intuitively, smaller k may not achieve significant reduction in clock cycles while a larger k reduces the opportunity to encounter consecutive zeros or ones. As a result, we can get an optimal k which leads to the smallest number of encoding clock cycles. In our experiments, the optimal k may vary for different programs, and it usually locates in a fixed range of [16–20]. Given the optimal k chosen for each program, Table 11.4 shows the clock cycles of encoding and decoding for different programs. We can see that the encoding

Table 11.3 Evaluation of area efficiency of PaCC architecture

Program	Optimal L_{th}	Compression ratio (%)	# of NV registers	Lower bound on L_{nv}	Area reduction ratio MCU only (%)	Area reduction ratio Total chip (%)
Fibonacci	9	23.7	381	357	26.2	17.3
Sorting	10	26.0	417	373	24.3	16.1
Sqrt	9	27.1	435	401	23.4	15.4
Rijndael	9	20.3	325	289	29.4	19.5
FFT	10	19.2	308	274	30.2	20.0
ZigBee MAC	10	25.2	405	381	25.0	16.5

Table 11.4 Evaluation of run-time of PaCC codec

Program	Optimal k	Clock cycles				Process time	
		Encode		Decode		On average (μs)	
		Mean	Std	Mean	Std	Encode	Decode
Fibonacci	19	243.2	21.2	97.8	3.3	24.3	9.8
Sorting	16	253.1	25.7	98.1	3.3	25.3	9.8
Sqrt	17	301.8	34.0	101.0	2.9	30.2	10.1
Rijndael	16	211.7	23.4	95.3	3.3	21.1	9.5
FFT	20	190.9	28.9	94.5	3.9	19.1	9.5
ZigBee MAC	20	279.5	42.5	98.5	2.2	27.0	9.9

Assuming the data encoding and decoding procedures runs at 10-MHz clock frequency, the clock cycle statistics is obtained from a circuit simulator
Mean means the average value, *Std* means the standard deviation

process needs extra 200–300 cycles to compress one vector, while the decoding one costs 90–100 cycles. Therefore, the time to store data takes less than 30 μs and the recall takes less than 10 μs at the 10-MHz clock frequency. It maintains the NVP's instant-on/instant-off features.

11.4.6 A Segment-Based Parallel Compression for Backup Acceleration in Nonvolatile Processors

In this section, we will introduce another compression structure referred to as SPaC: a segment-based parallel compression architecture. It achieves trade-offs between the compression time overheads in PaCC and the area overheads in a conventional NVP with full NVFF replacement.

11.4.6.1 SPaC Overview

We give the comparison of different NVP architectures in Fig. 11.17. As Fig. 11.17a shows, the conventional NVP connects each register with a nonvolatile cell. The backup process is totally parallel and fast but leads to nontrivial area overheads due

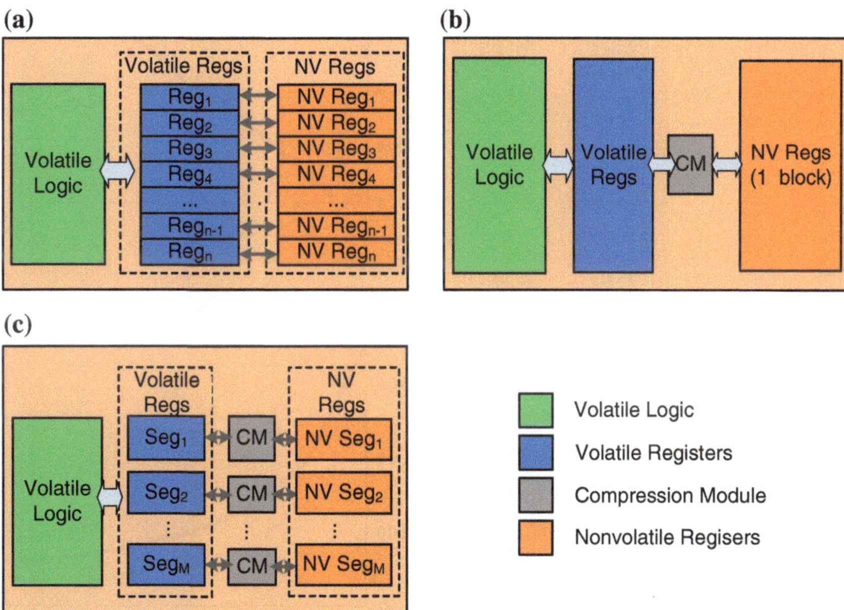

Fig. 11.17 Architecture comparison. **a** Full Replacement Architecture. **b** PaCC Architecture. **c** SPaC Architecture

to a large number of nonvolatile cells. PaCC in Fig. 11.17b uses a compression module (CM) to reduce the number of nonvolatile cells as well as the area. However, the CM compresses the data stream bit-by-bit causing longer backup time. Our proposed SPaC architecture in Fig. 11.17c partitions the registers into several segments and equips each segment with an individual CM. In SPaC, all segments are compressed in parallel to achieve faster backup speed against PaCC. Meanwhile, SPaC reduces the area against the full replacement approach.

Two key metrics of SPaC are the area and backup speed. We evaluate the metrics versus numbers of segments M to show their trends. The results in Figs. 11.18 and 11.19 are based on THU1010N (discussed in Sect. 11.3). Figure 11.18 shows the chip area normalized to the full replacement realization versus M. The area data are approximately linear to the number of segments, because the increasing area primarily comes from the additional CMs. Moreover, the total compression effect of PRLE algorithm degrades when the input vector is divided to more segments, which is another factor inducing the area increase. In our case, the area of SPaC cannot save area when $M > 7$. Figure 11.19 shows the compression speed under different M. Generally speaking, a larger M leads to deeper parallelism and fewer clock cycles. However, when $M > 6$, the speedup by further parallelism is trivial. We use three

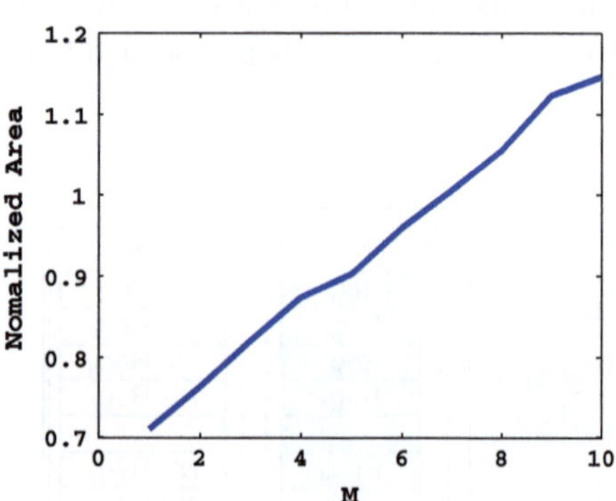

Fig. 11.18 Area evaluation

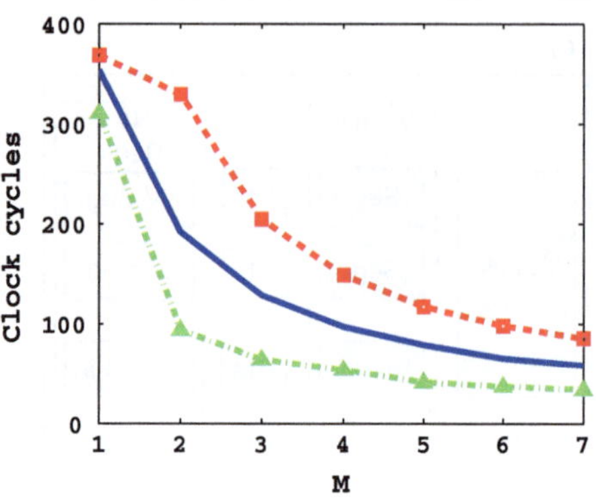

Fig. 11.19 Speed evaluation

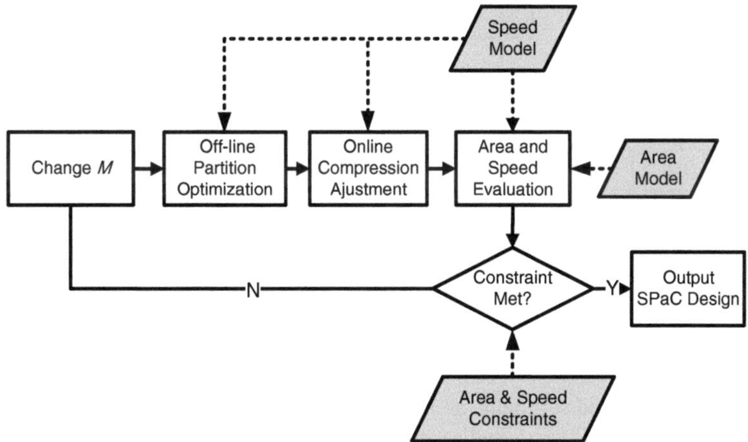

Fig. 11.20 Design flow of SPaC

curves to indicate the average value and the upper/lower bound in Fig. 11.19. The variations come from the input changes at different backup points. If the variation is large, it can significantly degrade the backup speed. This can be solved by an online scheduling controller in the next subsection. Considering both area efficiency and compression speed, the appropriate M is 2 or 3.

Although the data in Figs. 11.18 and 11.19 are based on a specific case, the trends of area and speed are common in other computer architectures (such as MIPS, X86). However, the demarcation point may be different in other processors, because they have different register numbers and architectures, and their area models may be different in other technology processes. Therefore, SPaC can be applied to other processors, but the number of segments should be determined according to the actual design and its requirements. For an actual processor, we propose the design flow for SPaC in Fig. 11.20. As Fig. 11.20 shows, different Ms are evaluated according to the area and speed constraints. We will change the value of M if the constraints are violated. Given a certain M, off-line partition optimization and online compression adjustment are introduced to minimize compression time.

11.4.6.2 SPaC Design

Figure 11.21 shows the detailed diagram of SPaC with M segments. The flip-flops are clustered into M segments denoted as $\{S_1, S_2, ...S_k, B_{k+1}, ..., B_M\}$ and each segment is connected to a CM module for parallel compression. Segments S_i usually have relative small workload variations and do not support compression reallocation. The determination of S_i is based on an off-line algorithm to balance the workloads on each CM. To support dynamic workload adjustment, we design a specific structure to allow the segments to share their CMs if some CMs are idle and others are busy.

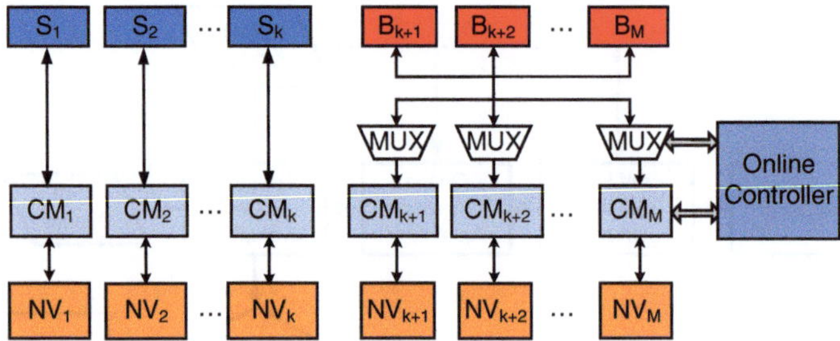

Fig. 11.21 SPaC structure

We denote such shared segments as B_i. To support the CM sharing among segments B_i, we use a set of MUXs to realize the switching operations.

First, we describe the off-line partition algorithm in the following.

Algorithm 2: Off-line Algorithm

Input: **V**, **M**, $\mathbf{Var}_{th}$, $\mathbf{loop}_{th}$
Output: $\mathbf{S} = (\mathbf{S}_1, \mathbf{S}_2, \ldots, \mathbf{S}_M)$
Variables: $std, time, step, loop$

$S_i = length(V)/M$ *for* $i = 1, 2, \ldots, M$; $std = S_{th}$;;
while $std \geq S_{th}$ *and* $loop \leq \mathbf{loop}_{th}$ **do**
  $time = CM(V, S)$;
  $std = STD(time)$;
  $step = ceil(std)$;
  $S(Indexof(\max(time))) = S(Indexof(\max(time))) - step$;
  $S(Indexof(\min(time))) = S(Indexof(\min(time))) - step$;
  $loop = loop + 1$;

Supposing that we partition the system-state vector V into M segments, S_{th} denotes the threshold of the standard deviation and $loop_{\text{th}}$ denotes the loop limitation. The output vector $\mathbf{S} = (\mathbf{S}_1, \mathbf{S}_2, \ldots, \mathbf{S}_M)$ represents the length of each segment. The variable std denotes the temporary standard deviation under the current partition S; $time = (t_1, t_2, , t_M)$ gives out the average clock cycles of all segments; $step$ is the max step value to change the vector length, and $loop$ is the iterating number. We use the equal partition as the initial S and set std to S_{th}. In each loop, we calculate the compressing time of each segment to get $time$ and its variation std. We find the segment with the maximum average clock cycles in $time$ and reduce its vector length by one step. Similarly, the opposite operation is performed to the segment with the minimum average clock cycles. We keep changing S until std is smaller than S_{th}, otherwise it will return an error message. In case of failures, we either reduce S_{th} or set a larger $loop_{\text{th}}$.

Furthermore, we illustrate the dynamic workload adjustment based on CM sharing. Each shared segment B_i is divided into two parts. One part is shared which is connected to MUXs. The remaining parts are directly connected to the segments' own CMs. During the compression, the online controller monitors the complete state of each segment. If one segment completes its compression and another is not, the online controller switches the shared part of the slowest segment to the CM of the fastest segment. To avoid area overheads of the multiplexing, the number of shared segments is small. The size of shared parts of each B_i is determined by the compression speed variations.

11.4.6.3 Evaluation Results

We compare metrics of an NVP using equal-size partition, off-line only partition and hybrid off-line/online partition under different Ms in Table 11.5. We can see that the off-line algorithm balances the workloads of different segments effectively while the hybrid algorithm further decreases the variations. As Table 11.5 shows, the off-line-only partition can improve the compression speed by 32 % compared to the equal-size partition. The hybrid strategy further reduces the variations by average 31.7 % and improves the overall speed performance by up to 10 %.

11.5 Nonvolatile Processor Applications

In this section, we describe two typical applications based on an NVP. The first one is a vehicle detection system. The second one is a self-powered sensor node aimed at body area monitoring. These two application systems have unique features differing from traditional sensor nodes, and we list them as follows:

1. Both systems are driven by immediate harvested energy without conventional energy storage devices, such as batteries.
2. Both systems work continuously under frequent power interruptions even using a square-wave power supply.

Those system features are attributed to the features of NVPs. The complexity to design a power system for a NVP-based sensor node can be significantly reduced without AC–DC regulators and energy storages. It implies the potential to reduce the cost and size of the total system.

11.5.1 Vehicle Detection System

The vehicle detection system is based on energy-driven nonvolatile sensor nodes. The whole system is depicted in Fig. 11.22. Each nonvolatile sensor node is equipped with a solar cell energy harvester with no batteries. The energy source can be sunlight outdoors or light sources indoors. Given an energy source, the sensor node continuously

Table 11.5 Compression speed comparison among equal partition, off-line method only and off-line + online method

M	Equally partition			Off-line only				Off-line + Online				
	Avg	3*Std	Sum	Avg	3*Std	Sum	Red. (%)	Avg	3*Std	Sum	Var. red. (%)	Overall red. (%)
2	288.9	31.8	320.7	199.2	22.5	221.7	30.8	203.8	15.6	219.4	30.6	1
3	211.1	24.9	236	130.6	25.4	156	33.8	130.2	17.7	147.9	30.3	5.1
4	155.4	25	180.4	97.2	24.9	122.1	32.3	100.2	17	117.2	31.7	4
5	124.4	28	152.4	80.2	23.1	103.3	32.2	82.3	16.1	98.4	30.3	4.7
6	101.8	18.4	120.2	67.7	22.7	90.4	24.7	67.5	13.6	81.1	40	10.2
7	96.1	24	120.1	58.1	15.9	74	38.3	59.5	11.5	71	27.6	4

Avg average value, *Std* standard deviation, *Var* variation, *Red* reduction

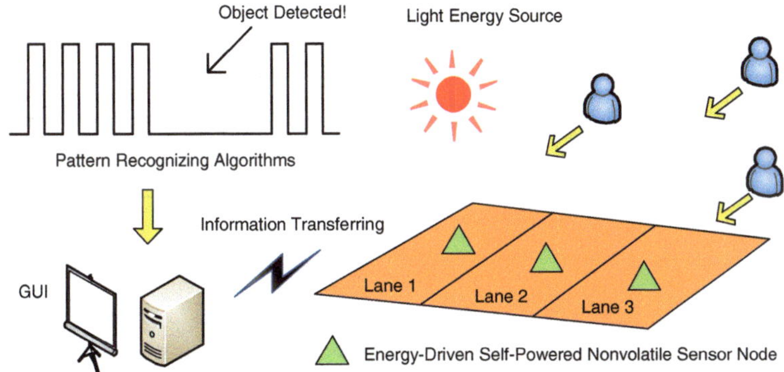

Fig. 11.22 The proposed moving vehicles detection system

counts the number. When a moving vehicle (e.g., a car) comes in between sunlight and sensor node, power to the sensor node is cut off. The nonvolatile sensor node will remember the current state and wait for the moving object to pass by. After that, the power supply is recovered, and the sensor node will continue to count. The counting number and related information can be stored in the local nonvolatile memory or wirelessly transferred to the remote data center. An object recognition algorithm is used in the data center to analyze the object occurring time and other information. A synchronization algorithm should be implemented among the collecting point and sensor nodes. After a certain period of time, the global time should be refreshed and synchronized with each node. A graphic user interface (GUI) is provided to show the real-time detecting results. The most novel technique used in this demo is the NVP-based energy-driven system. This system consists of the energy-harvesting module (solar cell), the power management unit (PMU), and the NVP, shown in Fig. 11.23. At several square centimeters in size, a solar cell is used to provide 6-V and more than 5-mW power supply under medium sunlight. The PMU realized the functions of energy detection and voltage regulation. It measures the energy stored on the capacitor and generates activation signals to the NVP as well as regulates the supply voltage.

To better describe the system working mechanism, we draw the signal timing diagram of the sleep and wake-up actions in Fig. 11.24. In the sleep action, when the PMU detects a power failure, it generates a sleep signal and maintains the power supply via the capacitor until the system state is stored in nonvolatile cells. In the wake-up action, the PMU detects the power recovery and provides power to the NVP until the voltage is stable. After that, it generates a wake-up signal to restart the NVP. According to the measured results, the wake-up action costs less than 100 μs and the sleep action takes around 50 μs, which enables our system to work under a frequently interrupted power supply.

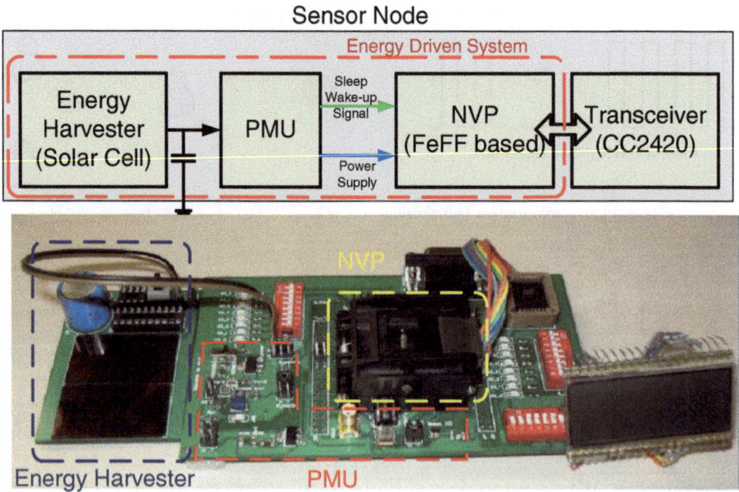

Fig. 11.23 Architecture and realization of energy-driven sensor node

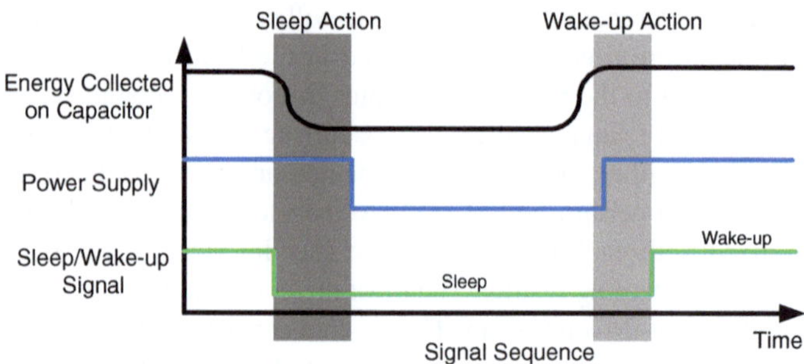

Fig. 11.24 Signal timing chart in sleep and wake-up actions

11.5.2 Self-Powered Body Sensors

Another NVP-based application focuses on the body health monitoring, which is significant to the personal life. Recently, many works have concentrated on the wireless body area network (WBAN) implementation. The body sensor nodes should achieve ultra-low power, low costs, small size, and high reliability. The NVP-based self-powered body sensors can be a promising candidate.

Figure 11.25a shows the block diagram of a self-powered body sensor. Generally, it consists of an energy-harvesting source (EH), a PMU, an NVP, and some peripherals. The NVP provides the node high robustness against power interruptions. The peripherals include several sensors and a RFID. The sensors are used to monitor the medical parameters of a human being and the RFID module enables the node to transmit those data to a sink in a wireless way. The actual sensor node is shown

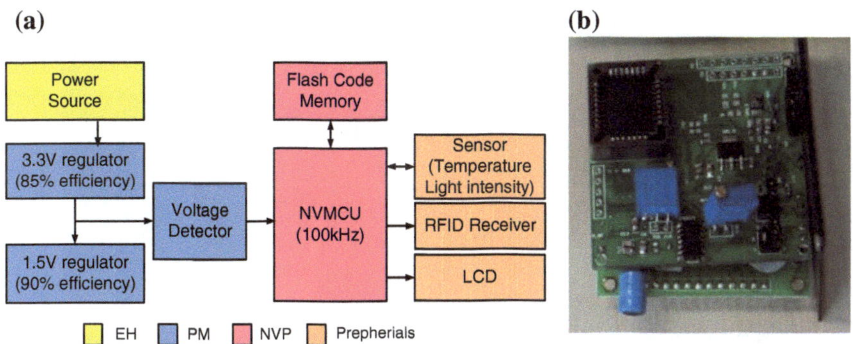

Fig. 11.25 Self-powered body sensor. **a** Block diagram of self-powered body sensor. **b** Actual self-powered body sensor

in Fig. 11.25b. The total size of the node is 50 × 50 × 27 mm. We adopt several power optimizing techniques to enable the sensor node to work under very limited power supply from small-size energy-harvesting devices. The node adopts a DC–DC converter with over 85 % energy-transforming efficiency and a ultra-low-power RFID module. By profiling the power consumption of the sensor node, we find that the Flash code memory contributes a more than 5 mA current under normal operations. As there is a large frequency gap between the NVP and the Flash memory, we design a specific program to power down the Flash memory when it is not read. With this technique, we reduce over 80 % power consumption of the Flash memory and the overall power of the node is reduced to 4 mW. The final demonstration (shown in Fig. 11.26) is a self-powered sensor node with harvested energy from sunlight or vibration. The sensor node monitors the temperature, sunlight duration, and intensity, and it transmits the data into a base station (PC). It mimics the working environment where a human being wears those sensor nodes outdoors or walking. It can collect those medical information reliably without a battery. The users can access those data via a RFID reader in a smart phone or specific devices.

11.6 Related Work

The NVP is a promising approach to realize a nonvolatile-memory-based computing system. Many researchers and companies have evaluated various ways to integrate nonvolatile memories in processors. Flash is a mature high-density nonvolatile memory and is widely used in the mainstream commercial microcontrollers [5, 6]. However, Flash is not suitable to implement distributed NVFFs, because it has drawbacks such as low endurance, slow writing speed, block erasing pattern, and high mask cost. Among existing nonvolatile memories [8], FeRAM and MRAM emerge as the most promising candidates for the NVP. Zwerg et al. [26] embedded a FeRAM into

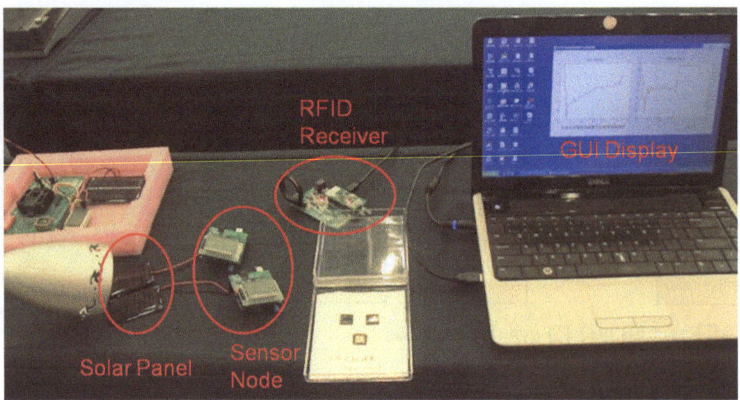

Fig. 11.26 Self-powered body sensor demonstration platform

a microcontroller for better tolerance to power failures. Xu et al. [27] had proposed to adopt STT-MRAM as the last-level on-chip cache in microprocessors. However, the centralized memory architecture cannot provide sufficient bandwidth and fast backup speed in accidental power failures.

In order to achieve faster sleep and wake-up features, some works had concentrated on the register-level nonvolatile memory implementation. Zhao et al. [4] employed MTJ-based flip-flops in FPGAs to achieve rapid start-up. Sakimura et al. [10] developed a magnetic flip-flop (MFF) library for systems-on-a-chip (SoC) design and tested the MFFs in a shifter circuit. Guo et al. [28] conducted an architectural analysis of a STT-MRAM-based processor, including the logic-in-memory, nonvolatile registers, and nonvolatile caches.

Recently, Rohm developed a lifetime-enhanced NVFF by adding a FeCap pair to a standard flip-flop and implemented a nonvolatile counter [1]. The hybrid flip-flop structure does not degrade the performance in the normal operations and prolongs the lifetime of the nonvolatile cells. Wang et al. [12] evaluated an NVP with ferroelectric flip-flops using a compare-and-write policy. Afterward, Yu et al. [2] proposed an evaluation of NVPs based on floating-gate technology. Their analysis demonstrated the performance, area, and power characteristics of an NVP based on the hybrid NVFFs. Simultaneously, Wang et al. [21] fabricated an actual NVP based on the ferroelectric flip-flops and obtained measured results on the sleep and wake-up properties. Most recently, Qazi et al. [29] provided an FIR filter based on ferroelectric flip-flops and demonstrated even faster sleep and wake-up speed.

To further improve the performance of an NVP, some design-optimizing methods are proposed. After observing large area overheads of the hybrid NVFFs, Wang et al. [30] presented a compare-and-compress architecture to reduce the NVP's area and Sheng et al. [31] reported a way to trade off the area overhead and the backup speed in an NVP.

In future, the NVP design may focus on the following aspects: high-speed and reliable NVFF design, hybrid nonvolatile memory architecture, and novel NVP applications.

11.7 Conclusion

In this chapter, we demonstrated the complete design flow to fabricate a ferroelectric NVP. Our experimental results show that the first fabricated NVP can achieve 7 μs sleep time and 3 μs wake-up time with zero standby power, which means over 30–100× speedup on the wake-up/sleep time and 70× energy savings on the backup and recall operations compared with the state-of-the-art industry microcontroller. Meanwhile, the ferroelectric NVP exhibits comparative performance and power consumption in normal operations. Furthermore, we design a PaCC and its variants SPaC to save up to 30 % silicon area in a ferroelectric NVP. Finally, we demonstrate two kinds of battery-less sensor nodes based on the NVP for the first time. They aimed at moving vehicle detection and body sensor applications.

Ferroelectric NVPs can realize energy-efficient computing systems with zero standby power, instant-on features, high resilience to power failures, and fine-grained power management. It has the potential to realize computing system powered by energy-harvesting devices, which eliminates the battery lifetime constraints and becomes a very promising solution for smart sensors and other applications.

Acknowledgments This work was supported in part by the NSFC under grant 60976032 and 61204032, High-Tech Research and Development (863) Program under contract 2013AA013201 and National Science and Technology Major Project under contract 2010ZX03006-003-01.

References

1. Nikkei Electronics Asia: Rohm Develops Non-Volatile Register; Slashes Dissipation. Website: http://techon.nikkeibp.co.jp/article/HONSHI/20080729/155646/.
2. Yu, W., Rajwade, S., Wang, S., Lian, B., Suh G.E., & Kan, E. (2011). A non-volatile microcontroller with integrated floating-gate transistors. In *DSN-W, 2011* (pp. 75–80).
3. Holland, C. First MRAM-based FPGA taped-out. Website: http://www.eetimes.com/General/DisplayPrintViewContent?contentItemId=4200035.
4. Zhao, W., Belhaire, E., Javerliac, V., Chappert, C., & Dieny, B. (2006). A nonvolatile flip-flop in magnetic FPGA chip. In *DTIS, 2006* (pp. 323–326).
5. Texas Instrument: Datasheet of MSP430F522X mixed signal microprocessors (2009).
6. Atmel: Datasheet of AT91SAM9G20-AT91 ARM thumb microcontrollers (2012).
7. Wu, X., Li, J., Zhang, L., Speight, E., Rajamony, R., & Xie, Y. (2010). Design exploration of hybrid caches with disparate memory technologies. *ACM TACO*, *7*(3), 15.
8. ITRS: Roadmap for Nonvolatile Memory. Website: http://www.itrs.net/.
9. Wang, P., Chen, X., Chen, Y., Li, H., Kang, S., Zhu, X. & Wu, W. (2011). A 1.0 V 45 nm nonvolatile magnetic latch design and its robustness analysis. In *CICC, 2011* (pp. 1–4).

10. Sakimura, N., Sugibayashi, T., Nebashi, R., & Kasai, N. (2008). Nonvolatile magnetic flip-flop for standby-power-free SOCs. In *CICC, 2008* (pp. 355–358).
11. Ueda, M., Otsuka, T., Toyoda, K., Morimoto, K., & Morita, K. (2002). A novel non-volatile flip-flop using a ferroelectric capacitor. In *ISAF, 2002* (pp. 155–158).
12. Wang, J., Liu, Y., Yang, H., & Wang, H. (2010). A compare-and-write ferroelectric nonvolatile flip-flop for energy-harvesting applications. In *ICGCS, 2010* (pp. 646–650).
13. Beeby, S. P., Tudor, M. J., & White, N. M. (2006). Energy harvesting vibration sources for microsystems applications. *Measurements of Science and Technology*, *17*(12), R175–R195.
14. Alippi, C., & Galperti, C. (2008). An adaptive system for optimal solar energy harvesting in wireless sensor network nodes. *IEEE Transactions on Circuits and Systems I: Regular Papers*, *55*(6), 1742–1750.
15. Lin, K., Yu, J., Hsu, J., Zahedi, S., Lee, D., Friedman, J., Kansal, A., Raghunathan, V., & Srivastava, M. (2005). Heliomote: Enabling long-lived sensor networks through solar energy harvesting. In *ACM SenSys, 2005* (pp. 309–309).
16. Sheikholeslami, A., & Glenn Gulak, P. (1997). A survey of behavioral modeling of ferroelectric capacitors. *IEEE Transactions on Ultrasonics Ferroelectrics and Frequency Control*, *44*, 917–924.
17. Dawber, M., Rabe, K. M., & Scott, J. F. (2005). Physics of thin-film ferroelectric oxides. *Reviews of Modern Physics*, *77*(4), 1083–1130.
18. Du, X. R., & Sheu, B. (2002). Modeling ferroelectric capacitors for memory applications. *IEEE Circuits and Devices Magazine*, *18*(6), 10–16.
19. Texas Instrument: datasheet of MSP430FR573X Mixed Signal Microcontrollers (2011).
20. Beach, R., Min, T., & Horng, C. (2008). A statistical study of magnetic tunnel junctions for high-density spin torque transfer-MRAM (STT-MRAM). In *IEDM, 2008* (pp. 1–4).
21. Wang, Y., Liu, Y., Li, S., & Sheng, X. (2012). A 3us wake-up time nonvolatile processor based on ferroelectric flip-flops. In *ESSCIRC, 2012* (pp. 149–152).
22. Beenker, G., & Immink, K. (1983). A generalized method for encoding and decoding run-length-limited binary sequences (corresp.). *IEEE Transactions on Information Theory, 29*(5), 751–754.
23. T. U. D. project: Benchmark Applications for Synthesizeable VHDL Model. (2006). Website: http://www.cs.ucr.edu/dalton.
24. Guthaus, M., Ringenberg, J., Ernst, D., Austin, T., Mudge, T., & Brown, R. (2001). Mibench: A free, commercially representative embedded benchmark suite. In *WWC, 2001* (pp. 3–14).
25. Texas Instrument: Z-stack - Zigbee Protocol Stack. (2009). Website: http://www.ti.com/tool/z-stack.
26. Zwerg, M., Baumann, A., Kuhn, R., Arnold, M., Nerlich, R., Herzog, M., Ledwa, R., Sichert, C., Rzehak, V., Thanigai, P., Eversmann, B.O. (2011). An 82 ua/Mhz microcontroller with embedded FeRAM for energy-harvesting applications. In *ISSCC, 2011* (pp. 334–336).
27. Xu, W., Sun, H., Wang, X., Chen, Y., & Zhang, T. (2011). Design of last-level on-chip cache using spin-torque transfer RAM (STT-RAM). *IEEE Transactions on VLSI System*, 483–493.
28. Guo, X., Ipek, E., & Soyata, T. (2010). Resistive computation: Avoiding the power wall with low-leakage, STT-MRAM based computing. In *ISCA, 2010* (pp. 371–382).
29. Qazi, M., Amerasekera, A., & Chandrakasan, A. P. (2013). A 3.4pJ FeRAM-enabled D flip-flop in 0.13um CMOS for nonvolatile processing in digital systems. *To appear in ISSCC 2013*.
30. Wang, Y., Liu, Y., Liu, Y., Zhang, D., Li, S., Sai, B., Chiang, M., & Yang, H. (2012). A compression-based area-efficient recovery architecture for nonvolatile processors. In *DATE, 2012* (pp. 1519–1524).
31. Sheng, X., Wang, Y., et al. (2013). SPaC: A segment-based parallel compression for backup acceleration in nonvolatile processors. In *DATE 2013* (pp. 865–868).

MIX
Papier aus verantwortungsvollen Quellen
Paper from responsible sources
FSC® C105338

If you have any concerns about our products,
you can contact us on
ProductSafety@springernature.com

In case Publisher is established outside the EU,
the EU authorized representative is:
Springer Nature Customer Service Center GmbH
Europaplatz 3, 69115 Heidelberg, Germany

Printed by Libri Plureos GmbH
in Hamburg, Germany